GRADEMAKER STUDY GUIDE AND WORKBOOK

Marketing

Fifth Edition

Charles W. Lamb, Jr.
M.J. Neeley Professor of Marketing,
M.J. Neeley School of Business,
Texas Christian University

Joseph F. Hair, Jr.
Alvin C. Copeland Endowed Chair of Franchising
and Director, Entrepreneurship Institute,
Louisiana State University

Carl McDaniel
Chairman, Department of Marketing,
College of Business Administration,
University of Texas at Arlington

Prepared by

Susan Peterson

Scottsdale Community College

Kathryn Dobie

University of Arkansas

Australia • Canada • Denmark • Japan • Mexico • New Zealand • Philippines
Puerto Rico • Singapore • South Africa • Spain • United Kingdom • United States

GradeMaker Study Guide and Workbook by Susan Peterson and Kathryn Dobie
Acquisitions Editor: Steve Scoble
Developmental Editor: Jamie Gleich Bryant
Marketing Manager: Sarah J. Woelfel
Production Editor: Anne Chimenti
Manufacturing Coordinator: Dana Began Schwartz
Cover Design: Michael H. Stratton
Printer: Mazer

Printed in the United States of America
2 3 4 5 02 01 00

For more information contact South-Western College Publishing, 5101 Madison Road, Cincinnati, Ohio, 45227 or find us on the Internet at http://www.swcollege.com

For permission to use material from this text or product, contact us by
- **telephone: 1-800-730-2214**
- **fax: 1-800-730-2215**
- **web: http://www.thomsonrights.com**

Library of Congress Cataloging-in-Publication Data

ISBN: 0-324-01440-6

This book is printed on acid-free paper.

Contents

SOLUTIONS

Introduction

This *Study Guide* was specifically written to accompany *Marketing*, Fifth Edition, by Charles W. Lamb, Jr., Joseph F. Hair, Jr., and Carl McDaniel. The purpose of this *Study Guide* is to help you review the material in the text, practice using the material, and check your understanding. The *Study Guide* is designed to help you get more out of reading and studying your text and to help you prepare for exams.

For each chapter in your text, this *Study Guide* provides the following materials:

1. **Responses to beginning-of-chapter learning objectives.** This section provides an overall perspective on the focus and goals of the text chapter.

2. **A one-page pretest.** This short test provides you a chance to test how much you understand the chapter material so that you can focus your study efforts on material that you have not yet grasped.

3. **An outline of the chapter materials.** The extensive outline sets forth the chapter structure, summarizes the major topics, and defines chapter key terms.

4. **Fill-in-the-blank vocabulary exercises.** This section provides an alphabetized list of key terms for your reference and allows you to test your recognition of key terms and to master important definitions.

5. **True/False questions.** These questions are designed to test your recall of key terms, marketing terminology, and marketing concepts.

6. **Agree/disagree questions.** These questions are designed to provide more in-depth critical thinking about topics covered in the chapter.

7. **Multiple-choice questions.** These questions are designed to test your ability to apply chapter concepts to marketing situations.

8. **Essay questions and problems.** Essay questions test your ability to integrate chapter information and to apply knowledge to broad concepts and relationships between marketing concepts. Problems that appear in selected chapters test your ability to apply analytical tools and technical marketing methods.

9. **Space** for note taking and for writing answers to chapter review questions and chapter case questions.

10. **Solutions to all exercises, questions, and problems.** These solutions are presented at the end of the *Study Guide*. Rationales are provided for false questions and all multiple-choice questions, and complete answers are provided for all pretest questions, essay and problem questions. Text page numbers are also provided for answer reference. The solutions are provided to help you evaluate your understanding of the material in the text chapter. You should refer to these answers only after you have completed the questions.

The outlines and all exercises are tied to the text through the learning objectives. Numbered learning objectives identify the material that covers each objective. If you are having trouble with a concept, review the material in the text identified by the same learning objective number.

Always read the text before using the *Study Guide*. This *Study Guide* is not a substitute for the text, but rather a tool that will enhance your understanding of text material. The material in the *Study Guide* is presented in the order that the topics are covered in each text chapter. This allows you to refer to the text for any sections that may pose difficulties. After reading your text, review the text material by reading through the learning objectives and chapter outline of this *Study Guide*. Then carefully work through each set of exercises and questions, making sure that you understand why the correct answers are correct. Answers to the essay questions are suggestions only and may not precisely match your responses.

This *Study Guide* can help you learn and evaluate your knowledge only to the extent that you use it correctly and regularly. Good luck with your course and the text. Marketing is an exciting field of study, and the intention of this *Study Guide* is to help you capture that excitement.

CHAPTER GUIDES and QUESTIONS

PART ONE

THE WORLD OF MARKETING

1 An Overview of Marketing

2 Strategic Planning: Developing and Implementing a Marketing Plan

3 The Marketing Environment and Marketing Ethics

4 Developing a Global Vision

CHAPTER 1 An Overview of Marketing

LEARNING OBJECTIVES

1 Define the term "marketing"

Marketing is the process of planning and executing the conception, pricing, promotion, and distribution of ideas, goods, and services to create exchanges that satisfy individual and organizational objectives.

2 Describe four marketing management philosophies

Four competing philosophies strongly influence the role of marketing and marketing activities within an organization. These philosophies are commonly referred to as production, sales, marketing, and societal marketing orientations.

The production orientation focuses on internal efficiency to achieve lower prices for consumers. It assumes that price is the critical variable in the purchase decision.

A sales orientation assumes that buyers resist purchasing items that are not essential and that consumers must be persuaded to buy. The marketing orientation is based on an understanding that a sale predominantly depends on the customer's decision to purchase a product and on the customer's perception of the value of that product. Responsiveness to customer wants is the central focus of the marketing orientation.

The societal marketing orientation holds that the firm should strive to satisfy customer needs and wants while meeting organizational objectives and preserving or enhancing both the individual's and society's long-term best interests.

3 Discuss the differences between sales and market orientations

Selling Orientation	Marketing Orientation
Organization's focus is inward on the firm's needs	Focus is outward on the wants and preferences of customers
Business is defined by its goods and services offered	Business is defined by benefits sought by customers
Product is directed to everybody	Product is directed to specific groups (target markets)
Primary goal is profit through maximum sales volume	Primary goal is profit through customer satisfaction
Goals are achieved primarily through intensive promotion	Goals are achieved through coordinated marketing

4 Describe the marketing process

Marketing managers are responsible for a variety of activities that together represent the marketing process. These include: matching the role of marketing with the firm's vision and mission, setting objectives, analyzing internal and external information, developing strategy, planning a marketing mix, implementing strategy, designing performance measures, and evaluating and adjusting strategy.

5 Describe several reasons for studying marketing

Marketing provides a delivery system for a standard of living, which is a monumental task in a society such as the United States, where a typical family consumes 2.5 tons of food per year.

No matter what an individual's area of concentration in business, the terminology and fundamentals of marketing are important for communicating with others in the firm.

Between one-fourth and one-third of the entire civilian work force in the United States performs marketing activities. Marketing offers career opportunities in areas such as professional selling, marketing research, advertising, retail buying, distribution management, product management, product development, and wholesaling.

As a consumer of goods and services, everyone participates in the marketing process every day. By understanding marketing, one can become a more sophisticated consumer.

PRETEST

Answer the following questions to see how well you understand the material. Re-take it after you review to check yourself.

1. Marketing is defined as:

2. What five conditions must be satisfied for any kind of exchange to take place?

3. The four variables of the marketing mix are:

4. Four marketing management philosophies are:

5. List five ways in which a marketing orientation is different from a sales orientation.

Sales Orientation	Marketing Orientation
________________	________________
________________	________________
________________	________________
________________	________________
________________	________________

6. List seven steps in the marketing process:

7. Name four reasons for studying marketing:

CHAPTER OUTLINE

1 Define the term "marketing"

I. What Is Marketing?

A. Marketing is not the same as selling or advertising.

B. Marketing includes selling, advertising, making products available in stores, arranging displays, maintaining inventories, and much more.

C. Marketing is a philosophy or a management orientation that stresses the importance of customer satisfaction, as well as the set of activities used to implement this philosophy.

D. The American Marketing Association definition of marketing:

Marketing is the process of planning and executing the conception, pricing, promotion, and distribution of ideas, goods, and services to create exchanges that satisfy individual and organizational goals.

1. The Concept of Exchange
The concept of **exchange** means that people give up something in order to receive something that they would rather have.

The usual medium of exchange is money. Exchange can also be fostered through barter or trade of items or services.

a. Five conditions must be satisfied for an exchange to take place:

1. There must be at least two parties.
2. Each party must have something the other party values.
3. Each party must be able to communicate with the other party and deliver the goods or services sought by the other trading party.
4. Each party must be free to accept or reject the other's offer.
5. Each party must want to deal with the other party.

Exchange may not take place even if all of these conditions exist, but these conditions are necessary for exchange to be possible.

2 Describe four marketing management philosophies

II. Marketing Management Philosophies

Four competing philosophies strongly influence an organization's marketing activities. These philosophies are commonly referred to as production, sales, marketing, and societal orientations.

A. Production Orientation

The **production orientation** focuses on internal capabilities of the firm rather than on the desires and needs of the marketplace. The firm is concerned with what it does best, based on its resources and experience, rather than with what consumers want.

B. Sales Orientation

A **sales orientation** assumes that more goods and services will be purchased if aggressive sales techniques are used and that high sales result in high profits.

C. The Marketing Concept and Market Orientation

1. This philosophy, called the **marketing concept**, states that the social and economic justification for an organization's existence is the satisfaction of customer wants and needs while meeting organizational objectives.

2. The marketing concept involves:

   a. Focusing on customer wants and needs so the organization can differentiate its product(s) from competitors' offerings

   b. Integrating all the organization's activities, including production, to satisfy these wants and needs

   c. Achieving long-term goals for the organization by satisfying customer wants and needs legally and responsibly

3. Firms that adopt and implement the marketing concept are said to be market oriented. **Market orientation** requires

   a. Top management leadership

   b. A customer focus

   c. Competitor intelligence

   d. Interfunctional coordination to meet customer wants and needs and deliver superior values

4. Understanding your competitive arena and competitor's strengths and weaknesses is a critical component of market orientation.
5. Market-oriented companies are successful in getting all business functions together to deliver customer value.

D. Societal Orientation

1. The philosophy called a **societal orientation** states that an organization exists not only to satisfy customer wants and needs and to meet organizational but also to preserve or enhance individual's and society's long-term best interests.

2. This orientation extends the marketing concept to serve three bodies rather than two: customers, the organization itself, and society as a whole.

3 Discuss the differences between sales and marketing orientations

III. Differences Between Contrasting Sales and Marketing Orientations

A. The Organization's Focus

1. Sales-oriented firms tend to be inward-looking. They focus on satisfying their own needs rather than those of customers.

2. Market-oriented firms derive their competitive advantage from an external focus. Departments in these firms coordinate their activities and focus on satisfying customers.

3. Customer Value

   a. **Customer value** is the ratio of benefits to the sacrifice necessary to obtain those benefits.

   b. Creating customer value is a core business strategy of many successful firms.

   c. Marketers interested in customer value

      1. Offer products that perform

      2. Give consumers more than they expect

      3. Avoid unrealistic pricing

      4. Give the buyer facts

      5. Offer organization wide commitment in service and after-sales support

4. Customer Satisfaction

   Customer satisfaction is the feeling that a product has met or exceeded the customer's expectations. The organizational culture focuses on delighting customers rather than on selling products.

5. Building Relationships

   Relationship marketing is a strategy that entails forging long-term partnerships with customers and contributing to their success.

   a. The Internet is an effective tool for generating relationships with customers.

   b. Customers benefit from stable relationships with suppliers.

   c. A sense of well-being occurs when one establishes an ongoing relationship with provider.

   d. Most successful relationship marketing strategies depend upon: customer-oriented personnel, effective training programs, employees with authority to make decisions and solve problems, and teamwork.

6. Customer-Oriented Personnel

   All employees in marketing-oriented firms must be customer-oriented for the customer satisfaction goals to be met. Often one employee is the only contact a customer has with a company.

7. The Role of Training

   Leading marketers recognize the role of employee training in customer service. In order to have full participation and understanding of the marketing philosophy, companies such as Walt Disney World, McDonald's, and American Express train all their employees to become customer-oriented.

8. Empowerment

Empowerment is the practice of giving employees expanded authority to solve customer problems as they arise. This technique improves customer service by improving responsiveness.

9. Teamwork

Teamwork entails collaborative efforts of people to accomplish common objectives.

B. The Firm's Business

1. A sales-oriented firm defines its business in terms of the goods and services it offers, like a encyclopedia publisher defining itself simply as a book publisher/seller.

2. A market orientated firm defines its business based on the benefits customers seek.

3. Why is this customer benefit definition so important?

   a. It ensures the firm keeps focusing on customers

   b. It encourages innovation and creativity by reminding people that there are many ways to satisfy customer wants.

   c. It stimulates an awareness of changes in customer desires and preferences.

4. Focusing on customer wants does not mean that customers will always receive the specific goods and services they want.

   a. Giving the customer exactly what he or she wants is not practical if doing so threatens the survival of the firm.

   b. Sound professional judgment must influence the decision about which goods or services should be offered. Perhaps the most prudent course will be to alter the way consumers perceive their needs and means of satisfaction.

C. Those To Whom the Product Is Directed

1. A sales-oriented organization targets its products at "everybody" or "the average customer." However, few "average" customers exist.

2. The market-oriented firm

   a. Recognizes that different customer groups have different wants

   b. Targets specific subgroups of customers

   c. Designs special products and marketing programs for these groups

D. The Firm's Primary Goal

1. The goal of a sales-oriented firm is profitability through sales volume. The focus is on making the sale rather than developing a long-term relationship with a customer.

2. The ultimate goal of most market-oriented organizations is to make a profit from satisfying customers. Superior customer service enables a firm to have large amounts of repeat business, customer loyalty, and higher profit margins.

E. Tools the Organization Uses to Achieve Its Goals

1. Sales-oriented firms seek to generate sales volume through intensive promotional activities, mainly personal selling and advertising.

2. Market-oriented organizations recognize that promotion is only one of the four basic tools that comprise the marketing mix.

   The tools are the marketing mix elements (the four P's): product, place (distribution), promotion, and price.

3. The important distinction is that market-oriented firms recognize that each of the four components of the marketing mix is of equal importance: sales-oriented organizations view promotion as the primary means of achieving their goals.

4 Explain how firms implement the marketing concept

IV. Implementation of the Marketing Concept

Some firms have trouble implementing the marketing philosophy. Changing to a customer-driven corporate culture must occur step by step.

A. Changes in Authority and Responsibility

1. Changing from a production or sales orientation to a marketing orientation may mean changes in the relationships of authority and responsibility within the firm.

2. Marketing personnel will be involved in product planning and production scheduling. The market research department may have much more authority than in the past.

3. Making changes slowly and involving all those who will be affected in the planning process will help to gain acceptance for the marketing concept.

B. Front-Line Experience for Management
Many organizations have a policy of sending executives into the field to actually work in customer-service, front-line jobs several days a year.

5 Describe the marketing process

V. The Marketing Process

A. Understanding the organization's mission and vision, and the role marketing plays.

B. Setting marketing objectives

C. Gathering, analyzing, and interpreting information about the organization's situation, including its strengths and weaknesses, as well as opportunities and threats in the environment.

D. Developing marketing strategy by deciding exactly which wants and whose wants the organization will try to satisfy, and by developing appropriate marketing activities to satisfy the desires of selected target markets.

E. Implementing the marketing strategy

F. Designing performance measures

G. Periodically evaluating marketing efforts and making changes if needed.

6 Describe several reasons for studying marketing

VI. Why Study Marketing?

A. Marketing Plays an Important Role in Society

Marketing provides a delivery system for a complex standard of living. The number of transactions needed everyday in order to feed, clothe, and shelter a population the size of the one in the United States is enormous and requires a sophisticated exchange mechanism.

B. Marketing Is Important to Businesses

Marketing provides the following vital business activities

1. Assessing the wants and satisfactions of present and potential customers

2. Designing and managing product offerings

3. Determining prices and pricing policies

4. Developing distribution strategies

5. Communicating with present and potential customers

C. Marketing Offers Outstanding Career Opportunities

1. Between one-fourth and one-third of the entire civilian work force in the United States performs marketing activities.

2. Marketing offers career opportunities in areas such as professional selling, marketing research, advertising, retail buying, distribution management, product management, product development, and wholesaling.

D. Marketing Affects Your Life Every Day

1. As consumers of goods and services, we participate in the marketing process every day.

2. Almost 50 cents of every dollar consumers spend goes to pay marketing costs such as market research, product research and development, packaging, transportation, storage, advertising, and sales-force expenses.

VOCABULARY PRACTICE

Fill in the blank(s) with the appropriate term or phrase from the alphabetized list of chapter key terms.

customer satisfaction
customer value
empowerment
marketing
marketing concept
marketing orientation
production orientation
relationship marketing
sales orientation
societal marketing concept
teamwork

1 Define the term "marketing"

1. The process of planning and executing the conception, pricing, promotion, and distribution of ideas, goods, and services to create exchanges that satisfy individual and organizational objectives defines ______________________________.

2 Describe four marketing management philosophies

2. There are four alternative marketing management philosophies. The first focuses on the firm's internal capabilities rather than the needs and desires of the marketplace; this is the ______________________________. The second philosophy is the ______________________________, which assumes that buyers resist purchasing nonessential items, so aggressive sales techniques should be used to sell more products. The third philosophy, which focuses on customer needs and wants, is the ______________________________. Finally, a firm that decides to preserve long-term best interests of its customers and society has adopted the ______________________________.

3. When a firm focuses on customer wants, integrates all firm activities to satisfy these wants, and achieves long-term goals by satisfying these wants legally and responsibly, the firm is using the ______________________________.

3 Discuss the differences between sales and market orientations

4. The ratio of the benefits to the sacrifice necessary to obtain those benefits is termed ______________________________.

5. When a marketer feels that a product meets or exceeds customers' expectations, then ______________________________ has been created.

6. A strategy of forging long-term partnerships with customers is called ______________________________. It means that firms can become part of their customer's organization and contribute to its success.

7. When people collaborate their efforts to accomplish common objectives, they are practicing ______________________________.

8. Marketing-oriented firms may give their employees expanded authority to solve customer problems immediately. This is known as ______________________________.

Check your answers to these questions before proceeding to the next section.

TRUE/FALSE QUESTIONS

Mark the statement **T** if it is true and **F** if it is false.

1 Define the term "marketing"

_____ 1. A marketing exchange cannot take place unless each party in the exchange has something that the other party values.

2 Describe four marketing management philosophies

_____ 2. The owners of the Plane Rubber and Tire Company are pleased with their low unit costs and high production volumes. Salespeople are unnecessary because buyers are always waiting for new tires to come off the assembly line. Plane currently has a production orientation.

_____ 3. The president of Hoppity Flea Collars does not find it necessary to conduct much marketing research because the telephone selling campaign has been such a successful marketing strategy. Hoppity has a marketing orientation.

_____ 4. Having a sales orientation is the same as having a market orientation since both have the ultimate goal of satisfying customer needs.

3 Discuss the differences between sales and marketing orientations

_____ 5. You are about to start manufacturing and selling ferret food. You have met with your board of directors and you all discussed the benefits and sacrifices regarding the purchase of your food. Knowing the ratio of benefits to sacrifices allows you to specify how much customer value you will achieve.

4 Describe the marketing process

_____ 6. The marketing mix variables are product, place, promotion, and price.

Check your answers to these questions before proceeding to the next section.

AGREE/DISAGREE QUESTIONS

For the following statements, indicate reasons why you may agree and disagree with the statement.

1. The marketing concept actually encompasses both the sales concept and the production concept.

   Reason(s) to agree:

   Reason(s) to disagree:

2. Marketing is the job of everyone in a business organization, not just the marketing department.

   Reason(s) to agree:

   Reason(s) to disagree:

3. Only students who are majoring in marketing should be required to take marketing courses.

   Reason(s) to agree:

   Reason(s) to disagree:

MULTIPLE-CHOICE QUESTIONS

Select the response that best answers the question, and write the corresponding letter in the space provided.

1 Define the term "marketing"

_____ 1. Which of the following is NOT true about marketing?
a. Marketing is a philosophy that stresses customer satisfaction.
b. Marketing is a process.
c. Marketing can involve any number of parties.
d. Marketing can be used for ideas, goods, or services.
e. Marketing involves products, pricing, promotion, and distribution.

_____ 2. In order for exchange to occur:
a. a complex societal system must be involved
b. each party must have something of value to the other party
c. a profit-oriented organization must be involved
d. money or other legal tender is required
e. organized marketing activities must also occur

_____ 3. If you were in the marketing consulting business which of the following clients could you not serve?
a. The Boston Museum of Science, which needs to determine what exhibits should it offer visitors
b. The State of Mississippi, which needs to attract tourists
c. Dr. Susan Scott, an orthopedic surgeon wishing to open a practice in your home town
d. The World Gym, which needs to determine where to locate its next outlet for customers
e. All of the above could be served by a marketing consultant

_____ 4. You are concerned with managing the exchange between the Red Cross and its blood donators. Which of the following costs would you have to be concerned about to create the ideal exchange?
a. The travel costs incurred by donators visiting the Red Cross blood donation sites.
b. The personal energy and time expended by the donator
c. The opportunity costs lost by not engaging in some other activity
d. All of the above are marketing costs that would be of concern to someone managing the exchange situation
e. None of the above are costs of exchange situations

2 Describe four marketing management philosophies

_____ 5. Fred Stone, the owner of Neanderthal Products, Inc. is production-oriented. If you were in charge of his marketing operations, which of the following statements might you use as a guiding principle if you wish to meet Mr. Stone's demand?
a. "I'm a customer and everyone is like me. I buy on price, therefore everyone does, as well."
b. "We need to buy the fastest production equipment as possible to raise productivity and keep prices at the lowest possible level."
c. "We produce the best widgits in the market place."
d. All of the above would be consistent with Mr. Stone's demands.
e. None of the above would be consistent, because all reflect a sales orientation.

_____ 6. Peter's company does an excellent and efficient job of churning thousands of Nit-Pickers off the assembly line every day. One problem with this ____________ approach to marketing is the failure to consider whether Nit-Pickers also meet the needs of the marketplace.
a. customer orientation
b. sales orientation
c. discount orientation
d. marketing orientation
e. production orientation

_____ 7. Jack Niven's company markets golf club polish. Jack knows that buyers may consider the product nonessential, and he assumes that if he hires a team of aggressive, persuasive salespeople, buyers will buy more of the polish. Jack has a:
   a. sales orientation
   b. production orientation
   c. promotion orientation
   d. marketing orientation
   e. customer orientation

_____ 8. Beth has noticed the lack of specialty recycling centers in her community, although local neighborhood clubs have repeatedly asked the city to provide such centers. Beth has decided to become certified in waste disposal and hopes to open a battery and motor oil recycling center next year. She hopes to include the innovative service of home pickup and delivery of recyclables. This business philosophy supports a(n) orientation.
   a. production
   b. sales
   c. retail
   d. marketing
   e. enterprise

_____ 9. The Ajax Insurance Company tells its salespeople to try to sell life insurance to everyone they meet or contact. In contrast, the Family Shelter Insurance Company concentrates on special insurance plans designed for single parents. Family Shelter is:
   a. missing out by not concentrating on the average customer
   b. a company that would state that they are in the business of selling insurance
   c. a selling-oriented company
   d. recognizing that different customer groups have different needs and wants
   e. aiming at a goal of profit through maximum sales volume

_____ 10. Bob & Gary's is a contemporary ice cream manufacturer that donates 10 percent of its earnings to the restoration of the Amazon rain forest. Bob & Gary's has which type of orientation?
   a. production
   b. sales
   c. promotion
   d. marketing
   e. societal marketing

3 Discuss the differences between sales and market orientations

_____ 11. A sales orientation __________, while a marketing orientation __________.
   a. achieves profit through customer satisfaction, achieves profit through maximum sales volume
   b. targets specific groups of people, targets everybody
   c. delivers superior customer value, focuses on selling goods and services
   d. has an "outward" focus, has an "inward" focus
   e. uses intensive promotion to maximize profits, uses coordinated interfunctional activities

4 Describe the marketing process

_____ 12. Which of the following is NOT a part of the marketing process?
   a. understanding the organization's mission
   b. developing performance appraisals for marketing personnel
   c. designing performance measures
   d. setting objectives
   e. determining target markets

_____ 13. The marketing manager for Oil of Olan, a skin care product, is working with an advertising agency to develop a new TV commercial targeting teen-agers. Which of the following marketing mix variables best describes this activity?
   a. product
   b. price
   c. target market
   d. distribution
   e. promotion

5 Describe several reasons for studying marketing

_____ 14. Jackie is a food science major at a state university and hopes to operate the family restaurant after graduation. Jackie has been advised to take a marketing course in the school of business as an elective, but she thinks this would be a waste of time. You are her friend and a marketing major. You advise that:
   a. marketing is not relevant for a business like a family restaurant
   b. Jackie declare a business minor because she needs a backup career
   c. more nutrition and gourmet cooking classes will be most useful for Jackie
   d. the main reason to take marketing is to teach Jackie how to advertise the restaurant
   e. marketing knowledge will help Jackie to understand how she can satisfy consumers' needs and wants

_____ 15. Jon owns a small laboratory that makes bifocal contact lenses. His company is growing fast, and there are many things he does not understand about his customers. Should Jon take a marketing course?
   a. Yes, because marketing is synonymous with selling, and Jon will want to learn aggressive sales techniques to continue the company's growth
   b. No, because he can hire an advertising firm and will not need further knowledge of marketing
   c. No, because marketing is a minor function in business
   d. Yes, because the concept of marketing will help Jon to better serve and satisfy his customers
   e. Yes, because marketing teaches businesses how to sell products that people don't need

Check your answers to these questions before proceeding to the next section.

ESSAY QUESTIONS

1. The concept of exchange is crucial to the definition of marketing. What are the five conditions that must be satisfied for an exchange to take place? Can marketing occur even if an exchange does not take place?

2. Assume you are a marketing manager. Describe the marketing strategy for each of the four orientations of marketing management philosophy.

3. Name and describe five key areas in which a market orientation differs from a sales orientation.

4. Briefly define the marketing process activities.

CHAPTER NOTES

Use this space to record notes on the topics you are having the most trouble understanding.

ANSWERS TO THE END-OF-CHAPTER DISCUSSION AND WRITING QUESTIONS

Use this space to work on the questions at the end of the chapter.

ANSWERS TO THE END-OF-CHAPTER CASE

Use this space to work on the questions at the end of the chapter.

CHAPTER 2 Strategic Planning: Developing and Implementing a Marketing Plan

LEARNING OBJECTIVES

1 Understand the importance of strategic marketing and know a basic outline for a marketing plan

Strategic marketing planning is the basis for all marketing strategies and decisions. The marketing plan is a written document that acts as a guidebook of marketing activities for the marketing manager. By specifying objectives and defining the actions required to attain them, a marketing plan provides the basis on which actual and expected performance can be compared. Creating a complete marketing plan is not a simple or quick effort. And the plan is only as good as the information it contains and the effort, creativity, and thought that went into its creation. Many of the elements in the plan are decided upon simultaneously and in conjunction with one another. Every marketing plan is unique to the firm for which is created.

Basic factors that should be covered include business mission, setting objectives, performing a situation analysis, selecting target markets, delineating a marketing mix, and establishing ways to implement, evaluate, and control the plan.

2 Define an appropriate business mission statement

The mission statement is based on a careful analysis of benefits sought by present and potential customers and analysis of existing and anticipated environmental conditions. The firm's long-term vision, embodied in the mission statement, establishes boundaries for all subsequent decisions, objectives, and strategies. A mission statement should focus on the market or markets the organization is attempting to serve rather than on the good or service offered.

3 Know the criteria for stating good marketing objectives

Objectives should be realistic, measurable, and time specific. Objectives must also be consistent, and indicate the priorities of the organization.

4 Explain the components of a situation analysis

The situation analysis is sometimes call a SWOT analysis, because firms identify their strengths, weaknesses, opportunities, and threats. This analysis may help the firm discover a strategic window of opportunity, or a differential advantage. When examining external opportunities and threats, marketing managers must analyze aspects of the marketing environment in a process called environmental scanning. The six most often studied macroenvironmental forces are: social, demographic, economic, technological, political and legal, and competitive forces.

5 Identify strategic alternatives and describe tools used to help select alternatives

Firms can use the strategic opportunity matrix to allow firms to explore four options: market penetration, market development, product development, and diversification. Firms select the alternative that best helps them reach their overall strategic goal: market share or profit. Corporate culture plays a large role in the selection process.

There are several major techniques for selecting alternatives. The portfolio matrix is a method of determining the profit potential and investment requirements of a firm's SBUs by classifying each as a star, cash cow, dog, or problem child and then determining appropriate resource allocations for each. A more detailed alternative to the portfolio matrix is the market attractiveness/company strength matrix, which measures company and market viability.

6 Discuss target market strategies

The target market strategy identifies which market segment or segments to focus on. The process begins with a market opportunity analysis, or MOA, which describes and estimates the size and sales potential of market segments that are of interest to the firm. In addition, an assessment of key competitors in these market segments is performed. After the market segments are described, one or more may be targeted by the firm.

The three strategies for selecting target markets are appealing to the entire market with one marketing mix, concentrating on one segment, or appealing to multiple market segments using multiple marketing mixes.

7 Describe elements of the marketing mix

The term **marketing mix** refers to a unique blend of product, distribution, promotion, and pricing strategies designed to produce mutually satisfying exchanges with a target market. Distribution is often referred to as place, thus giving the "four P's" of marketing: product, place, promotion, and price. Products can be tangible goods, ideas, or services. Distribution strategies are concerned with making products available when and where customers want them. Promotion includes personal selling, advertising, sales promotion, and public relations. Price is what the buyer must give up to obtain a product.

8 Understand why implementation, evaluation, and control of the marketing plan is necessary

After selecting strategic alternatives, plans should be implemented, that is, put into action. The plan should be evaluated to see if it has achieved its objectives. The final step in the strategic planning process, control, is the alteration of plans, if necessary. A marketing control system ensures that marketing goals are achieved within guidelines.

9 Identify several techniques that help make strategic planning effective

Effective strategic planning should be treated as an ongoing process, not an annual exercise. Effective planning requires creativity and should challenge existing assumptions about the firm and the environment. Perhaps the most critical element is the support and participation of top management. Their involvement in planning must be sincere and ongoing.

PRETEST

Answer the following questions to see how well you understand the material. Re-take it after you review to check yourself.

1. List and briefly describe six major elements of a marketing plan.

2. Name and briefly describe four strategic alternatives that can be used by a firm.

3. List and briefly describe two tools that can be used to select a strategic alternative.

4. What is a marketing strategy?

5. List and briefly describe the four Ps ("marketing mix").

CHAPTER OUTLINE

1 Understand the importance of strategic marketing and know a basic outline for a marketing plan

I. The Nature of Strategic Planning

1. Marketing managers must plan, organize, and control marketing activities. They must develop both long-range (strategic) and short-range (tactical) plans.

    1. **Strategic planning** is the managerial process of creating and maintaining a fit between the organization's objectives and resources and evolving market opportunities.

    2. Strategic decisions require long-term resource commitments with major financial consequences. A good strategic plan can help to protect a firm's resources against competitive onslaughts.

    3. Strategic marketing management addresses two questions: What is the organization's main activity at a particular time? And how will it reach its goals?

B. What is a marketing plan?

1. Planning is the process of anticipating future events and determining strategies to achieve organizational objectives in the future.

2. Marketing planning involves designing activities relating to marketing objectives and the changing marketing environment.

3. Issues such as product lines, distribution channels, marketing communications, and pricing are all delineated in the marketing plan.

C. Why write a marketing plan?

1. The marketing plan serves as a reference point for the success of future activities.

2. The marketing plan allows you to examine the external marketing environment and the internal business, allowing the firm to enter the marketplace with an awareness of possibilities and problems.

D. Marketing Plan Elements

There are elements common to all marketing plans, such as defining the business mission and objectives, performing a situation analysis, delineating a target market, and establishing components of the marketing mix.

2 Define an appropriate business mission statement

II. Define the Business Mission

A. The firm's **mission statement** is the answer to the question "What business are we in and where are we going?" A mission statement should focus on the market or markets the organization is attempting to serve rather than on the good or service offered.

B. Defining the business in terms of goods and services rather than in terms of the benefits customers seek is sometimes called **marketing myopia**. In this context, the term **myopia** means narrow, short-term thinking, which can threaten an organization's survival.

C. The organization may need to define a mission statement and objectives for a **strategic business unit (SBU)**, which is a subgroup of a business.

Strategic business units will have the following characteristics:

1. A distinct mission and a specific target market
2. Control over their resources
3. Their own competitors
4. A single business or a collection of related businesses
5. Plans independent of the other businesses in the organization

3 Know the criteria for stating good marketing objectives

III. Set Marketing Plan Objectives

A **marketing objective** is a statement of what is to be accomplished through marketing activities.

A. Marketing objectives should be consistent with organization objectives, should be measurable, and should specify the time frame during which they are to be achieved.

B. Objectives communicate marketing management philosophies, provide direction, serve as motivators, are a basis for control, and force executives to clarify their thinking.

4 Explain the components of a situation analysis

IV. Before specific marketing activities can be defined, marketers must understand the current and potential environment that the product or service will be marketed in.

A. Conduct a Situation Analysis

1. A situation analysis is sometimes referred to as a **SWOT analysis**. That is, the firm should identify its internal strengths (S) and weaknesses (W) and also examine external opportunities (O) and threats (T).

2. **Environmental scanning** is the collection and interpretation of information about forces, events, and relationships in the external environments that may affect the future of the organization.

3. Environmental scanning helps identify market opportunities and threats and provides guidelines for the design of marketing strategy.

4. The six most often studied macroenvironmental forces are social, demographic, economic, technological, political/legal, and competitive.

5. Corporate culture is a pattern of basic assumptions that an organization has accepted to cope with the firm's internal environment and the changing external environment.

   These assumptions have worked in the past and are taught to new members of the firm as the correct way to perceive, think, and feel about the firm's environments.

6. Corporate culture affects how a firm reacts to the problems and opportunities of the external environment. A firm can usually be categorized as a prospector, reactor, defender, or analyzer.

   a. *Prospectors* focus on identifying and capitalizing on emerging market opportunities.

   b. *Reactors* respond to environmental pressures and are followers. They tend to maintain the status quo and not take risks.

   c. *Defenders* have a specific market domain and do not search outside that domain for opportunities. Defenders emphasize operating efficiency.

   d. *Analyzers* tend to be both conservative and aggressive. They defend successful markets and try to identify emerging opportunities. Analyzers are usually "second in" to new product markets.

7. A **strategic window** is the limited period of time during which the "fit" between the key requirements of a market and the particular competencies of a firm are at an optimum.

5 Identify strategic alternatives and describe tools used to help select alternatives

V. Develop Strategic Alternatives

1. To discover a marketing opportunity or strategic window, management must know how to identify the alternatives. One method is the strategic opportunity matrix.

   1. A firm following a **market penetration** alternative would try to increase market share among existing customers.

   2. **Market development** is a strategic alternative that attracts new customers to existing products, perhaps by expanding the target market or expanding geographically.

      The ideal solution is finding new uses for old products that will stimulate additional sales among existing customers while also bringing in new buyers.

3. A product development strategy entails the creation of new products for present markets. Advantages of this strategy are current knowledge of the target market and established distribution channels.

4. **Diversification** refers to a strategy of increasing sales by introducing new products into new markets. This strategy can be very risky when a firm is entering unfamiliar markets.

B. Select a Strategic Alternative

Several tools aid corporate decision makers in selecting a strategic alternative.

It is also important to recognize several factors that affect the selection, including corporate philosophy and culture:

1. Most companies have a philosophical stance on the issue of immediate profit versus market share. In the long run these are compatible goals, but in the short run there may be a conflict.

2. The portfolio matrix specifies four share/growth categories for SBUs:

   a. **Stars** are market leaders and growing fast. Stars have large reported profits but require a lot of cash to finance the rapid growth.

   b. A **cash cow** usually generates more cash than is required to maintain its market share. It is in a low-growth market but has a dominant market share.

   c. **Problem children**, also called **question marks**, exhibit rapid growth but poor profit margins. They have a low market share in a high-growth industry.

   Problem children require a tremendous amount of cash to obtain better market share.

   d. A **dog** has low growth potential and a small market share.

   Most dogs eventually leave the marketplace. The firm often harvests them by cutting all support costs to a bare minimum.

3. After classifying the various SBUs into the matrix, the next step is to allocate future resources for each.

   a. If an SBU is believed to have the potential to be a star, the organization may decide to follow a build strategy, providing the financial resources necessary to achieve this objective.

   b. If an SBU is a very successful cash cow, a key objective would be to hold or preserve market share so the organization can take advantage of the very positive cash flow.

   c. The harvest objective is appropriate for all SBUs except those classified as stars. The basic objective is to increase the short-term cash return without regard for long-run impact.

   d. Question marks and dogs are often divested, or removed from the company portfolio. SBUs with low shares of low-growth markets are usually the targets of this strategy.

4. The General Electric model for selecting strategic alternatives is known as the **market attractiveness/company strength matrix**. These dimensions are:

   a. richer and more complete than the portfolio matrix

   b. much harder to quantify

3. Differential Advantage

1. A differential advantage is one or more unique aspects of an organization that cause target consumers to patronize that firm rather than competitors. A differential advantage may exist in the firm's image or in any element of the marketing mix.

2. Sources of advantage are superior skills or superior resources. An organization with a differential advantage can deliver superior customer value or attain lower relative costs.

6 Discuss target market strategies

VI. Describe the Marketing Strategy

Marketing strategy involves three activities: selecting one or more target markets, setting marketing objectives, and developing and maintaining a marketing mix that will produce mutually satisfying exchanges with target markets.

A. Target Market Strategy

Target market(s) can be selected by appealing to the entire market, concentrating on one segment, or appealing to multiple market segments using multiple marketing mixes.

B. Marketing Objectives

A **marketing objective** is a statement of what is to be accomplished through marketing activities. Marketing objectives should be consistent with organization objectives, should be measurable, and should specify the time frame during which they are to be achieved.

7 Describe elements of the marketing mix

VII. The Marketing Mix

The term **marketing mix** refers to a unique blend of product, distribution, promotion, and pricing strategies designed to produce mutually satisfying exchanges with a target market.

Distribution is sometimes referred to as place, thus giving us the "four P's" of the marketing mix: product, place, promotion, and price.

A. Product Strategies

The heart of the marketing mix, the starting point, is the product offering and product strategy. The product includes its package, warranty, after-sale service, brand name, company image, and many other factors.

B. Distribution Strategies

Distribution strategies, which usually involve wholesalers and retailers, are concerned with making products available when and where customers want them. Physical distribution also involves all the business activities that are concerned with storing and transporting raw materials or finished products.

C. Promotion Strategies

Promotion includes personal selling, advertising, sales promotion, and public relations. Promotion's role in the marketing mix is to inform, educate, persuade, and remind target markets about the benefits of an organization or a product.

D. Pricing Strategies

Price is often the most flexible of the four marketing mix elements, the quickest element to change. Price is a very important competitive weapon and very important to the organization, because price multiplied by the number of units sold equals total revenue for the firm.

8 Understand why implementation, evaluation, and control of the marketing plan is necessary

VIII. Implementation

A. **Implementation** is the process of gaining the organizational compliance required to put marketing strategies into action. Brilliant marketing strategies are doomed to fail if they are not properly implemented.

B. **Evaluation** entails gauging the extent to which marketing objectives have been achieved during a specified time period.

C. Control provides the mechanisms for evaluating marketing results in light of the strategic plan and for correcting actions that do not help the organization reach those goals within budget guidelines.

D. After an organization implements its strategic plan, it must track results and monitor the external environment. A key tool is the marketing audit.

1. A **marketing audit** is a thorough, systematic, periodic evaluation of the goals, strategies, structure, and performance of the marketing organization.

2. The marketing audit has four characteristics

   - The marketing audit covers all the major marketing issues facing an organization and not just trouble spots.
   - The marketing audit takes place in an orderly sequence and covers the organization's marketing environment, internal marketing system, and specific marketing activities. The diagnosis is followed

by an action plan with both short-run and long-run proposals for improving overall marketing effectiveness.

- The marketing audit is normally conducted by an inside or outside party who is independent enough to have top management's confidence and to be objective.
- The marketing audit should be carried out on a regular schedule instead of only in a crisis.

9 Identify several techniques that help make strategic planning effective

IX. Effective Strategic Planning

Effective strategic planning requires continual attention, creativity, and management commitment.

1. It is not an annual process but an ongoing process.
2. Sound planning involves creativity. The firm needs to challenge existing assumptions.
3. Perhaps the most critical element is the support and participation of top management.

VOCABULARY PRACTICE

Fill in the blank(s) with the appropriate term or phrase from the alphabetized list of chapter key terms.

cash cow
control
differential advantage
diversification
dog
environmental scanning
evaluation
four P's
implementation
market attractiveness/company strength matrix
market development
marketing audit
marketing planning
marketing mix
marketing myopia
marketing objective
marketing plan
marketing strategy
market opportunity analysis
market penetration
mission statement
planning
portfolio matrix
problem child (question mark)
star
strategic business unit (SBU)
strategic planning
strategic window
sustainable competitive advantage
SWOT analysis

1 Understand the importance of strategic marketing and know a basic outline for a marketing plan

1. The managerial process of creating and maintaining a fit between the organization's objectives and resources and evolving market opportunities is ________________________________.

2. The process of anticipating events and determining strategies to achieve organizational objectives is ________________________________, while designing activities relating to marketing objectives and the changing marketing environment is ________________________________.

3. The written document that acts as a guidebook of activities in the areas of product, place, promotion, and price is known as the ______________________________.

2 Define an appropriate business mission statement

4. The foundation of any marketing plan is the answer to the question, "What business are we in and where are we going?" The answer is the firm's ______________________________.

5. Defining a business in terms of goods and services rather than in terms of the benefits customers seek is called ______________________________, meaning narrow, short-term thinking.

6. After defining its mission, an organization may need to divide its firm into subgroups that have their own planning and operations. One of these sections would be called a(n) ______________________________.

3 Know the criteria for stating good marketing objectives

7. A statement of what is to be accomplished with marketing activities is a(n) ______________________________.

4 Explain the components of a situation analysis

8. A situation analysis that allows the company to determine its present status, its current capabilities, and its future expectations is the ______________________________.

9. The identification of market opportunities and threats to provide guidelines for the design of marketing strategy is known as ______________________________. Forces identified with this process include social, demographic, economic, technological, political/legal, and competitive forces.

10. Another technique for examining opportunities is to find the limited time during which the fit between key market requirements and the firm's competencies are optimized. This time is called a(n) ______________________________.

5 Identify strategic alternatives and describe tools used to help select alternatives

11. Firms can conceptualize strategic alternatives with the strategic opportunity matrix, which describes four possible growth opportunities. If a firm tries to increase market share through present products marketed to existing customers, it is following the ______________________________ alternative. If the firm is attracting new customers to its existing products, ______________________________ is taking place. If the firm creates new products for present markets, it is engaging in ______________________________. Finally, if the firm increases sales by introducing new products into new markets, the firm is opting for ______________________________.

12. To determine SBU cash contributions and requirements, managers can use a framework that classifies each SBU by growth and market share. This is the ______________________________, which describes four categories

according to growth and market share dominance. The category that describes a fast-growing market leader is a(n) ____________________. The category that would describe high market share SBUs that are in low-growth markets is the ____________________. SBUs that exhibit rapid growth but poor profit margins are ____________________. Finally, an SBU with a low-growth potential and a small market share would be called a(n) ____________________. A modified version of this framework was developed by GE, and is known as the ____________________.

13. One or more unique aspects of an organization that cause target consumers to patronize that firm instead of a competitor is a(n) ____________________. To make sure this unique aspect can be held on a long-term basis, the firm should make sure it is a(n) ____________________.

6 Discuss target market strategies

14. The activities of selecting and describing one or more target markets and developing and maintaining a marketing mix to satisfy these markets is known as a(n) ____________________.

15. When marketers want to estimate the size and sales potential of market segments, as well as assess key competitors in these market segments, they conduct a(n) ____________________.

7 Describe elements of the marketing mix

16. The unique blend of product, distribution, promotion, and pricing strategies designed to produce mutually satisfying exchanges with a target market is the ____________________. These elements are also known as the ____________________.

8 Understand why implementation, evaluation, and control of the marketing plan is necessary

17. The phase of the marketing process in which marketers gain the organizational compliance required to put marketing strategies into action is termed ____________________. The phase in which marketers gauge the extent to which objectives have been achieved during a specified time period is the process of ____________________. Finally, providing mechanisms for correcting actions is called ____________________.

18. A thorough, systematic, periodic evaluation of the goals, strategies, structure, and performance of the marketing organization is a(n) ____________________.

Check your answers to these questions before proceeding to the next section.

TRUE/FALSE QUESTIONS

Mark the statement **T** if it is true and **F** if it is false.

1 Understand the importance of strategic marketing and know a basic outline for a marketing plan

_____ 1. The owner of the Ace Auto Store is considering the permanent price reduction of all inventory to position the store as the lowest-priced auto parts store in town. This is an example of strategic decision making.

_____ 2. The first step in developing a marketing plan is the creating a SWOT analysis.

2 Define an appropriate business mission statement

_____ 3. Mike's Motos manufactures and sells mopeds, scooters, and other small motorcycles. Management does not define the business in terms of offered products; instead, the business is defined as "serving transportation needs for students." Mike's Motos suffers from marketing myopia because it has ignored the fact that students can get transportation from cars, buses, and bicycles.

3 Know the criteria for stating good marketing objectives

_____ 4. Pets Market has the marketing objective of being the best retailer of pet food and supplies in the country. This would be considered a useful objective.

4 Explain the components of a situation analysis

_____ 5. An example of a "threat" which a tobacco manufacturer could use in its SWOT analysis could include impending legislation that would lessen the amount of advertising tobacco companies can do.

_____ 6. You have been hired to collect and analyze information about factors that may affect your new company. You are also responsible for identifying market opportunities and threats. Your job can be summarized as "environmental scanning."

_____ 7. The Futuro firm is continually developing new products to meet the needs of existing and new markets. This firm is a defender.

5 Identify strategic alternatives and describe tools used to help select alternatives

_____ 8. Twinky Tins has been producing the same assortment of cookie tins for the life of the company. The company has decided that its products, targeted at bakeries, would also fulfill the needs of household consumers. Pursuing this new market would be an example of market development.

_____ 9. Finkle's Fishing Rods has just developed a new line of tackle boxes that further meet the needs of Finkle's fishing customers. This is an example of product development.

6 Discuss target market strategies

_____ 10. Your school of business is engaged in describing and estimating the size and sales potential of market segments (such as traditional students, executives, and the local community). In addition, the school is assessing major competitors (such as other colleges and educational programs). Your school is conducting a market opportunity analysis.

7 Describe elements of the marketing mix

_____ 11. A small, independent motion picture studio decides to use theaters to advertise a new artistic movie release. This is an example of using "place" in the marketing mix.

8 Understand why implementation, evaluation, and control of the marketing plan is necessary

_____ 12. The control process actually begins while planning is taking place.

_____ 13. A marketing audit is a control device used primarily by large corporations to study past performance.

Check your answers to these questions before proceeding to the next section.

AGREE/DISAGREE QUESTIONS

For the following statements, indicate reasons why you may agree and disagree with the statement.

1. Every organization—profit or non-profit—can use the same basic outline for a marketing plan.

   Reason(s) to agree:

   Reason(s) to disagree:

2. Since the business environment is ever-changing, it is a waste of time to do a situation analysis.

   Reason(s) to agree:

   Reason(s) to disagree:

3. The tools used to help select strategic alternatives—such as the portfolio matrix—are only meant for large firms.

   Reason(s) to agree:

   Reason(s) to disagree:

4. Implementation is secondary; strategy is more important.

Reason(s) to agree:

Reason(s) to disagree:

MULTIPLE-CHOICE QUESTIONS

Select the response that best answers the question, and write the corresponding letter in the space provided.

1 Understand the importance of strategic marketing and know a basic outline for a marketing plan

____ 1. If Plan-O-Co is proceeding through the decision-making phases involved in the strategic planning process, which of the following activities is NOT being performed?
- a. defining the business mission
- b. conducting a situation analysis
- c. developing strategic alternatives
- d. designing a marketing information system to solve specific problems
- e. establishing strategic business units

____ 2. Production Aid Consultants, Inc., has recently been engaged in several special meetings where issues such as business mission, situation analysis, market and growth alternatives, and implementation approaches have been discussed. Production Aid is apparently engaged in:
- a. target market planning
- b. writing the mission statement
- c. the strategic planning processes
- d. business analysis
- e. strategic contingency planning

2 Define an appropriate business mission statement

____ 3. Frito-Lay defines its business as "snack-food" rather than just "corn chips." This is an example of:
- a. a marketing mix strategy
- b. a mission statement
- c. quantifiable goals
- d. a financial statement
- e. organizational accomplishment

____ 4. Sharon Mauser, the senior vice president of Progressive Products, Ltd. is in the process of developing Progressive's organizational mission statement. Which of the following factors should she consider when developing such a mission statement?
- a. What benefits do present customer need and want
- b. What benefits will potential customers need and want
- c. What existing environmental conditions will influence the choices made in the future by Progressive Products, Inc.
- d. What anticipated environmental conditions will influence the choices made in the future by Progressive Products, Inc.
- e. All of the above should be considered by Ms. Mauser

____ 5. By defining its business as "running department stores" instead of "providing a range of products and services that deliver value to U.S. families," Sears Roebuck was engaging in:
a. market segmentation
b. marketing myopia
c. market development
d. strategic planning
e. product differentiation

____ 6. A high tech company has the business mission of "providing high quality products at a fair price to customers." This mission statement
a. is an example of marketing myopia.
b. is too broad of a statement to be of use in serving customers.
c. could stifle creativity in discovering opportunities to serve customers.
d. is not sincere enough.
e. does not meet customer needs.

____ 7. AT&T sells many products and services in addition to telephones. Some of these include computers, modems, fax machines, and a variety of business and home long-distance telephone services. How would this be best justified in the strategic planning process?
a. New business units are needed to continue growth.
b. Local delivery has high market attractiveness.
c. Diversification is needed to survive.
d. This mission statement recognizes the firm as a communications company, not just a telephone manufacturer.
e. It is more efficient to produce a wide variety of products.

____ 8. Pepsico is a large conglomerate that has separate subsidiaries called Pepsi-Cola (soft drinks), Tropicana (juices), Pepsi Bottling, and Frito-Lay (snack foods). Each of these subsidiaries has its own functional departments, its own planning, its own financial goals, and its own target markets. These subsidiaries may also be called
a. product market niches
b. diversified divisions
c. strategic business units
d. strategic alliances
e. heterogeneous elements

3 Know the criteria for stating good marketing objectives

____ 9. As the marketing director for a new pharmaceutical that prevents baldness, you have been asked by the CEO to provide marketing objectives for the next year. Your product is the second one of its type in the market. Which of the following is the most appropriate objective?
a. To generate sales of $20 million during the first year of the launch.
b. To attain a market share of 100 percent in the first year.
c. To significantly increase the company's sales.
d. To be number one in the market for baldness-prevention products.
e. To be known as the best product in the baldness-prevention market.

4 Explain the components of a situation analysis

____ 10. Gabble's Granola has set up a committee to formally study its current status and capabilities and its future expectations. Gabble's Granola is conducting a(n):
a. marketing audit
b. profit and loss statement
c. environmental scan
d. situation analysis
e. strategic window search

____ 11. When Sam joined the Dale Corporation, he noticed some interesting characteristics about the company. Instead of being encouraged to take risks, employees were chastised for making mistakes. The marketing department was not used to taking the leadership in pricing; it simply reacted to what competition was doing. This scenario illustrates the company's:
a. marketing plans
b. strategic alternatives
c. business mission
d. corporate culture
e. response functions

____ 12. Though a small company in the industry, Software, Inc., constantly monitors market needs and develops new software products that meet those needs. The company spends much money on market research and on product development to maintain its reputation for innovation. This company is a(n):
a. reactor
b. analyzer
c. defender
d. offender
e. prospector

5 Identify strategic alternatives and describe tools used to help select alternatives

____ 13. When Disney started targeting its theme parks toward adults rather than just children, it was selecting which strategic alternative?
a. product development
b. market development
c. market penetration
d. product penetration
e. diversification

____ 14. Starbucks, the giant gourmet coffee retailer, started selling its own branded ice cream to its current coffee customers. This is an example of a strategy called:
a. market development
b. market penetration
c. product penetration
d. product development
e. diversification

____ 15. Licoh, a manufacturer of printing equipment, recently bought a local baseball team to increase its overall sales. Licoh is following the strategy of:
a. market penetration
b. product development
c. product penetration
d. market development
e. diversification

____ 16. The Formula Foods Corporation has used a system to classify its various subsidiaries or divisions to determine the future expected cash contributions and the future expected cash requirements for each division. The divisions have been designated as cash cows, dogs, stars, and problem children. Formula Foods is using:
a. the portfolio matrix
b. market share analysis
c. response functions
d. the market attractiveness matrix
e. the company strength matrix

____ 17. The baking products division of Basic Foods, Inc., is the market leader in a mature and low-growth market. The baking products division generates more dollars than is required in order to maintain market share, and in portfolio matrix terms it is known as Basic Food's:
a. green light
b. cash cow
c. star
d. dog
e. spotlight

____ 18. In its portfolio of business subsidiaries, Tupple Toys has three units that are showing rapid growth but poor profit margins. Apparently, everything about these units demands more and more additional cash. In the Boston Consulting Group's portfolio matrix, these units would be classified as:
a. dogs
b. cash cows
c. problem children
d. stars
e. leeches

____ 19. A local private college has been offering an accounting program for several years. The program has a large but declining enrollment, and the program represents the largest income earner for the business division. The recommended strategic option is to:
a. cultivate
b. divest
c. hold
d. harvest
e. build

____ 20. The Kandy Korner has a moderate market share with little growth opportunity in a price-competitive market. Its financial resources are adequate, and it is returning a medium-level profit. In the market attractiveness/company strength matrix approach, the business should:
a. be maintained unless attractiveness begins to slip
b. bypass the largest competitors
c. be divested
d. require that market attractiveness be expressed quantitatively
e. make a large investment in the business

____ 21. Chipsters, Inc., has attracted and employed some of the most creative and experienced computer chip designers in the industry. This has led Chipsters to a position of great strength, attributed to:
a. company strengths
b. superior skills
c. superior resources
d. strategic windows
e. differential advantages

6 Discuss target market strategies

____ 22. A power tool company has done consumer research and describes its customers as being male, 28-55 years of age, married, 62 percent high school graduates, and earners of below-average income. This is the group of consumers most likely to buy the tools and is referred to as the:
a. social responsibility group
b. exchangers
c. target market
d. advertisees
e. scanned environment members

7 Describe elements of the marketing mix

____ 23. John Porter is the new vice president of marketing and is designing the marketing mix for his company. The starting point of Mr. Porter's marketing mix will be the:
a. analysis of what production equipment is available and owned by the company
b. design of the promotion campaign to be used for the product
c. selection of the places through which the good or service will be sold
d. development of the good or service to be sold
e. determination of the price to be charged for the good or service, enabling future revenues and budgets to be estimated

____ 24. Barak Austin is thinking about opening a new hardware and home improvement store in a community that research has shown needs more outlets. On which area of the marketing mix should he focus his attention?
a. customer needs/wants
b. production
c. product
d. promotion
e. distribution

____ 25. Diana's job is to decide whether her company's advertising money will be spent on television, radio, newspaper, or direct mail. To which of the four P's do Diana's duties relate?
a. publicity
b. price
c. promotion
d. place
e. product

____ 26. Albert must make a quick change to his marketing plan to boost sales, so he should change the ____________ element of the marketing mix?
a. product
b. place
c. price
d. promotion
e. publicity

8 Understand why implementation, evaluation, and control of the marketing plan is necessary

____ 27. Becky's Bolt Supply has set up a management committee that is arranging the delegation of authority and responsibility for marketing strategies, determining a time frame for completion of tasks, and overseeing resource allocation. The task of this committee is:
a. market planning
b. strategic analysis
c. alternative control
d. implementation
e. strategic planning

____ 28. The managers at Unicorn Research realize that implemented plans do not always lead to the desired results. The managers feel they should develop a mechanism for correcting actions that are not efficiently aiding the firm in reaching its objectives. Unicorn's concerns center around:
a. control
b. implementation
c. objectives
d. planning
e. budgeting

_____ 29. A planning manager from corporate headquarters finds that his eastern region has no effective method of allocating resources or evaluating goals and performance of the marketing organization. He suggests that the region should prepare a:
a. service audit
b. tactical plan
c. marketing audit
d. market share analysis
e. response function

9 Identify several techniques that help make strategic planning effective

_____ 30. Effective strategic planning requires:
a. a General Electric planning matrix
b. management to stay out of the process
c. a designated single time frame during the year
d. the planner to challenge existing company assumptions
e. a stringent and narrow mission statement for guidelines

_____ 31. Ethan, the owner of a hobby store, is concerned that his employees are not seriously embracing the marketing concept. You are his friend and a marketing expert. You would suggest all of the following ideas EXCEPT:
a. a clear declaration of intent and support for the new policies by Ethan
b. an increased authority for individual employees to make customer-related decisions as problems occur
c. getting everyone who will be affected by changes in policies and procedures involved in the planning process
d. slowly implementing new policies that encourage a focus on the customer
e. concentrating on creating a specific marketing department focused on this problem

Check your answers to these questions before proceeding to the next section.

ESSAY QUESTIONS

1. What are the elements of the marketing plan?

2. What is a marketing objective? State four criteria that marketing objectives should have, and write a good example of an objective that has these criteria.

3. What is a target market? List three general strategies for selecting target markets and give an example for each one.

4. The Boston Consulting Group (BCG) portfolio matrix is a technique that managers can use to classify SBUs. Using the following matrix, label the horizontal and vertical axes with the SBU classifications used by the BCG and label each quadrant. Then describe each quadrant and include the basic tactic followed in each category.

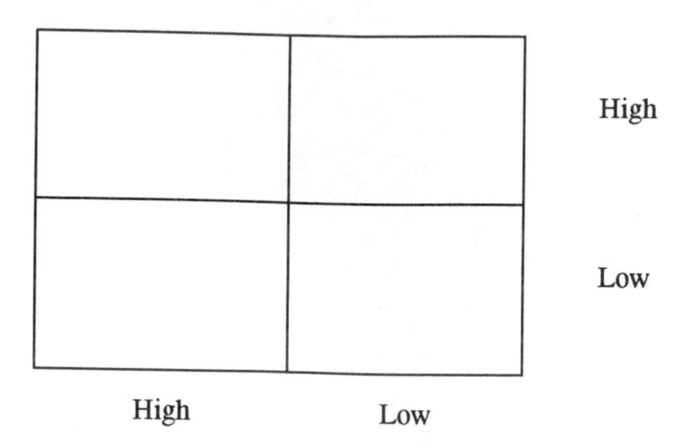

5. The marketing mix refers to a unique blend of marketing variables known as the four P's. Using a new breakfast cereal as an example, name and briefly describe each of the four P's.

CHAPTER NOTES

Use this space to record notes on the topics you are having the most trouble understanding.

ANSWERS TO THE END-OF-CHAPTER DISCUSSION AND WRITING QUESTIONS

Use this space to work on the questions at the end of the chapter.

ANSWERS TO THE END-OF-CHAPTER CASE

Use this space to work on the questions at the end of the chapter.

CHAPTER 3 The Marketing Environment and Marketing Ethics

LEARNING OBJECTIVES

1 Discuss the external environment of marketing and explain how it affects a firm

The external environmental variables of marketing are the uncontrollable factors that continually mold and reshape the target market. The firm and its customers are affected by social, demographic, economic, technological, political and legal, and competitive factors. The monitoring and evaluating of these environments is a task called environmental scanning.

2 Describe the social factors that affect marketing

The social factors, which include the values and lifestyles of a population, strongly influence the attitudes and desires of target markets. These are perhaps the most difficult variables for marketers to forecast or influence.

3 Explain the importance to marketing managers of current demographic trends

Demographic trends are trends in the vital statistics of a population such as birth rates, death rates, age group totals, and locations. These trends are important for marketers as they attempt to forecast demand and changes in the composition of their markets.

4 Explain the importance to marketing managers of multiculturalism and growing ethnic markets

Multiculturalism occurs when all major ethnic groups in an area are roughly equally represented. Growing multiculturalism makes the marketers tasks more challenging. Niches within ethnic markets may require micro marketing strategies. An alternative to a niche strategy is maintaining a core brand identity while straddling different languages, cultures, ages and incomes with different promotional campaigns. A third strategy is to seek common interests, motivations or needs across ethnic groups.

5 Identify consumer and marketer reactions to the state of the economy

Economic factors in the environment affect consumer demand. Marketers follow trends in consumer income to forecast what type of products will be demanded. Inflation, a period of rising prices, tends to make consumers less brand loyal and more likely to stock up on sale and coupon items.

6 Identify the impact of technology on a firm

Changes in technological and resource factors can have a momentous effect on an entire industry. The use of a new technology can assist a firm in coping with the detrimental effects of many of the other environmental factors. Not taking advantage of a new technology may cause customers to move to more innovative products and firms.

7 Discuss the political and legal environment of marketing

Federal and state legislatures establish many rules for operating a business. These regulations affect every aspect of the marketing mix. The two most directly involved federal agencies are the Consumer Product Safety Commission (prescribes safety standards for consumer goods) and the Federal Trade Commission (prevents unfair methods of competition).

8 Explain the basics of foreign and domestic competition

The competitive environment refers to the number of competitors a firm must face, the relative size of the competitors, and the degree of interdependence within the industry. The nature of a firm's competition is beyond the control of management, yet the marketing mix, particularly pricing, is clearly affected by the type and amount of competition facing the firm. International competition is a growing concern for most businesses in the United States for two reasons: numerous U.S. firms must consider going international as a growth strategy, and more and more foreign firms, entering the United States to do business, compete on quality as well as price.

9 Describe the nature of ethical decision making

Business ethics may be viewed as a subset of the values of society as a whole. As members of society, marketing managers are morally obligated to consider the ethical implications of their decisions. Many companies develop codes of ethics as a guideline for their employees.

10 Discuss corporate social responsibility

Social responsibility in business refers to a firm's concern for the way its decisions affect society. Some people argue that a business has an obligation to society because of the social costs of economic development. Another argument is that it is in the firm's best interest to better the environment in which it operates, and that firms can avoid restrictive regulations by responding to societal needs. Finally, some would say that a firm has the assets and resources to help solve social problems and is therefore morally obligated to do so.

In contrast, critics argue that businesses do not have the ability to prioritize social needs, are interfering with government responsibilities when they respond to societal needs, and should not be distracted from the obligation to earn profits for shareholders.

PRETEST

Answer the following questions to see how well you understand the material. Re-take it after you review to check yourself.

1. What is a target market?

2. Give three examples of social factors that affect marketing.

3. What is demography?

4. Name four generations that are often used as target markets. Describe key characteristics of each generation.

5. List and briefly describe at least five types of legislation that impact marketing.

CHAPTER OUTLINE

1 Discuss the external environment of marketing and explain how it affects a firm

I. The External Marketing Environment

A. The defined group of consumers that managers feel is most likely to buy a firm's product is the **target market**, and it is this group for which the marketing mix is designed.

The external environment continually molds and reshapes the target market. Purchasers are influenced by the marketing mix: product, place, promotion, and price, all of which are controlled by the marketing manager. Uncontrollable external elements also influence purchasers.

External environmental factors are:

1. Social
2. Demographic
3. Economic
4. Technological
5. Political and legal
6. Competitive

B. Understanding the External Environment

A critical task for the firm is the process of **environmental scanning**, the collection and interpretation of information about forces, events, and relationships that may affect the future of an organization. An organization's environmental scanning is performed by a group of specialists who collect and evaluate environmental data on a continuing basis.

C. Environmental Management

Although marketing managers consider the external environment to be uncontrollable, they may be able to influence certain aspects. A company may attempt to influence evolving external factors with **environmental management**, the implementation of strategies that attempt to shape the external environment within which a firm operates.

2 Describe the social factors that affect marketing

II. Social Factors

A. Social factors include the attitudes, values and lifestyles of a population. These are perhaps the most difficult external variables for marketing managers to forecast, influence, or integrate into marketing plans.

B. Marketing-oriented values of the late 1990s:

1. Nearly one-fourth of American adults are interested in new kinds of products and services. The cultural creative scan, an information source efficiently, seizes what they are interested in, and explores the topic in depth.

2. Other values evident in the late 1990's include an emphasis on:

   a. Environmentally safe and responsible products

   b. Nonmaterial accomplishments

c. Devices that save time and tools to aid in relaxation and sleep, due to a **poverty of time**

C. The Growth of Component Lifestyles

Another trend in the United States is component lifestyles. Rather than conforming to traditional stereotypes, which are often associated with one's job, Americans are choosing products to meet a variety of needs and interests.

D. The Changing Role of Families and Working Women

One of the changes with the most dramatic impact on consumers has been the increase in working women and dual-career families.

1. Approximately 58 percent of all females between 16 and 65 are in the work force.

2. Two-career families have greater household incomes but less available time for family activities. Their purchase roles are changing, as well as their purchase patterns. Women are the principle buyers for 45 percent of all cars and trucks sold in the U.S.

3 Explain the importance to marketing managers of current demographic trends

III. Demographic Factors

A. **Demography** is the study of a population's vital statistics such as ages, births, deaths, and locations. Demographic factors are an extremely important indicator of composition and change in a market.

Demographic characteristics are strongly related to consumer buyer behavior in the marketplace and are good predictors of how the target market will respond to a specific marketing mix.

B. Generation Y: Born to Shop

The "Y" generation is composed of about 58 million Americans under the age of 16. This has had a great impact on companies selling products from toys to Club Med family vacations. Environmental changes affecting Generation Y include:

1. Nearly 60% of children under age 6 have working mothers.

2. 61% of children aged three to five are attending preschool.

3. Nearly 60% of households with children aged 7 or younger have personal computers.

4. More than one-third of elementary school students nationwide are Black or Hispanic.

5. In recent years approximately 15% of U.S. births were to foreign-born mothers.

6. Nearly one in three births in the early 1990's was to an unmarried woman.

7. One-quarter of children under age 6 are living in poverty.

C. Generation X: Savvy and Cynical

1. In 1998, approximately 48 million consumers were between the ages of 17 and 29, the group labeled **generation X**.

2. This group of consumers is willing to spend on movies, eating out, clothing, and electronic items, but they are critical and suspicious of most marketers' attempts to target them.

D. Baby Boomers: America's Mass Market

One of the most important demographic changes is the increase of middle-aged consumers as the baby boomers (born 1946-1964) move into their 40s and 50s.

1. The importance of individualism among baby boomers has led the United States to become a personalized economy that delivers products at a good value on demand.
2. Successful products in a personalized economy share three characteristics: custom-designed for small target markets, immediacy of delivery, and value.

3. This group focuses on family, health, convenience, finances, and reading materials.

E. Older Consumers: Not Just Grandparents

The 50-plus consumers are healthier, wealthier, and better educated that earlier generations. This group represents a tremendous marketing opportunity.

This group is only 26 percent of the population, but it buys

1. Half of all domestic cars
2. Half of all silverware
3. Nearly half of all home remodeling
4. 80% of all travel dollars

F. On the Move

The average U.S. citizen moves every six years. The greatest gains in population during the past decade have been in the Southeast and the Far West.

4 Explain the importance to marketing managers of multiculturalism and growing ethnic markets

IV. Growing Ethnic Markets

The United States is shifting from a society dominated by whites and rooted in Western culture to a society blended with three large racial and ethnic minorities: African-Americans, U.S. Hispanics, and Asian-Americans.

A. Ethnic and Cultural Diversity

Multiculturalism occurs when all major ethnic groups in an area are roughly equally represented.

1. Growing multiculturalism makes the marketers tasks more challenging. Niches within ethnic markets may require micro marketing strategies.

2. An alternative to a niche strategy is maintaining a core brand identity while straddling different languages, cultures, ages and incomes.

3. A third strategy is to seek common interests, motivations, or needs across ethnic groups. This is sometimes called **stitching niches**, combining ethnic, age, income, and lifestyle markets to form a larger market.

5 Identify consumer and marketer reactions to the state of the economy

V. Economic Factors

A. Rising Incomes

1. Two-thirds of all American households earn a "middle-class" income, between $18,000 and $75,000. In 1998, almost half the households were in the upper end of that income range, as opposed to a quarter in that range in 1980.

2. Today over 8 percent of households earn over $75,000, a large increase over 1980 when only 2.6 percent of the households earned over this amount. The trend toward dual-income families contributes heavily to this phenomenon.

3. The rise in consumer income provides more **discretionary income** (income after taxes and necessities) for higher-quality, higher-priced goods and services.

B. Inflation

Inflation is a general rise in prices resulting in decreased purchasing power. Inflation generally causes consumers to do two things:

1. Decrease their brand loyalty, as they search for the lowest prices

2. Take advantage of coupons and sales to stock up on items

C. Recession

A **recession** is a period of economic activity when income, production, and employment tend to fall--all of which reduce demand. The effects of reduced demand can be countered by

1. Improving existing products and introducing new ones.

2. Maintain and expand customer services.

3. Emphasize top-of-the-line products and promote product value.

6 Identify the impact of technology on a firm

VI. Technology and Resource Factors

Changes in technological and resource factors can have a momentous effect on an entire industry.

New technology can assist a firm in coping with many of the other environmental factors. For example, new processes can reduce production costs and help a firm fight inflation and recession.

A. The United States is not creating new technology as fast as it has in the past. The U.S. share of new patents, research and development funding, and the focus of R&D expenditures (defense versus nondefense spending) have fallen behind some other countries.

1. The federal government spends about $75 billion a year on R&D, private industry spends another $120 billion.

2. **Basic** or **pure research** attempts to expand the frontiers of knowledge. U.S. companies seem to have trouble transforming the results of pure research into goods and services.

3. **Applied research** attempts to develop new or improved products.

B. Many U.S. companies have fallen into the habit of seeking short-term profits and taking minimal risk. This approach leads to the development of line extensions (slight variations on existing products) rather than real innovations.

C. Several political and cultural alterations could encourage innovation, such as tax incentives, changes in organizational structure, and company encouragement of risk taking.

7 Discuss the political and legal environment of marketing

VII. Political and Legal Factors

Government establishes many operating rules for businesses, and legal rules and restrictions affect every aspect of the marketing mix. Business needs government regulations to protect the interests of society, protect businesses from each other, and protect consumers. Government needs business to generate revenue.

A. Federal legislation has made an impact on marketing in several categories:

1. The **Sherman Act**, the **Clayton Act**, the **Federal Trade Commission Act**, the **Celler-Kefauver Antimerger Act**, and the **Hart-Scott-Rodino Act** were passed to regulate the competitive environment.

2. The **Robinson-Patman Act** was designed to regulate pricing practices.

3. The **Wheeler-Lea Act** was created to control false advertising.

B. State laws also supervise business practices, and many states have additional regulations affecting marketers directly.

C. The three regulatory agencies most directly involved in marketing affairs are the **Consumer Product Safety Commission**, the **Federal Trade Commission**, and the **Food and Drug Administration**.

1. The Consumer Product Safety Commission (CPSC) has the power to prescribe mandatory safety standards for almost all products that consumers use. It can ban products from the marketplace, fine offending firms, and sentence their officers to up to a year in prison.
2. The Federal Trade Commission (FTC) is empowered to prevent persons or corporations from using unfair methods of competition in commerce and has a vast array of regulatory powers.

3. The powers of the FTC include

a. Cease-and-desist order: A final order is issued to cease an illegal practice. These orders are often challenged in the courts.

b. Consent decree: A business consents to stop the questionable practice without admitting illegality.

c. Affirmative disclosure: An advertiser is required to provide additional information about products in its advertisements.

d. Corrective advertising: New advertising is ordered to correct the past effects of misleading advertising.

e. Restitution: The company must provide refunds to consumers misled by deceptive advertising (for practices carried out after the issuance of a cease-and-desist order).

f. Counteradvertising: The FTC has proposed that the Federal Communications Commission permit advertisements in broadcast media to counteract advertising claims (free under certain conditions).

4. The FTC's powers were slightly weakened by the FTC Improvement Act of 1980, which removed the use of "unfairness" as a standard for industrywide rules against advertising and mandated congressional hearings on the FTC's decisions every six months.

5. The Food and Drug Administration (FDA) is charged with enforcing regulations against selling and distributing adulterated, misbranded, or hazardous food and drug products.

8 Explain the basics of foreign and domestic competition

VIII. Competitive Factors

The competitive environment refers to the number of competitors a firm must face, the relative size of the competitors, and the degree of interdependence within the industry. The nature of a firm's competition is beyond the control of management, yet the marketing mix, particularly pricing, depends on the type and amount of competition facing the firm.

A. Competition for Market Share

1. In the past, foreign firms entered U.S. markets by emphasizing price, but today the accent is on quality.

2. U.S. companies often battle one another in global markets just as intensively as they battle in the domestic market.

Regardless of the form of the competitive market, as population growth slows, costs rise, and available resources tighten, firms find they must work harder to maintain their profits and market share.

B. Global Competition

Global competition is a growing facet of U.S. business. Not only do U.S. firms consider going international as a growth strategy, but more and more foreign firms are entering the United States to do business.

9 Describe the nature of ethical decision making

IX. Ethical Behavior in Business

Ethics refers to moral principles or values that generally govern the conduct of an individual or a group.

Morals are rules or habits, typically stated as good or bad, that people develop as a result of cultural values and norms.

Business ethics are actually a subset of the values held by society as a whole. These values are acquired through family, and through educational and religious institutions.

Ethics can be very situation-specific and time-oriented.

Approaches for developing a personal set of ethics includes

- Examining the consequences of a particular act

- Stressing the importance of rules
- Developing a moral character

A. Morality and Business Ethics

There are three levels of morality:

1. **Preconventional morality** is childlike in nature, calculating, self-centered, and selfish.
2. **Conventional morality** is concerned with the expectations of society. Loyalty and obedience are paramount.
3. **Postconventional morality** represents the morality of the mature adult. People at this level are concerned with how they see and judge themselves and their acts over the long run.

B. Ethical Decision Making

The following factors tend to influence ethical decision making and judgments:

- The extent of ethical problems within the organization
- Top-management actions on ethics
- Potential magnitude of the consequences
- Social consensus within managerial peers
- Probability of a harmful outcome
- Length of time between the decision and the onset of consequences
- Number of people to be affected

C. Developing Ethical Guidelines

1. Many firms have developed a specific **code of ethics** to help employees make better decisions. The American Marketing Association has such a code.

   An explicit code helps

   a. Identify acceptable business practices

   b. Internally control behavior

   c. Reduce employee confusion in decision making

   d. Facilitate discussion about right and wrong in issues that may arise

2. Although many companies have issued policies on ethical behavior, marketing managers must put them into effect. These managers must address the "matter of degree" issue in many situations.

10 Discuss corporate social responsibility

X. Corporate Social Responsibility

A. **Corporate social responsibility** is business's concern for society's welfare. Specifically, this concern is demonstrated by managers who consider the long-range best interests of the company and the company's relationship to the society within which it operates.

B. Despite arguments to the contrary, most large corporations feel that their social responsibility extends beyond simply earning profits.

1. Today a firm must also develop environmental controls, provide equal employment opportunities, create a safe workplace, produce safe products, and do much more.

2. Four types of social responsibility make up the **pyramid of corporate social responsibility**:

   a. Economic
   b. Legal
   c. Ethical
   d. Philanthropic

Multinational companies also have important social responsibilities and in some cases can be a dynamic force for social change in host countries.

VOCABULARY PRACTICE

Fill in the blank(s) with the appropriate term or phrase from the alphabetized list of chapter key terms.

applied research
baby boomers
basic research
code of ethics
component lifestyle
Consumer Product Safety Commission (CPSC)
corporate social responsibility
demography
discretionary income
environmental management
ethics
Federal Trade Commission (FTC)
Food and Drug Administration (FDA)
Generation X
inflation
morals
multiculturalism
personalized economy
poverty of time
pyramid of corporate social responsibility
recession
stitching niches
target market

1 Discuss the external environment of marketing and explain how it affects a firm

1. The marketing mix is designed by marketing managers to appeal to a specific group of potential buyers, known as the ______________________________. This group is most likely to buy the firm's product.

2. When a team of specialists is created to collect and evaluate environmental information, the process is called environmental scanning. After this process, a company may decide to implement strategies that attempt to shape the external environment within which it operates. This is known as ______________________________.

2 Describe the social factors that affect marketing

3. There has been a shift away from traditional values in the United States. Part of this is due to dual career families who can do nothing but work, eat, and sleep. These people suffer from ______________________________. In addition, many people choose goods and services to meet diverse needs and interests rather than conforming to a single way of life. This is known as piecing together a ______________________________.

3 Explain the importance to marketing managers of current demographic trends

4. The study of people's vital statistics such as their ages, births, deaths, and locations is known as ______________________________. One item studies might be the amount of money people have beyond necessities and taxes, known as ______________________________.

5. The group of people born between 1966 and 1977 are termed ______________________________. The group of people born between 1946 and 1964, known as ______________________________, value home life and convenience. This latter group insists that marketers deliver goods and services at a good value upon demand. This has created a ______________________________.

4 Explain the importance to marketing managers of multiculturalism and growing ethnic markets

6. When all ethnic groups in an area are roughly equally represented, such as in a city, county, or census tract, ______________________________ occurs. This has not occurred equally in the United States.

7. One strategy for targeting multicultural markets is termed ______________________________, which involves combining ethnic, age, income, and lifestyle markets on some common basis to form a mass market.

5 Identify consumer and marketer reactions to the state of the economy

8. Economic factors can affect the amount of discretionary or disposable income available to purchase goods and services. A general rise in prices, called ______________________________, results in decreased purchasing power. A period of economic activity when income, production, and employment all tend to fall is called a(n) ______________________________ and is characterized by a reduction in demand.

6 Identify the impact of technology on a firm

9. The United States often engages in the type of research and development that attempts to expand the frontiers of knowledge. This is known as pure or ______________________________, which does not emphasize commercial viability. In contrast, ______________________________ attempts to specifically develop new or improved products.

7 Discuss the political and legal environment of marketing

10. There are three federal agencies most directly involved in marketing affairs. The purpose of the ______________________________ is to protect the health and safety of consumers in an around their homes. The ______________________________ is empowered to prevent unfair methods of competition. Finally, the ______________________________ enforces regulations against selling and distributing harmful food and drug products.

9 Describe the nature of ethical decision-making

11. Moral principles or values that generally govern the conduct of an individual or a group refers to ________________________________. Rules or habits typically stated as good or bad are ________________________________.

12. Many firms have developed an explicit set of guidelines to help employees make better decisions. This set of guidelines is a(n) ________________________________.

10 Discuss corporate social responsibility

13. The concern of a business for the long-range best interests of the company and its relationships to the society within which it operates is termed ________________________________.

14. One theory defines four kinds of social responsibilities: economic, legal, ethical, and philanthropic. This theory is known as the ________________________________.

Check your answers to these questions before proceeding to the next section.

TRUE/FALSE QUESTIONS

Mark the statement **T** if it is true and **F** if it is false.

1 Discuss the external environment of marketing and explain how it affects a firm

_____ 1. A firm's target market is the geographic location where the firm intends to focus its marketing efforts.

_____ 2. To be successful, marketers should control the elements in the external environment and the four P's of the marketing mix.

2 Describe the social factors that affect marketing

_____ 3. Most Americans hold traditionalist values; that is, they hold conservative political and social views, have traditional religious beliefs, and believe in traditional division of labor between men and women.

3 Explain the importance to marketing managers of current demographic trends

_____ 4. Demography is the study of people's vital statistics, such as their ages and locations.

_____ 5. The generation that followed the Baby Boomers is called Generation Y.

_____ 6. People born between 1946 and 1964 who are married, employed, and have children are called baby boomers.

4 Explain the importance to marketing managers of multiculturalism and growing ethnic markets

_____ 7. A multinational food company has created several cross-cultural frozen food products such as taco pizzas, rice-filled burritos, and Cuban-Asian barbecue sauces. This is an example of "stitching niches."

5 Identify consumer and marketer reactions to the state of the economy

_____ 8. U.S. incomes have continued to decline after several periods of recession and inflation.

7 Discuss the political and legal environment of marketing

_____ 9. Ricky Simons thinks that a competing firm may be engaging in unfair competition. The federal agency that would address this charge would probably be the Federal Trade Commission.

8 Explain the basics of foreign and domestic competition

_____ 10. Marketing managers should be concerned with competition because most marketing firms exist in a purely competitive environment.

9 Describe the nature of ethical decision making

_____ 11. The business ethics we see today shape the values held by society.

10 Discuss corporate social responsibility

_____ 12. Phillips Petroleum contributes its resources to local communities to help preserve nature and wildlife. By being a good corporate citizen, Phillips is fulfilling its philanthropic responsibilities on the pyramid of corporate social responsibility.

_____ 13. Avoidance of existing government regulations is an example of socially responsible marketing.

Check your answers to these questions before proceeding to the next section.

AGREE/DISAGREE QUESTIONS

For the following statements, indicate reasons why you may agree and disagree with the statement.

1. There is really no difference in attitudes between Baby Boomers and Generation X; when Baby Boomers were younger, they had the same attitudes and opinions that Generation Xers have today.

   Reason(s) to agree:

   Reason(s) to disagree:

2. The United States is a "melting pot" of cultures.

   Reason(s) to agree:

Reason(s) to disagree:

3. The most important environmental factors that impact marketing efforts are economic factors.

   Reason(s) to agree:

   Reason(s) to disagree:

4. Most corporations practice conventional morality; that is, they make ethical decisions only because it may translate into more profits in the long-term.

   Reason(s) to agree:

   Reason(s) to disagree:

MULTIPLE-CHOICE QUESTIONS

Select the response that best answers the question, and write the corresponding letter in the space provided.

1 Discuss the external environment of marketing and explain how it affects a firm

_____ 1. Resorts International, Inc., focuses much of its marketing effort on Baby Boomers who have large disposable incomes and young children. The company offers a large variety of services, including the traditional sporting and leisure services as well as creative child care services. Baby Boomers would be considered Resorts International's:
   a. market niche
   b. environment
   c. target market
   d. market segmentation
   e. external environment

_____ 2. The Mondo Moped Company spends much money trying to influence the passage of bills in Congress that would limit the amount of foreign-made mopeds that can be sold in the United States. The Mondo Moped Company is engaging in:
   a. defensive marketing
   b. pure competition
   c. corporate espionage
   d. legislative planning
   e. environmental management

2 Describe the social factors that affect marketing

_____ 3. Tina Mathews is the Director of Marketing Planning for Visionary Electronics Systems. She has been examining the social factors that may affect the way in which Visionary Electronics does business with consumers in the next decade. Which of the following social factors might affect the way her firm does business?
- a. The population is growing older
- b. The income levels will rise in the next decade
- c. More products will be made with high-tech ceramics
- d. Today's shoppers are environmentalists
- e. Fewer babies are being born

_____ 4. The manufacturer of New Rage Clothing has a marketing department that keeps track of current external environmental changes. Although some information is relatively easy to collect, analyzing and forecasting ____________ trends is extremely difficult, yet these trends may be the most important factor for this company.
- a. social
- b. economic
- c. legislative
- d. technological
- e. demographic

_____ 5. The Dispozall Company produces disposable toiletries (razors, combs, curlers, and so on) designed for the busy person who wants simplicity, ease of use, and the convenience of disposable products. Which current environmental trend(s) may pose a major new threat to this company?
- a. social factors that are increasing are an interest in the environment and recycling
- b. technological factors that are automating production
- c. demographic factors that demonstrate a decrease in the birthrate
- d. many large competitors in the same industry
- e. product safety legislation currently under review by the Consumer Product Safety Commission

_____ 6. Ricky is a helicopter pilot for the forestry service. He is also a computer programmer, stock portfolio manager, and breeder of pedigreed sable ferrets. His other interests include reading, woodworking, and jazz. Marketers would describe Ricky's needs for goods and services as:
- a. fitting into a well-educated segment
- b. a component lifestyle
- c. inappropriate in all segments
- d. the typical pilot target segment
- e. a conforming lifestyle

3 Explain the importance to marketing managers of current demographic trends

_____ 7. Georgetta Wilson is a marketer of consumer packaged goods that can be purchased in grocery stores nationwide. Which of the following demographic changes should she be concerned about as impacting the sales of her firm's products?
- a. The population growth rate of less than one percent and a rate of new household formation under two percent is the slowest U.S. growth rate since the Great Depression.
- b. The 1990 census shows that about 23 million people in the United States are single.
- c. Today's fifty-plus consumers are wealthier, healthier, and better educated than earlier generations of older Americans.
- d. All of the above demographic changes could affect the sales of packaged goods produced by Ms. Wilson's firm.
- e. None of the above are true demographic changes that have occurred. Further, even if they have occurred they would have little if any impact on consumer packaged goods sales in general, and the products sold by Ms. Wilson's firm.

_____ 8. This demographic target market was "born to shop." Entire industries, such as children's software, children's versions of adult clothing (such as Baby Gap), and creative educational products have spawned to target this group of people. The target market described above is:
a. Generation X
b. Generation Y
c. grandparents
d. Baby Boomers
e. Generation A

_____ 9. You will be marketing to "Generation X" and know that all of the following facts about this group will affect your marketing plan EXCEPT:
a. this group is materialistic
b. Generation Xers spend lots of money on eating out and electronic items
c. this group is the first generation of latchkey children because of dual career households
d. Generation Xers are the most likely group to be successfully employed
e. members of Generation X do not mind indulging themselves with movies, clothing, and alcoholic beverages

_____ 10. The number one trait that parents of baby boomers wanted to drive into their children was:
a. optimism
b. "me first"
c. conformity
d. to think for themselves
e. materialism

4 Explain the importance to marketing managers of multiculturalism and growing ethnic markets

_____ 11. Which of the following statements about multiculturalism in the United States is false?
a. Today, roughly 75% of the U.S. is composed by whites.
b. The Hispanic population grew faster than any other ethnic group in the 1980s.
c. Asian-Americans are expected to represent about 7% of the total U.S. population in the year 2023.
d. Minority ethnic groups are expected to grow more than the white population in the future.
e. The nation's most ethnically diverse county is San Francisco county.

_____ 12. It is the year 2023, and you are the CEO of a multinational consulting firm. Your client would like to target the largest segment of the United States population, so you advise your client to target:
a. Asian-Americans
b. African-Americans
c. Hispanic-Americans
d. whites
e. minorities

_____ 13. The Cool Clothing Company has just hired you to develop a marketing strategy for their new line of casual wear. The firm wants to target several different ethnic groups, so you suggest a "stitching niches" approach. Which of the following would you use to illustrate your suggestion?
a. create a different product for each of the niche markets the firm is targeting
b. position the products so they simultaneously meet the needs of the different ethnic groups
c. maintain Cool's core identity by creating different promotional campaigns and brand extensions for the various markets served
d. use the same marketing mix that has worked so well for the firm throughout the Midwestern states of Iowa, Nebraska, and Minnesota
e. gain market share by developing an entirely distinct marketing mix for the largest niche market

5 Identify consumer and marketer reactions to the state of the economy

_____ 14. Jorge is a marketing manager faced with planning marketing strategies during times of inflation. He should be aware that inflation causes consumers to:
a. purchase more goods and services to support their psychological well-being such as counseling and stress management training
b. decrease their brand loyalty to products they have traditionally used
c. buy in small quantities until inflation is over
d. consume more meals away from home
e. put more money into savings accounts because prices are too high

_____ 15. Harold is the owner of a luxury car dealership. He is very worried about his business because sales of his luxury cars have declined dramatically. He has read that unemployment is up, that consumer spending is down, and consumer income is also down. Which of the following best describes the current economy?
a. recession
b. inflation
c. stagnation
d. price escalation
e. depression

6 Identify the impact of technology on a firm

_____ 16. Your company has just allotted $25 million to research and development, with a goal of solving the specific problem of removing fats from meat products while keeping taste. This way, "fat-free" burgers can be marketed in fast-food restaurants. Your company is engaging in:
a. basic research
b. product introduction
c. pure research
d. fundamental development
e. applied research

7 Discuss the political and legal environment of marketing

_____ 17. Ms. Ashley Timber is Vice President of Marketing for Grand Appliances, Inc. She wants her firm to develop newer and better consumer appliances, but knows they must conform to federal laws and regulatory statutes. Which of the following federal agencies could most likely provide information which would affect how her firm develops new appliances?
a. Labor Department
b. Food and Drug Administration
c. Federal Trade Commission
d. Department of Commerce
e. Consumer Product Safety Commission

_____ 18. For several years, an unethical pharmaceutical company has been advertising that its major product can get help overweight consumers lose 20 pounds per week. There are two problems to this product claim: the only consumers who may lose that much weight in such a period of time are extremely obese, and using the pharmaceutical over a long period of time causes severe side effects. The FTC has finally caught up to this company and has ordered it to discontinue its current advertising and to run advertisements that state that the original advertisement was not accurate. This last action is an example of:
a. a Cease and Desist Order
b. corrective advertising
c. affirmative disclosure
d. counteradvertising
e. restitution

8 Explain the basics of foreign and domestic competition

_____ 19. Danny finds that his startup company, a manufacturer of cordless telephones, has been seriously affected by the type and number of competitors in this area. In particular, one area of his marketing mix, ____________, is dependent upon the competition.
a. product
b. price
c. promotion
d. distribution
e. production

9 Describe the nature of ethical decision making

_____ 20. Badd's Butcher Shop sells meat that has not been inspected by USDA inspectors. This meat could possibly carry harmful diseases. However, Badd's can buy the meat extremely cheaply from an out-of-state wholesaler, and the meat will help bring in a larger profit. Badd's Butcher Shop is operating at a level of:
a. conventional morality
b. preconventional morality
c. social responsibility
d. unconventional morality
e. postconventional morality

_____ 21. Recently, a local beer distributor has been concerned with whether it is ethical to run an ad in the local college newspaper. Community attitudes toward alcohol have been changing, and it may not be acceptable to advertise to young people. What level of morality is the beer distributor adopting?
a. unconventional morality
b. conventional morality
c. social responsibility
d. preconventional morality
e. postconventional morality

_____ 22. The pesticides company that questions the long-term effects of its products on the environment is operating at what level of morality?
a. unconventional
b. preconventional
c. postconventional
d. conventional
e. disconventional

10 Discuss corporate social responsibility

_____ 23. Lighthouse Industries has a policy of hiring blind people whenever possible. Not only is this good business, it is also:
a. mandated by Supreme Court rulings
b. environmental marketing
c. a moral obligation
d. an ethical job orientation
e. socially responsible

_____ 24. Like most other large corporations, the Bizby Corporation feels that its social responsibility:
a. is the province of marketing managers
b. may encourage more government regulation
c. is secondary to earning
d. is inversely proportional to profits
e. extends beyond simply earning profits

_____ 25. The pyramid of corporate social responsibility assumes that ____________ responsibility is the foundation, without which the other responsibilities could not exist.
a. economic
b. ethical
c. moral
d. philanthropic
e. legal

Check your answers to these questions before proceeding to the next section.

ESSAY QUESTIONS

1. While marketing managers can control the marketing mix, they cannot control the elements in the external environment that continually mold and reshape the target market. Assume you are a cigarette manufacturer. Name five elements of the external environment, and briefly describe how each factor might affect the marketing of cigarettes.

2. List three demographic trends apparent in the United States in the 1990s. For each trend, describe one way that the trend affects marketing.

3. Ethical development can be thought of as having three levels of morality. Name and briefly define the three levels. Assume that you are a salesperson for Bolt, a highly addictive beverage that, if consumed regularly, can cause death in children, brain damage in teenagers, and mere disorientation for adults. The product sells well and is extremely profitable. Taking on the role of this salesperson, describe the types of thoughts that might occur at each level of morality.

4. What is the pyramid of corporate social responsibility? What are the different levels that should be achieved?

CHAPTER NOTES

Use this space to record notes on the topics you are having the most trouble understanding.

ANSWERS TO THE END-OF-CHAPTER DISCUSSION AND WRITING QUESTIONS

Use this space to work on the questions at the end of the chapter.

ANSWERS TO THE END-OF-CHAPTER CASE

Use this space to work on the questions at the end of the chapter.

CHAPTER 4 Developing a Global Vision

LEARNING OBJECTIVES

1 Discuss the importance of global marketing

As world trade barriers are removed and domestic competition increases, global marketing is becoming a viable and important opportunity for U.S. businesses. However, the United States' ability to compete in global markets has significantly declined in the last four decades, and many domestic firms are losing market share to foreign businesses.

2 Discuss the impact of multinational firms on the world economy

Because of their vast size and financial, technological, and material resources, multinational corporations (MNCs) have a tremendous influence on the world economy. These firms have the potential to solve complex social, economic, and environmental problems in both developed and undeveloped nations. But if these corporations misuse their power, they can have a devastating effect on economic conditions in the world.

3 Describe the external environment facing global marketers

Global marketers face the same environmental factors domestically or internationally: culture, level of economic development, political and legal structure, demographics, and natural resources. However, several of these factors have a greater impact and are more difficult to understand in the global environment. Cultural considerations include the language, societal values, attitudes and beliefs, and customary business practices. Level of economic development may vary widely among countries. The political and legal structure may include tariffs, quotas, boycotts, exchange controls, trade agreements, and market groupings. In short, all the external environment factors may be different and more complicated in global marketing.

4 Identify the various ways of entering the global marketplace

Firms use these strategies to enter global markets in descending order of risk and profit: foreign direct investment, joint venture, contract manufacturing, licensing, and exporting.

5 List the basic elements involved in developing a global marketing mix

In developing a global marketing mix, a firm's major consideration is how much it will adjust the four P's (product, place, promotion, and price) within each country. Using a global marketing approach, a firm makes few or no adjustments to product and promotion strategies. However, national differences in distribution channels and economic conditions usually require such firms to adjust place and price.

PRETEST

Answer the following questions to see how well you understand the material. Re-take it after you review to check yourself.

1. What is global marketing?

2. Name five advantages that multinational firms enjoy over domestic firms.

3. Name and briefly describe five external environments that global marketers face.

4. Name and briefly describe five methods for entering the global marketplace. What are the risk factors for each method?

5. List and briefly describe four product/promotion options when entering global markets.

CHAPTER OUTLINE

1 Discuss the importance of global marketing

I. The Rewards of Global Marketing

Global marketing, marketing to target markets throughout the world, is an imperative for almost every business. Managers must recognize and react to international opportunities as well as to foreign competition via imports, which is found in almost every industry.

In addition, managers need to understand how customer and supplier networks operate worldwide.

A. The word *global* has assumed a new meaning, referring to a boundless mobility and competition in social, business, and intellectual arenas.

B. **Global vision** means recognizing and reacting to international marketing opportunities, being aware of threats from foreign competitors in all markets, and effectively using international distribution networks.

C. The Importance of Global Marketing to the United States

1. The United States exports about a fifth of its industrial production and a third of its farm products.

2. One out every sixteen jobs in the United States is directly or indirectly supported by exports.

3. Almost one-third of corporate profits is derived from international trade and foreign investment.

4. About 85 percent of all U.S. exports of manufactured goods are made by 250 companies, with less than 10 percent of all manufacturing businesses, or around 25,000 companies, exporting their goods on a regular basis. Most small and medium-size firms are essentially nonparticipants in global trade and marketing.

5. Men and women have differing perceptions of the benefits of globalization.

2 Discuss the impact of multinational firms on the world economy

II. The Multinational Firm

The **multinational corporation (MNC)** is a company that is heavily engaged in international trade. Multinational corporations move resources, goods, services, and skills across national boundaries without regard to the country in which the headquarters is located.

A. Most multinational corporations are enormous, with sales larger than the gross domestic product of most countries of the world.

B. Large multinational corporations have advantages over other companies:

1. They can often overcome trade problems.

2. They may be able to sidestep regulatory problems in certain countries by establishing divisions that operate as local firms.

3. They may save a great deal in labor costs by moving some operations to low-cost countries.

4. They can shift costs from one plant to another.

5. They can tap new technology from around the world.

C. Global Marketing Standardization

1. **Global marketing** is defined as individuals and organizations using a global vision to effectively market goods and services across national boundaries.

2. **Global market standardization** presumes that the markets throughout the world are becoming more alike and that globally standardized products can be sold the same way all over the world.

3. Some markets and products, such as McDonald's, Pepsi, and Coca-Cola, do seem to fit global market standardization concepts. Industrial and high-tech products are also likely to be marketed globally in a standardized manner.

4. Most firms try to identify, yet minimize, needed product modifications from country to country. Advertising usually differs in each market.

3 Describe the external environment facing global marketers

III. The External Environment Facing Global Marketers

A. Culture

Central to any society is its culture, the common set of values shared by its citizens that determine what is socially acceptable. Culture forms the basis for the family, educational, religious, and social class systems.

1. Without understanding a country's culture, a firm has little chance of effectively penetrating the market.

2. Important aspects to consider are the meaning and significance of language and slang terms, difficulties in translation, etiquette, and negotiation procedures.

3. Not understanding a culture's notion of time can sometimes lead to situations that are awkward and embarrassing.

B. Economic and Technological Development

In general, complex and sophisticated industries and higher family incomes are found in developed countries, and more basic industries and lower incomes are found in less-developed nations. Five stages of economic growth and technological development exist:

1. The **traditional society** is largely agricultural, with a social structure and a value system that provide little opportunity for upward mobility. It is custom-bound, and its economy typically operates at the subsistence level because of primitive or no technology.

2. The **preindustrial society** is characterized by economic and social change and the emergence of a rising middle class with an entrepreneurial spirit. Nationalism may also emerge, hindering the entrance of multinational corporations.

3. The **takeoff economy** is the period of transition from a developing to a developed nation. New industries arise, and a generally healthy social and political climate emerges.

4. The **industrializing society** is characterized by the spread of technology to most sectors of the economy. Capital goods and consumer goods begin to be produced. A large middle class arises, and the demand for luxuries and services grows.

5. The **fully industrialized society** is an exporter of manufactured products, many of which are based on advanced technology. The wealth of industrialized nations creates tremendous market potential.

C. Political Structure

Government policies in other countries run the gamut, from no private ownership and minimal individual freedom to little central government and maximum personal freedom.

1. As rights of private property increase, government-owned industries and centralized planning tend to decrease.

2. The early 1990s were characterized by a large group of countries in Eastern Europe and the former Soviet Union moving from centralized planning to market economies, opening up market opportunities for many firms.

3. A threat to some firms engaged in global marketing is the growth of nationalist sentiment in countries where citizens have strong loyalties and devotion to their country. Foreign firms can lose the right to conduct business in those countries.

4. Legal structures, often closely intertwined with the political environment, are designed to either encourage or limit trade.

   a. A **tariff** is a tax levied on the goods entering a country.

   b. A **quota** is a limit on the amount of a specific product that can enter a country.

   c. A **boycott** is an exclusion of all products from certain countries or companies, usually used by a government to exclude business from countries with which it has a political difference.

   d. An **exchange control** is a law compelling a company earning foreign exchange from exports to sell the foreign currency to a control agency, usually a central bank.

   e. A **market grouping**, or common trade alliance, occurs when several countries agree to work together to form a common trade area that enhances trade opportunities.

      The best-known market grouping is the European Union (an outgrowth of the European Community, a group of twelve European countries).

   f. A **trade agreement** is an agreement to stimulate international trade.

      1). The **Uruguay Round** of trade negotiations, which created the World Trade Organization, to lower trade barriers worldwide.

      2). Signed by 117 nations, the agreement reduces tariffs by one-third world-wide.

      3). The agreement makes several major changes in patents, copyrights and trademarks; financial, legal, and accounting services; agriculture; limits on importing textiles and apparel; and establishes the World Trade

Organization.

5. The trend toward globalization has brought to the fore several examples of the influence of political structure and legal considerations.

    a. The Japanese Keiretsu

        1). The **keiretsu**, or societies of business, are entities that grew out of Japanese nationalism. They take two primary forms:

            (a) Bank-centered keiretsu are massive industrial combines of twenty to forty-five core companies organized around a bank, for investment purposes.

            (b) Supply keiretsu are groups of companies dominated by the major manufacturer they provide with supplies.

        2). The keiretsu have blocked U.S. companies from the Japanese market. It is estimated that up to 90 percent of domestic Japanese transactions are among parties organized into some kind of long-term relationship.

    b. North American Free Trade Agreement (NAFTA)

        1). The **North American Free Trade Agreement (NAFTA)** is the world's largest free-trade zone. It includes Canada, the United States, and Mexico, a combined population of 360 million and economy of $6 trillion.

        2). When NAFTA went into effect, tariffs on many items disappeared, and a web of licensing requirements and quotas were lifted.

        3). The first industries to benefit were autos, textiles, capital goods, financial services, construction equipment, electronics, telecommunications, and petrochemicals.

        Countries south of the U.S. border have been forming their own trade agreements. The largest new trade agreement is **Mercosur**, which includes Brazil, Argentina, Uruguay, and Paraguay.

    c. The European Union

        1). The European Union (an outgrowth of the European Community, often called the Common Market), is a group of twelve countries that have lifted most internal trade barriers. Its members are Belgium, France, Germany, Italy, Luxembourg, the Netherlands, Denmark, Ireland, Spain, the United Kingdom, Portugal, and Greece.

        2). The European Union has been evolving for nearly four decades.

        3). The EU is an attractive market, with 320 million consumers and purchasing power almost equal to the United States. However, it is full of diversity.

        4). The Maastricht Treaty of 1993 proposes to take the EU further toward economic, monetary, and political union. The treaty standardized trade rules and taxes, health and safety standards, duties, customs procedures, and product testing.

5). One problem facing U.S. marketers is the possibility of a protectionist movement by the European Union against outsiders. Rivalry is very intense in the EU.

D. Demographic Makeup

1. Population is an important factor to marketers.

a. The three most densely populated nations in the world are China, India, and Indonesia.

b. It is also important to consider whether the population is urban or rural. Urban populations are usually more attractive to marketers, because marketers have easier access to these populations.

2. The amount and distribution of income within a country is also an important indicator of its market potential. The wealthiest nations include Japan, the United States, Switzerland, Sweden, Canada, Germany, and several of the Arab oil-producing nations.

3. Even countries with very low per capita incomes have pockets of extremely affluent consumers.

4. The largest population increases are in the developing nations, not the industrialized countries.

E. Natural Resources

The abundance of natural resources in some countries can create huge amounts of personal wealth.

1. Many oil-rich nations experienced an influx of wealth during the 1970s, but their economies suffered in the 1980s, when the price of oil fell.

2. Other countries are hampered by their dependence on imports for many raw metals, timber, agricultural products, and other raw materials.

4 Identify the various ways of entering the global marketplace

IV. Global Marketing By The Individual Firm

A company should consider moving to the global arena only after its management has a solid grasp of the global environment.

A firm has five basic methods for entering the global marketplace.

A. Export

Exporting is selling domestically produced products in another country. This is usually the least complicated and least risky alternative for entering the global market.

1. A company can sell directly to foreign importers or buyers, or it may decide to sell to independent exporting intermediaries located in its domestic market.

2. The most common intermediary is the export merchant, also known as the **buyer for export**, who is usually treated like a domestic customer by the domestic manufacturer. The buyer for export assumes all risks and sells internationally for its own account.

3. Another type of intermediary is the export agent or broker who plays the traditional broker's role by bringing the buyer and seller together without taking title to or possession of the goods.

   a. An **export broker** operates primarily in agriculture and raw materials.

   b. An **export agent** may be a foreign sales agent-distributor, similar to a domestic manufacturer's agent but living in the market country, or a resident agent in the manufacturer's country for foreign buyers.

B. Licensing

Licensing is a legal process whereby a licensor agrees to let another firm use its manufacturing process, trademarks, patents, trade secrets, or other proprietary knowledge.

1. The licensee agrees to pay a fee or royalty.

2. Licensing is less risky than direct manufacturing in the foreign country.

3. The licensor must make certain it can exercise the control over the licensee's activities needed to ensure proper quality, pricing, distribution, and so on.

4. Franchising is one form of licensing that has grown rapidly in recent years. Over 350 U.S. franchisors operate more than 32,000 outlets in foreign countries.

C. Contract Manufacturing

Some firms become heavily involved in global marketing by subcontracting the manufacturing to a firm in the foreign country.

1. **Contract manufacturing** is simply private-label manufacturing by a foreign company that affixes the parent company's name to the goods.

2. The parent company markets the goods.

D. Joint Venture

1. In a **joint venture**, the domestic firm buys part of a foreign company or joins with a foreign company to create a new entity.
2. Joint ventures, although financially risky, give management a voice in company affairs that it might not have under licensing. The partner that learns the fastest comes to dominate the relationship and can then rewrite its terms. This is a new form of competition and a chance to learn.

E. Direct Investment

Direct foreign investment, the active ownership of a foreign company or of oversees manufacturing or marketing facilities, offers the greatest potential rewards and the greatest risks.

1. The large investment may be a tremendous financial risk.

2. A firm may want to acquire an existing foreign firm in order to acquire personnel or some other valuable resource.

3. The United States is a popular place for direct investment by foreign companies.

5 List the basic elements involved in developing a global marketing mix

V. Developing a Global Marketing Mix

A. The first step in creating a marketing mix is developing a thorough understanding of the global target market. Often this understanding can be accomplished with global marketing research.

1. Global marketing research uses the same tools as domestic research. However, obtaining a sample can be difficult in developing countries, where telephone ownership is rare and mail service is slow or sporadic.

2. Differences in culture may also affect the ability of the researcher to collect data.

B. Product and Promotion

1. Some companies have a global marketing strategy of **one product, one message**. This strategy, essentially the global standardization strategy, fits certain products. A slight alteration of advertising or certain changes in packaging and labeling may still take place.

2. In the context of global marketing, **product invention** can be taken to mean either creating a new product for a market or drastically changing an existing product.

3. Many firms use a strategy of extending or adapting products or messages (**message adaptation**).

a. Some firms use the same promotional strategy worldwide but alter products to fit each market.

b. The promotional strategy for a standardized product may stress different customer benefits in different markets or use varying approaches that take into account cultural differences.

Language barriers, translation problems, and cultural differences have generated numerous headaches for international marketing managers.

4. **Product adaptation** is another alternative for global marketers, to meet local conditions. Flavors may be changed, or colors or packaging.

C. Meeting ISO 9000 Standards

ISO 9000 is a standard of quality management that was created in the late 1980s by the International Organization for Standardization. ISO 9000 is designed to certify that a firm is operating with sound quality procedures.

1. It is an internationally recognized system, with over 30,000 certificates issued to companies worldwide.

2. It is extremely popular and well known in Europe.

3. More and more large firms are insisting that their suppliers be certified under ISO 9000. The certificate is a guarantee that the supplier will be a quality vendor partner.

D. Pricing

1. Pricing in the international market must take into account the extra transportation costs, insurance costs, taxes, and tariffs.

2. Because many people in developing nations lack purchasing power, the global firm must decide if it can cut costs and still produce a product that the market will buy.

3. Some companies overproduce certain items and end up dumping them in the international market.

    a. **Dumping** occurs when products are sold either below cost or below their selling price in their domestic market.

    b. Dumping is specifically defined by the World Trade Organization. The Uruguay Round rewrote the international law on dumping.

4. **Countertrade**, a form of trade in which all or part of the payment for goods or services is in the form of other goods or services, is also very popular in international trade. It is estimated that almost 30 percent of global trade is countertrade.

E. Distribution

1. Some foreign distribution systems offer U.S. firms many challenges.

    The Japanese system is considered the most complicated in the world. Imported goods wend their way through layers of agents, wholesalers, and retailers decreed by historical patterns and social relationships.

2. Retail institutions overseas may also differ from what a company is accustomed to in its domestic market. U.S.-type retail outlets may be unavailable or impractical in developing countries. Customers may have different shopping habits.

3. Many developing nations' distribution channels and physical infrastructure are inadequate.

VOCABULARY PRACTICE

Fill in the blank(s) with the appropriate term or phrase from the alphabetized list of chapter key terms.

buyer for export
contract manufacturing
countertrade
direct foreign investment
dumping
export agent
export broker
exporting
fully industrialized society
General Agreement on Tariffs and Trade (GATT)
global marketing
global marketing standardization
global vision
industrializing society
joint venture
keiretsu
licensing
Maastricht Treaty
Mercosur
multinational corporation (MNC)
North American Free Trade Agreement (NAFTA)
preindustrial society
takeoff economy
traditional society
Uruguay Round
World Trade Organization (WTO)

1 Discuss the importance of global marketing

1. Individuals and organizations utilizing a certain vision to effectively market goods and services across national boundaries are practicing ______________________________. Having a(n) ______________________________ means that management recognizes and reactions to international marketing opportunities, is aware of threats from foreign competitors in all marketing, and effectively utilizes international distribution networks.

2 Discuss the impact of multinational firms on the world economy

2. A company heavily engaged in international trade that moves resources, goods, services, and skills across national boundaries without regard to the country in which the headquarters is located is a(n) ______________________________.

3. An international marketing strategy that presumes that the markets throughout the world are becoming more alike is called ______________________________.

3 Describe the external environment facing global marketers

4. A major environmental factor facing the global marketer is the level of economic development of a country. The earliest form of society is largely agricultural, with a social structure and a value system that provide little opportunity for upward mobility. This is the ______________________________.

   The second stage of economic development involves economic and social change and the emergence of a rising middle class. This is the ______________________________. A period of transition from a developing to a developed nation is the ______________________________. The fourth phase of economic development is characterized by spreading technologies and economic growth. This is the ______________________________. Finally, the fifth stage of economic development is characterized by technologically advanced nations engaged in the export of manufactured products. This is the ______________________________.

5 An example of a legal agreement used to stimulate international exchange is the ______________________ of trade negotiations adopted in 1994 by 117 nations. This agreement created ______________________, a new trade organization that replaced the old ______________________, created in 1948.

6. Entities that grew out of Japanese nationalism are the ______________________, or societies of business, which take two primary forms: bank-centered and supply.

7. The world's largest free-trade zone includes Canada, the United States, and Mexico and is made possible by the ______________________. Another large trade agreement includes Brazil, Argentina, and Paraguay, and is called ______________________. The most common market grouping, the European Union, has removed many internal trade barriers on its way to a United Europe. This was accomplished with the signing of the ______________________.

4 Identify the various ways of entering the global marketplace

8. Selling domestically produced products in another country is ______________________. The most common intermediary for international trade is the export merchant, also known as a(n) ______________________. This intermediary assumes all risks and sells internationally for its own account. A second type of intermediary brings buyers and sellers together. This intermediary is called a(n) ______________________ and deals primarily in agriculture and raw materials. A foreign sales agent-distributor or a hired purchasing agent of foreign customers is a(n) ______________________.

9. A legal process whereby one firm agrees to let another firm use its manufacturing process, trademarks, patents, trade secrets, or other proprietary knowledge is called ______________________. This is an effective way for a firm to move into the international market with relatively low risk.

10. If a firm does not want to become involved in licensing arrangements it may engage in private-label manufacturing by a foreign company. This strategy is called ______________________. Another alternative is for a domestic firm to join with a foreign firm to form a new entity. This strategy is called a(n) ______________________. Finally, a domestic firm could have active, majority ownership is a foreign company. This alternative is known as ______________________.

5 List the basic elements involved in developing a global marketing mix

11. Selling products either below cost or below their sale price in their domestic market is the practice of ______________________. This practice is illegal in the United States and Europe.

12. International trade does not always involve cash. If all or part of the payment for goods or services is in the form of other goods or services, ______________________ is being practiced, a common form of which is barter.

Check your answers to these questions before proceeding to the next section.

TRUE/FALSE QUESTIONS

Mark the statement **T** if it is true and **F** if it is false.

2 Discuss the impact of multinational firms on the world economy

_____ 1. A major advantage of a multinational corporation is its ability to sidestep regulatory problems.

_____ 2. Global marketing, like domestic mass marketing, assumes that markets throughout the world are becoming more alike.

3 Describe the external environment facing global marketers

_____ 3. A society in the earliest stage of economic development is called a preindustrial society.

_____ 4. The government of Bolimia has announced that it will no longer permit U.S. cars to be exported into its country. This is an example of a quota.

_____ 5. GATT, NAFTA, and the EU are all examples of market groupings.

_____ 6. Some countries in Africa are highly populated. However, the size of the population alone does not tell an exporter enough about the country's potential as a target market.

4 Identify the various ways of entering the global marketplace

_____ 7. The Crafty Corporation produces arts and crafts materials at its U.S. manufacturing plant. Crafty then sells the products to markets in Germany, Japan, and Mexico. This is an example of exporting.

_____ 8. Athena Athletic shoe company hired a private-label manufacturer in China to produce shoes that Athena plans to sell to the Chinese market. Athena is engaged in licensing.

_____ 9. Active ownership of a foreign company or overseas manufacturing or marketing facilities is direct foreign investment.

5 List the basic elements involved in developing a global marketing mix

_____ 10. Wave surfboards are produced in Australia and sold both in Australia and in the United States. The price in the United States for the surfboards is below the price charged for the same product in its country of origin. This is an example of dumping.

Check your answers to these questions before proceeding to the next section.

AGREE/DISAGREE QUESTIONS

For the following statements, indicate reasons why you may agree and disagree with the statement.

1. Every firm—both domestic and global—needs to develop a global vision.

   Reason(s) to agree:

   Reason(s) to disagree:

2. The most important environmental factor that impacts global marketing efforts is culture.

   Reason(s) to agree:

   Reason(s) to disagree:

3. When marketing to other countries, firms should always adapt both their product and promotion.

   Reason(s) to agree:

   Reason(s) to disagree:

4. When entering foreign markets for the first time, firms should consider exporting first.

   Reason(s) to agree:

   Reason(s) to disagree:

MULTIPLE-CHOICE QUESTIONS

Select the response that best answers the question, and write the corresponding letter in the space provided.

1 Discuss the importance of global marketing

_____ 1. Your firm, a maker of industrial parts, has never engaged in global marketing. You have just completed an examination of the firm's capabilities and the global environment. Which of the following reasons might you employ to convince your boss to think globally?
- a. To remain competitive domestically, the firm has to be competitive internationally.
- b. A global vision enables managers to understand that customer and distribution networks operate worldwide.
- c. The toughest domestic competition is increasingly coming from foreign competition.
- d. Global marketing has become imperative for business because many marketing opportunities exist internationally.
- e. All of the above arguments can and should be offered to your boss.

2 Discuss the impact of multinational firms on the world economy

_____ 2. World Wide Widgets, Inc., is a "member" of several nations. It pays taxes, produces goods, and hires employees in each country yet still has stockholders and corporation management in the United States. WWW would be best described as a(n):
- a. multinational corporation
- b. domestic trader
- c. industrial exporter
- d. export agent
- e. global enterprise

_____ 3. McDonald's offers the same Big Mac in many different countries throughout the world. This is the:
- a. combination market approach
- b. global marketing standardization approach
- c. product extension approach
- d. traditional marketing concept
- e. cultural marketing strategy

3 Describe the external environment facing global marketers

_____ 4. Ferretware is considering expanding its business and selling globally. Its first step should be:
- a. setting up a joint venture
- b. examining the external environments of target countries
- c. determining estimated risks and returns
- d. selling directly to foreign importers
- e. finding a plant location in a foreign country

_____ 5. Management at SoSugar decided to change all its packaging worldwide to a single color and design. In certain countries sales declined after the change, and it was discovered that the color had negative connotations in those areas. Management overlooked the importance of the environmental factor of:
- a. demographics
- b. culture
- c. natural resources
- d. economic development
- e. political structure

_____ 6. In the tiny country of Praire, most people survive by farming beans and grazing herds of llamas. The nation has little or no technology and is bound by customs that have been largely unchanged for centuries. This country is at the ____________ stage of development.
a. traditional
b. preindustrial
c. takeoff economy
d. industrializing
e. fully industrialized

_____ 7. In the republic of Noveland, the rate of economic growth has been one of the fastest in the world, and investors worldwide have been attracted to the emerging industries and generally stable sociopolitical climate. Noveland is a(n) ____________ society.
a. industrializing
b. traditional
c. preindustrial
d. fully industrialized
e. takeoff economy

_____ 8. In the country of Akire, the success of key industries has led to a strong market for support industries that produce component parts and services. This has also led to a growth of the middle class and a demand for more goods and services. Akire is in the ____________ stage of economic development.
a. preindustrial
b. traditional
c. industrializing
d. takeoff economy
e. fully industrialized

_____ 9. The nation of Zembia has a small middle class which dominates business in major urban areas. Though Marketing in Zembia is still difficult since it lacks modern distribution and communication systems. Zembia is a(n) ____________ society.
a. fully industrialized
b. traditional
c. preindustrial
d. industrializing
e. takeoff economy

_____ 10. The U.S. government allows Japan-based Nippon Products to sell only five million toasters in the United States. What kind of trade barrier has the U.S. government imposed on Nippon Products?
a. a boycott
b. a quota
c. an exchange control
d. a tariff
e. a license

_____ 11. Rosarita's company became excited about selling blenders in Indonesia when the country began encouraging foreign investors and imports. Because of Indonesia's enormous population, Rosarita felt the potential for annual sales was staggering and began manufacturing blenders for export. Yet sales the first two years were so small that the venture lost money. Rosarita's company probably overlooked ____________ factors.
a. cultural
b. political
c. demographic
d. educational
e. country resource

4 Identify the various ways of entering the global marketplace

_____ 12. You have started a company that matches U.S. electronics manufacturers with buyers in southeast Asia. You are otherwise known as a(n):
a. buyer for export
b. licensor
c. export agent
d. export broker
e. joint venture partner

_____ 13. You are a small business marketer that specializes in fruit juices. You would like to quickly enter the international marketing arena by exporting. Given your limited resources, which of the following marketing actions might you employ in your exporting effort?
a. sell a new fruit juice created specifically for each country
b. use a dumping strategy to unload your production output to the nations who buy your exports
c. establish a manufacturing plant in a foreign country
d. enter a joint venture agreement with a foreign firm in each foreign country targeted
e. attempt to market the same juices internationally, but tailor the promotion to each country's cultural environment

_____ 14. The Blazer Lazer corporation is concerned about having control over the firm in Spain that is licensed for the next three years to produce Blazer's laser scalpels. Blazer wants to ensure that the licensee will not suddenly decide to set up a separate business as competition using the knowledge it has gained. To prevent this, Blazer has decided to:
a. have the Spanish firm sign a legal, formal licensing agreement
b. use capital-intensive equipment
c. register the patent and trademark in the United States
d. send all the profits home rather than reinvest
e. ship a critical component from the United States rather than producing it in Spain

_____ 15. Farah's Faucet Fixtures (FFF), based in Texas, has decided to find a firm in Asia to do contract manufacturing for the firm. As FFF plans its Asian operations, it will:
a. allow the Asian firm to begin local promotional plans
b. need to set up a marketing division in Asia
c. begin to build the plant in Asia
d. plan the pricing for the contract manufacturer to use
e. do marketing research to determine where to best locate the plant

_____ 16. Disney sells the rights for an investment company to run a Disneyland theme park in Tokyo. The investment company gains most of the profits from the enterprise while paying Disney a percentage in royalties. This is an example of:
a. a joint venture
b. exporting
c. direct investment
d. licensing
e. contract manufacturing

_____ 17. The Bullhide Boot Company wants to expand to South America. The company feels it is important to retain strict quality control over the production process and wants this global expansion to yield a high return. Bullhide Boot should consider:
a. a joint venture
b. licensing
c. direct investment
d. contract manufacturing
e. exporting

5 List the basic elements involved in developing a global marketing mix

_____ 18. Your firm is entering the market in mainland China. You need to conduct market research for the new line of food products made specifically for Chinese. Which of the following issues is a problem your firm might have in conducting and using such research?
a. Questioning a Chinese citizen about private matters may not elicit the answers needed.
b. Obtaining population parameters to produce representative samples might be difficult due to a limited amount of secondary market data available
c. The secondary data available from China may be quite outdated.
d. Using telephone surveying methods would bias the sample because many Chinese do not own or use a telephone
e. All of the above would be problems that your firm might encounter

_____ 19. US-based Harley Davidson offers the same basic motorcycles in Japan as it does in the US. However, the company found that its American advertising theme, "One steady constant in an increasingly screwed-up world" would not appeal to the Japanese market. Harley-Davidson should use a(n) ________ strategy in Japan.
a. one product, one message
b. invariant message
c. product adaptation
d. product and message extension
e. message adaptation

_____ 20. Pandora's Personal Care Company has developed different personal care products for its global markets, which are not at all similar to the features of the product mix offered in the United States. One likely reason for this product invention is that:
a. U.S. manufacturers cannot protect product designs with patents in other nations
b. global pricing policies are the same as those used at home
c. no countries have similar cultural backgrounds
d. consumers in different countries use products differently
e. various cultures respond to advertising messages differently

_____ 21. PepsiCo has entered the market in many developing nations. It has discovered that these markets present many special pricing problems because:
a. the rate of capital accumulation exceeds the rate of population growth
b. exchange rates fluctuate
c. there is a lack of mass-purchasing power
d. nationalism is rising
e. no government regulations exist

_____ 22. Akia Electronics is trying to gain fast market share in a foreign market and sets its prices in this market well below those in its own domestic market. Akia is ________ its products in the foreign market.
a. licensing
b. offloading
c. boycotting
d. repatriating
e. dumping

_____ 23. Due to the exchange difficulties with a small South American country, Leever's Jeans is accepting cases of rum and cigars as payment. This is an example of:
a. barter
b. price controls
c. a tariff
d. exchange controls
e. a joint venture

Check your answers to these questions before proceeding to the next section.

ESSAY QUESTIONS

1. Assume you are the president of Techtronix stereo corporation and are considering marketing your digital tapedecks globally. List and explain five important external environmental factors that you should examine for each country before embarking on your global venture.

2. One major factor in the external environment facing the global marketer is the level of economic development. List and briefly describe the five stages of economic growth and technical development from the lowest to the highest level.

3. Global legal structures are designed to either encourage or limit trade. Define and differentiate five of these structures, and give one specific example of each.

4. What is the difference between a keiretsu, NAFTA, and the European Union? Describe each fully.

5. Assume you are a global marketing consultant and have been asked to name the available options or methods of entry into the global marketplace. Name and describe the five methods of entry in the order of high risk/high return to low risk/low return.

6. What is a joint venture? Name and describe three advantages and three disadvantages of joint ventures to the parties involved.

7. For each of the four P's of the marketing mix, discuss considerations unique to global marketing.

CHAPTER NOTES

Use this space to record notes on the topics you are having the most trouble understanding.

ANSWERS TO THE END-OF-CHAPTER DISCUSSION AND WRITING QUESTIONS

Use this space to work on the questions at the end of the chapter.

ANSWERS TO THE END-OF-CHAPTER CASE

Use this space to work on the questions at the end of the chapter.

CHAPTER GUIDES and QUESTIONS

PART TWO

ANALYZING MARKETING OPPORTUNITIES

CHAPTER 5 Consumer Decision Making

LEARNING OBJECTIVES

1 Explain why marketing managers should understand consumer behavior

An understanding of consumer behavior is essential to the marketer who endeavors to satisfy the needs and wants of his or her customers and wants to communicate effectively with them. Understanding consumer behavior reduces uncertainty when creating the marketing mix.

2 Analyze the components of the consumer decision-making process

The consumer decision making process begins with a stimulus that triggers problem recognition, revealing an unmet need or want. The next step is to determine whether additional information is needed to make the decision. Next, the alternatives are evaluated, and purchase decision rules are established. A purchase decision is then made. Postpurchase evaluation, including the important concept of cognitive dissonance, is based on the evaluation of the outcomes.

3 Explain the consumer's postpurchase evaluation process

Postpurchase evaluation weighs a consumer's satisfaction with the purchase. The outcome of a purchase may be positive, negative, or neutral. Cognitive dissonance, which is an internal tension that the consumer experiences due to doubts about the decision, is an extremely important concept to marketers who try to provide customer satisfaction.

4 Identify the types of consumer buying decisions and discuss the significance of consumer involvement

Consumers face three basic categories of decision making: (1) routine response behavior, used for frequently-purchased, low cost items that require very little decision effort; (2) limited decision making, used for products that are purchased occasionally; and (3) extensive decision making, used for products that are unfamiliar and expensive or infrequently bought.

The levels of consumer involvement in the purchase task signify the economic and social importance of the purchase to the consumer. Depending on the level of consumer involvement, the extensiveness of the purchase process will vary greatly.

5 Identify and understand the cultural factors that affect consumer buying decisions

Culture is the essential character of a society that distinguishes it from other cultural groups. cultural influences on consumer buying decisions include culture and values, language, myths, customs, rituals, laws, and the artifacts, or products. Subcultures are also important parts of a culture. Subcultures can be based on demographic characteristics (such as age), geographic regions, political beliefs, and national and ethnic backgrounds.

6 Identify and understand the social factors that affect consumer buying decisions

Social factors include family, reference groups, opinion leaders, social class, life cycle, culture, and subculture. Consumers may use products or brands to identify with or become a member of a reference group. Opinion leaders reference groups members who influence others' purchase decisions. Family members also influence purchase decisions; children tend to shop in patterns like their parents'. Marketers often define their tar markets in terms of consumers' life cycle stage, social class, culture, and subculture; consumers with simi characteristics generally have similar consumption patterns. Because all consumer behavior is shaped individual and social factors, the main goal of marketing strategy is to understand and influence them.

7 Identify and understand the individual factors that affect consumer buying decisions

Individual factors that affect consumer buying decisions include gender, age and family life cycle, and personality, self-concept, and lifestyle. The gender or age of a consumer affects how the consumer may make purchasing decisions. Marketers also often define their target markets in terms of consumers' life cycle, following changes in consumers' attitudes and behavioral tendencies as they mature.

8 Identify and understand the psychological factors that affect consumer buying decisions

Psychological factors include perception, motivation, learning, and values, beliefs, attitudes, and values. Perception allows consumers to recognize their consumption problems. Motivation is what drives consumers take actions to satisfy specific consumption needs. Almost all consumer behavior results from learning, which the process that creates changes in behavior through experience. Consumers with similar values, beliefs, and attitudes tend to react alike to marketing-related inducements.

PRETEST

Answer the following questions to see how well you understand the material. Re-take it after you review to check yourself.

1. List and briefly describe the five steps of the consumer decision-making process.

2. List five factors influence the involvement with which a consumer has in the decision-making process.

3. List and briefly describe three cultural factors that influence consumer decision-making.

4. List and briefly describe three social factors that influence consumer decision-making.

5. List and briefly describe three individual factors that influence consumer decision-making.

6. List and briefly describe four psychological factors that influence consumer decision-making.

CHAPTER OUTLINE

1 Explain why marketing managers should understand consumer behavior

I. The Importance of Understanding Consumer Behavior

Consumer behavior describes how consumers make purchase decisions, and how they use and dispose of the purchased goods or services.

Understanding consumer behavior reduces uncertainty when creating the marketing mix.

2 Analyze the components of the consumer decision-making process

II. The Consumer Decision-Making Process

There are five steps in the **consumer decision-making process**: need recognition, information search, evaluation of alternatives, purchase, and postpurchase behavior.

A. **Need recognition** occurs when consumers are faced with an imbalance between actual and desired states.

1. A **stimulus** is any unit of input affecting the five senses. Stimuli can be either internal or external.

    a. **Internal stimuli** are a person's normal needs.

    b. **External stimuli** stem from sources outside one's self.

2. A **want** exists when someone has an unfulfilled need and has determined which product will satisfy it.

3. Consumers recognize unfulfilled wants in a number of ways. Recognition occurs when the consumer

    a. Has a current product that isn't performing properly

    b. Is about to run out of a product that is generally kept on hand

    c. Sees a product that appears to be superior to the one currently being used

B. Information search can occur internally, externally, or both.

1. **Internal information search** is the process of recalling information stored in the memory.

2. **External information search** seeks information in the outside environment.

    a. **Nonmarketing-controlled information sources**, such as personal experience, personal sources such as friends, and public sources (Underwriters Laboratories, Consumer Reports), are product information sources that are not associated with advertising or promotion.

b. **Marketing-controlled information sources**, such as mass media advertising, sales promotion, salespeople, and product labels and packaging, are biased toward a specific product because they originate with marketers promoting the product.

3. The factors influencing an external search are the consumer's perceived risk, knowledge, prior experience, and interest level in the good or service.

4. The search yields an **evoked set** (also called a consideration set), a group of brands from which a buyer can choose.

C. Evaluation of Alternatives and Purchase

Evaluation involves the development of a set of criteria. These standards help the consumer evaluate and compare alternatives.

1. Consumers often set minimum or maximum levels of an attribute (cutoffs) that determine whether a product will be considered as a viable choice.

2. Adding new brands to an evoked set affects the consumer's evaluation of the existing brands in that set.

3. The goal of the marketing manager is to determine which attributes are most important in influencing a consumer's choice.

3 Explain the consumer's postpurchase evaluation process

III. Postpurchase Behavior

Expectations of value affect satisfaction.

A. When buying products, consumers expect certain outcomes from the purchase. How well these expectations are met determines whether the consumer is satisfied or dissatisfied with the purchase.

B. **Cognitive dissonance** is a state of inner tension felt when consumers are aware of a lack of consistency between their values or opinions and their behavior.

Consumers can reduce cognitive dissonance by:

1. Finding new information that reinforces positive ideas about the purchase

2. Avoiding information that contradicts their decision

3. Revoking the original decision by returning the product

4 Identify the types of consumer buying decisions and discuss the significance of consumer involvement

IV. Types of Consumer Decision Behavior and Consumer Involvement

A. **Involvement** refers to the amount of time and effort a buyer invests in the search, evaluation, and decision processes of consumer behavior.

B. **Routine response behavior** is used for frequently purchased, low-cost goods and services that require very little decision effort. These can also be called low-involvement products.

C. **Limited decision making** typically occurs when a consumer has previous product experience, but is unfamiliar with the current brands available. It is also associated with lower levels of involvement.

D. **Extensive decision making** is used for an unfamiliar, expensive, or infrequently bought item. It is the most complex type of consumer buying decision and is associated with high involvement on the part of the consumer.

E. Factors Determining the Level of Consumer Involvement

Five factors affect involvement level:

1. Previous experience: Previous experience tends to lead to low involvement.

2. Interest: A high level of interest in a product leads to high-involvement decision making.

3. Perceived risk of negative consequences: The higher the perceived risk of making the decision, the more involved the consumer will be.

4. Situation: The circumstances of the situation may change a low-involvement situation into a high-involvement situation because of increased risks.

5. Social visibility: Involvement increases as the social visibility of the product increases.

F. Involvement Implications for the Marketing Manager

1. Marketing managers must offer extensive and informative promotion for high-involvement products.

2. In-store promotion is important for low-involvement products.

3. Linking a low-involvement product to a higher-involvement issue is another tactic that can increase sales.

5 Identify and understand the cultural factors that affect consumer buying decisions

V. Cultural Influences on Consumer Buying Decisions

Cultural factors exert the broadest and deepest influence over a person's consumer behavior and decision making.

A. Culture and Values

1. **Culture** is the essential character of a society that distinguishes it from other cultural groups.

2. Elements of culture are values, language, myths, customs, rituals, and laws as well as artifacts or products.

3. Culture is pervasive, functional, learned, and dynamic.

4. **Values** are the enduring beliefs shared by a society that a specific mode of conduct is personally or socially preferable to another mode of conduct. Values represent what is most important in people's lives.

5. The personal values of consumers have important implications for marketers as they seek to target their message more effectively.

B. Understanding Culture Differences

1. Underlying core values can vary.

2. Products have cultural values and rules that influence their perception and use. Elements such as the meaning of colors and language can impact the perceptions of a product.

C. A **subculture** is a homogeneous group of people who share elements of the overall culture as well as cultural elements unique to their own group.

1. Subcultures may be geographically concentrated such as Cajuns in the bayou regions of southern Louisiana.

2. Subcultures may be geographically dispersed such as Harley-Davidson owners (HOGs)

D. A **social class** is a group of people who

1. Are nearly equal in status of community esteem

2. Regularly socialize among themselves

3. Share behavioral norms

E. Social class is typically measured as a combination of occupation, income, education, wealth, and other variables.

6 Identify and understand the social factors that affect consumer buying decisions

VI. Social Influences on Consumer Buying Decisions

Social factors include all effects on buyer behavior that result from interactions between a consumer and the external environment.

A. Reference Groups

Reference groups are all of the formal and informal groups in society that influence an individual's purchasing behavior.

1. Reference groups are an important concept to marketers because consumers may use products to establish identity with a group or to gain membership into it. Reference groups are also an information source for consumption behavior.

2. Direct membership groups are face-to-face membership groups that touch people's lives directly.

a. **Primary membership groups** are reference groups with which people interact regularly in an informal, face-to-face manner, such as family, friends, and co-workers.

b. **Secondary membership groups** are reference groups with which people associate less consistently and more formally, such as clubs, professional groups, and religious groups.

3. Indirect membership groups are groups of which one is not a member.

a. **Aspirational group** are groups one would like join. To gain membership one must conform to group **norms** (the attitudes and values deemed acceptable by the group).

b. **Nonaspirational reference groups** (dissociative groups) are groups with which an individual does not want to associate or be identified.

4. An **opinion leader** is an individual who influences the opinion of others.

a. An opinion leader may not be influential across product categories.

b. The product endorsement of an opinion leader is most likely to succeed if an association between the spokesperson and the product can be established.

Opinion leaders are often the first to try new products and services.

B. The Family

The family is the most important social institution for many consumers; it strongly influences our values, attitudes,self-concept, and socialization process (the passing down of cultural values and norms to children).

Not all family decisions are jointly made. Some decision roles are specific to a product or user. Trends in the American family are having a strong influence on decision-making roles.

1. Purchase and information roles within the family:

a. Instigators: suggest, initiate, or plant the seed for the purchase process

b. Influencers: provide valued opinions

c. Decision makers

d. Purchasers

e. Consumers: the users

2. Children today have a great influence over the purchase decisions of their parents.

7 Identify and understand the individual factors that affect consumer buying decisions

VII. Individual Influences on Consumer Buying Decisions

A person's buying decisions are also influenced by personal characteristics that are unique to each individual.

A. Gender

1. Gender differences include the obvious physiological differences, and distinct cultural, social and economic roles.

B. Age and Family Life Cycle Stage

1. Consumer tastes in food, clothing, cars, furniture and recreation are often age related.

2. The *family life cycle* defines an orderly series of stages through which consumer's attitudes and behavioral tendencies evolve. The family life cycle is often used as an indicator of consumer purchase priorities.

C. Personality, Self-Concept, and Lifestyle

1. **Personality** is a way of organizing and grouping the consistencies of an individual's reactions to situations.

2. Personality traits are the characteristics used to describe a personality, such as adaptability, self-confidence, sociability, and so on.

3. **Self-concept** is how a consumer perceives himself or herself in terms of attitudes, perceptions, beliefs, and self-evaluations. Self-concept provides for consistent and coherent behavior.

   a. The **ideal self-image** is the way an individual would like to be.

   b. The **real self-image** is the way an individual actually perceives himself or herself.

   c. Another important component of self-concept is *body image*, the perception of one's own physical features.

4. **Lifestyle** is a person's mode of living as identified by activities, interests, and opinions.

   a. *Psychographics* is the analysis technique used to examine consumer lifestyles and to categorize consumers.

   b. Lifestyle analysis has proved valuable in segmenting and targeting consumers.

8 Identify and understand the psychological factors that affect consumer buying decisions

VIII. Individual Factors Influencing Consumer Buying Decisions

1. **Selective exposure** is the process whereby a consumer notices certain stimuli (such as advertisements) and ignores other stimuli.

2. **Selective distortion** is the process whereby a consumer changes or distorts information that conflicts with his or her feelings or beliefs.

3. **Selective retention** is the process whereby a consumer remembers only the information that supports his or her personal feelings or beliefs.

4. Marketers must recognize the importance of cues in consumers' perception of products; such cues as package design or brand name may affect consumer perception of product quality.

5. The threshold level of perception is referred to by experts as the level of change required to make a "just noticeable difference" in consumer perception. Experts estimate that at least a 20 percent change in the stimulus is required.

6. Sending messages subconsciously to consumers is what is known as **subliminal perception**.

B. Motivation

Motives are the driving forces that cause a person to take action to satisfy specific needs.

Maslow's hierarchy of needs is a method of classifying human needs and motivations into five categories (in ascending order of importance):

1. Physiological needs: food, water, shelter

2. Safety needs: security, freedom from pain and discomfort

3. Social needs: love, sense of belonging

4. Self-esteem needs: self-respect, accomplishment, prestige, fame, recognition of one's accomplishments

5. Self-actualization: self-fulfillment, self-expression

C. Learning

Learning is the process that creates changes in behavior, immediate or expected, through experience and practice.

1. Experiential learning is learning by doing.

2. Conceptual learning is learning by applying previously learned concepts to a new situation.

3. Learning occurs faster and is retained longer

   a. The more reinforcement that is received during learning

   b. The more important the material is to be learned

   c. The greater the stimulus repetition is

4. Behavioral learning process:

   Stimulus ---> Response ---> Reward

5. Repetition, in which advertising messages are spread over time rather than clustered at one time in order to increase learning, is a key strategy in promotional campaigns.

6. **Stimulus generalization** occurs when one response is extended to a second stimulus similar to the first.

7. **Stimulus discrimination** is the learned ability to differentiate among stimuli.

8. *Product differentiation* is a marketing tactic designed to distinguish one product from another.

D. Beliefs and Attitudes

1. Beliefs

a. **Beliefs** are organized patterns of knowledge that an individual holds as true about his or her world.

b. Consumers tend to develop a set of beliefs about a product's attributes and through these beliefs, form a **brand image**, a set of beliefs about a particular brand.

2. Attitude

An **attitude** is a learned tendency to respond consistently toward a given object. Beliefs help form the basis for attitudes, as do values.

Often the marketer's goal is to change attitudes toward a brand. This goal might be accomplished in three ways:

a. Changing beliefs about the brand's attributes

b. Changing the relative importance of these beliefs

c. Adding new beliefs

VOCABULARY PRACTICE

Fill in the blank(s) with the appropriate term or phrase from the alphabetized list of chapter key terms.

aspirational reference group
attitude
belief
cognitive dissonance
consumer behavior
consumer decision-making process
culture
evoked set
extensive decision making
external information search
ideal self-image
internal information search
involvement
learning
lifestyle
limited decision making
marketing-controlled information source
Maslow's hierarchy of needs
motive
need recognition
nonaspirational reference group
nonmarketing-controlled information source
norm
opinion leader
perception
personality
primary membership group
real self-image
reference group
routine response behavior
secondary membership group
selective distortion
selective exposure
selective retention
self-concept
social class
socialization process
stimulus
stimulus discrimination
stimulus generalization
subculture
value
want

1 Explain why marketing managers should understand consumer behavior

1. Marketers study the processes the final consumer uses to make purchase decisions, as well as the use and disposal of the purchased product or service. This is known as the study of ______________________, which also includes the analysis of factors that influence purchase decisions and product usage.

2 Analyze the components of the consumer decision-making process

2. When buying products, consumers follow a series of steps called the ______________________. The first step in the process is when the consumer is faced with an imbalance between actual and desired states. A unit of input that affects any of the five senses is known as a(n) ______________________ and starts these conditions under which the consumer experiences ______________________.

3. A marketing manager cannot create a need, but tries to create a(n) ______________________ during the problem recognition stage.

4. A consumer can explore alternative sources of information during the information search stage. This information search should ultimately yield a group of brands called the buyer's ______________________, or consideration set. When a person tries to recall past information stored in his or her memory, the person is conducting a(n) ______________________. Alternatively, when a

person consults a friend or an advertisement, a(n) ______________________________ is being used. Getting information from an advertisement would be considered using a(n) ______________________________, while getting the information from a friend would be using a(n) ______________________________.

3 Explain the consumer's postpurchase evaluation process

5. After buying a product, a consumer may start to feel an inner tension due to an inconsistency among values, opinions, and behavior. This phenomenon is known as ______________________________.

4 Identify the types of consumer buying decisions and discuss the significance of consumer involvement

6. The amount of time and effort a consumer gives to the search, evaluation, and decision process is known as ______________________________.

7. Consumers have different levels of decision making. When buying frequently purchased, low cost products, a consumer generally uses ______________________________. When a consumer acquires information about an unfamiliar brand in a familiar product category, the consumer is using ______________________________. Finally, when the consumer is buying an unfamiliar, expensive product, or an infrequently bought item, ______________________________ is used.

5 Identify and understand the cultural factors that affect consumer buying decisions

8. The set of values, norms, and attitudes that shape human behavior is called ______________________________. This can be divided into smaller sets on the basis of unique elements. This smaller set is ______________________________.

6 Identify and understand the social factors that affect consumer buying decisions

9. A group in society that influences the purchasing behavior of an individual is a(n) ______________________________. Direct groups can have two forms: The group of people with which we interact in a regular, informal, and face face manner is the ______________________________. The group of people we interact with on a less consistent and more formal basis is the ______________________________. Indirect, nonmembership groups can also have two forms: A group that one would like to be a member of is a(n) ______________________________, while a group that one would like to avoid is a(n) ______________________________. A value or attitude deemed acceptable by a group is a(n) ______________________________. Any of these groups contain an important individual who is influential in convincing others, called a(n) ______________________________. Marketers try to persuade these individuals to buy products and services.

10. The family strongly influences the consumer by passing down cultural values and norms to each generation. This is known as the ______________________________.

11. A group of people who are considered nearly equal in community esteem and share behavioral norms is called a(n) ______________________________.

7 Identify and understand the individual factors that affect consumer buying decision

12. A broad concept that can be thought of as a way of organizing and grouping the consistencies of an individual's reactions to situations is called ______________________________. An individual's attitudes, perceptions, beliefs, and evaluations about himself or herself as a person make up that individual's ______________________________, which is actually a combination of two facets. The way a person would like to be is the ______________________________, while how a person actually perceives himself or herself is the ______________________________.

13. A person's mode of living as identified by activities, interests, and opinions is called a(n) ______________________________. The analysis technique used to measure this is called psychographics.

8 Identify and understand the psychological factors that affect consumer buying decisions

14. The process by which we select, organize, and interpret stimuli is called ______________________________. Consumers decide which stimuli to notice or ignore. This is known as ______________________________. Two concepts are related to this phenomenon. First, when consumers change information that is in conflict with their feelings or beliefs, ______________________________ takes place. Second, when a consumer remembers only that information that supports personal feelings or beliefs, ______________________________ happens.

15. A driving force that causes a person to take action to satisfy specific needs is a(n) ______________________________. These forces are described as physiological, safety, social, esteem, and self actualization needs in ______________________________. The process that creates changes in behavior through experience and practice is known as ______________________________. A key strategy used in promotion to increase the amount of this process is repetition. A related concept is when one response is extended to a second similar stimulus, known as ______________________________. Alternatively, some marketers strive for differentiation between stimuli, known as ______________________________.

16. An enduring belief that a specific mode of conduct is personally or socially preferable to an alternative mode of conduct is a(n) ______________________________. An organized pattern of knowledge that an individual holds to be true about the world is a(n) ______________________________. A learned tendency to respond in a consistently favorable or unfavorable manner toward an object is a(n) ______________________________.

Check your answers to these questions before proceeding to the next section.

TRUE/FALSE QUESTIONS

Mark the statement T if it is true and F if it is false.

2 Analyze the components of the consumer decision-making process

_____ 1. Jerry noticed that the day before the big Super Bowl telecast, his T.V.broke. This scenario is an example of need recognition.

_____ 2. After searching for information on computer diskettes, Erika is about to choose one brand: Sony, Verbatim, 3M, or Maxell. This group of diskette brands makes up Erika's evoked set.

4 Identify the types of consumer buying decisions and discuss the significance of consumer involvement

_____ 3. In high involvement buying situations, consumers learn in an almost random fashion by buying products first and then evaluating them.

_____ 4. Ken is familiar with 35mm cameras and photography accessories. While shopping for a new camera bag, Ken came across a brand he had never heard of or seen before. Acquiring information about this unfamiliar brand could be called extensive decision making on Ken's part.

_____ 5. Bertha perceives the purchase of athletic shoes to be a socially risky decision because she thinks people will judge her by the shoes she wears. Violetta, however, does not perceive shoes as a particularly risky behavior. As a result, buying athletic shoes will be a high involvement activity for Bertha, but not for Violetta.

_____ 6. Detailed, informative advertisements are most effective for high involvement products because consumers actively search for additional information on these products prior to making their purchase decision.

5 Identify and understand the cultural factors that affect consumer buying decisions

_____ 7. The entire United States would be an example of a global subculture.

_____ 8. The United States, like most other countries, has a social class structure with clear delineations of who fits into upper, middle, and lower classes.

6 Identify and understand the social factors that affect consumer buying decisions

_____ 9. Cindy follows all the latest fashion trends at her Beverly Hills high school. Her friends greatly influence the ways she dresses. Cindy's friends are an example of her primary membership group.

8 Identify and understand the psychological factors that affect consumer buying decisions

_____ 10. Jon Austin is a staunch Republican. He was just given a pamphlet about the positive aspects of the Democratic party. Jon reads this pamphlet and then throws it away. By the next day, he has forgotten the points made in the pamphlet. This is an example of selective distortion.

_____ 11. Maslow's hierarchy of needs categorizes human needs into five levels: physiological needs, safety needs, social needs, esteem needs, and self actualization needs.

_____ 12. Values, beliefs, and attitudes are closely interrelated concepts, although beliefs are often about entire brands, while attitudes are often concerned with specific attributes.

Check your answers to these questions before proceeding to the next section.

AGREE/DISAGREE QUESTIONS

For the following statements, indicate reasons why you may agree and disagree with the statement.

1. No matter how important or trivial the decision, consumers go through all five steps of the consumer decision making process.

   Reason(s) to agree:

   Reason(s) to disagree:

2. The same basic cultural values cut across all social classes.

   Reason(s) to agree:

   Reason(s) to disagree:

3. Most firms use aspirational reference group association to get consumers to buy their products.

   Reason(s) to agree:

Reason(s) to disagree:

4. Perception is reality. The way a consumer perceives something is more important than what is the truth.

   Reason(s) to agree:

   Reason(s) to disagree:

MULTIPLE CHOICE QUESTIONS

Select the response that best answers the question, and write the corresponding letter in the space provided.

2 Analyze the components of the consumer decision-making process

_____ 1. Brenda watches a TV commercial that promotes a new fat-free line of snack products and decides that she will buy a few the next time she goes to the grocery store. The TV commercial is a(n):
   a. problem recognition
   b. internal stimulus
   c. external stimulus
   d. purchase outcome
   e. cognitive dissonance

_____ 2. Dan is a prestige oriented shopper and will only buy and wear clothing that have a little embroidered polo pony. This illustrates the:
   a. physiological drive
   b. satisfaction of a need
   c. satisfaction of a belief
   d. satisfaction of a want
   e. need motivator

_____ 3. David is trying to decide what kind of new car he is going to buy. He relies on Consumer Reports, other car magazines, and the advice of car mechanics. David is using:
   a. marketing controlled information sources
   b. nonmarketing controlled information sources
   c. demographic information sources
   d. secondary data sources
   e. internal search sources

3 Explain the consumer's postpurchase evaluation process

_____ 4. If you were attempting to reduce postpurchase anxiety for your brand of large home appliances, which of the following methods might you attempt to use?
a. send letters to each buyer thanking them for their purchase
b. produce products that positively emphasize the attributes customers want from larger home appliances
c. provide warranties to assure customers of the appliances' reliability
d. conduct follow-up phone calls from the appliance manufacturer assuring customers that if they are not completely satisfied, the manufacturer will do what is necessary to make the customer happy
e. all of the above could be employed to reduce cognitive dissonance

4 Identify the types of consumer buying decisions and discuss the significance of consumer involvement

_____ 5. When Steve goes to the grocery store every other week, he buys the same brands of coffee, milk, cereal, and dog food. This type of buying behavior is called:
a. routine response behavior
b. extensive decision making
c. motivational response
d. limited decision making
e. situation convenience

_____ 6. When Jill went to purchase nail polish to wear at her wedding, she went to four stores, spent three hours, and looked at over 200 color shades before selecting the perfect one. This nail polish (which cost $1.25) is properly designated a high involvement product because of:
a. brand loyalty
b. trial investment
c. financial risk
d. cognitive dissonance
e. situational factors

_____ 7. You are the brand manager of a new candy bar. Marketing research shows that consumers do not usually make planned purchases for candy bars and buy on impulse instead. Given the research, which of the following strategies would be most appropriate?
a. Run numerous TV commercials to "pull" consumers into the grocery stores.
b. Price the candy very low to encourage children to buy the product.
c. Determine the important choice criteria the customer uses to shop for candy purchases and then appeal to those criteria when the consumer undertakes alternative evaluation.
d. Place the candy bars at the point-of-purchase within good view of shoppers.
e. Match the lifestyles of the target market to the messages used in the advertising.

5 Identify and understand the cultural factors that affect consumer buying decisions

_____ 8. Which of the following statements about Asian cultures is NOT true?
a. Asians value social harmony.
b. Asians de-emphasize individuality.
c. Asians have a strong work ethic.
d. Asians respect someone who is honest.
e. Asians place a high value on freedom of expression.

_____ 9. Joe has a bachelor of electrical engineering degree from an Ivy League university and works as a hardware specialist in Silicon Valley. Joe is most likely a part of the:
a. capitalist class
b. upper middle class
c. middle class
d. working class
e. working poor

6 Identify and understand the social factors that affect consumer buying decisions

_____ 10. Erika competes on the college swim team and is working hard to qualify for her first college conference competition. She subscribes to all the swimming magazines and reads them as soon as they arrive each month. Past Olympic champion swimmers and divers are often used in the advertisements in the magazines. The ads are quite effective because these champions are a(n) ___________ group for Erika.
a. secondary reference
b. primary reference
c. direct reference
d. dissociative
e. aspirational

_____ 11. Donna feels that the only type of consumer who would wear a fur coat is wasteful and materialistic, and she would never consider owning one herself. The people who typically buy fur coats are in Donna's ___________ group for that type of purchase.
a. out reference
b. nonaspirational reference
c. ex membership
d. integrated
e. low aspiration

_____ 12. Reference group influence would be weakest for determining which brand of ___________ a person buys.
a. car
b. clothing
c. frozen corn
d. beer
e. cigarettes

_____ 13. Ethan tends to buy the same brands of mouthwash and toilet paper as his ___________, which is often the strongest source of group influence upon the individual for many product purchases.
a. family
b. social class
c. psychographic group
d. subculture
e. dissociative group

_____ 14. A baby food manufacturer has spent a large amount of money on packaging, advertising, and store displays. After a successful introduction with sales higher than expected, sales suddenly dropped off dramatically. Subsequent research revealed that many babies refused to eat it. The baby food manufacturer forgot that babies also play an important role in the family decision process as a(n):
a. instigator
b. consumer
c. decision maker
d. purchaser
e. selector

7 Identify and understand the individual factors that affect consumer buying decision

_____ 15. Jill, a new junior executive, feels that she is a trendy, upwardly mobile professional woman and wants to project an impression of competence and independence. She carefully shops for suits like the kind worn by the two women vice presidents at her firm. Jill admires the vice presidents and strives to be like them. She is dressing to fit her:
a. status seeking image
b. social compliant orientation
c. real self image
d. ideal self image
e. personality

_____ 16. John's purchase behavior is influenced by his hobbies of antique firearms, working out, and computers; his interest in scuba diving, music, and swimming; and his deeply held political and cause related opinions. All of these things are part of the personal influence factor called:
a. lifestyle
b. personality
c. beliefs
d. attitude
e. values

8 Identify and understand the psychological factors that affect consumer buying decisions

_____ 17. Julie, an accounting major, read an article that states that accounting majors receive the highest starting alary offers for business majors. The article also states that marketing majors start with lower salaries but surpass all other majors' salaries within ten years. Julie doesn't remember reading this last part of the article, just the first part. This is an example of:
a. selective distortion
b. selective exposure
c. selective retention
d. perception retention
e. reinforcement

_____ 18. Universal Alarm Systems uses advertisements that depict a young mother with her baby at home alone at night. A prowler has been stalking the house, but he suddenly leaves when he notices the presence of the alarm system. These advertisements are designed to appeal to the consumers':
a. self esteem needs
b. safety needs
c. economic needs
d. physiological needs
e. social needs

_____ 19. When Lever Brothers introduced the new Lever 2000 body soap, they gave away over one million bars of soap for consumers to try in order encourage:
a. selective perception
b. learning
c. consumer needs
d. psychographics
e. problem recognition

_____ 20. Leonard has used and liked Colgate toothpaste for years, so when the company introduced Colgate mouthwash, he bought some. This is an example of:

a. stimulus discrimination
b. selective retention
c. product reinforcement
d. social learning
e. stimulus generalization

_____ 21. Jesefe has certain opinions about personal computers. She thinks that they are complex and expensive, but are high-quality, reliable products. This is a description of one consumer's ____________ about a certain class of product.

a. values
b. attitude
c. belief
d. facts
e. motives

_____ 22. Ralph Creamden is the brand manager for Top Stuff Clothing Products. His marketing research shows that his targeted audience does not hold favorable attitudes towards his products. If you were an advisor to Mr. Creamden, which of the following actions might you suggest?

a. changing the belief(s) about the brand attributes
b. changing the relative importance of these beliefs
c. adding new beliefs to the ones already possessed by consumers
d any of the above might stimulate attitude change towards his brand
e. none of the above, because attitudes are all but impossible to change

_____ 23. When a European cereal company began exporting its product to the United States, managers discovered that many consumers believed that the cereal tasted bad. The company began a promotion emphasizing the nutrition and unique nutty taste of the cereal. This is an illustration of efforts to:

a. change beliefs about attributes
b. change the importance of beliefs
c. add new beliefs
d. reinforce current beliefs
e. discover consumer needs

Check your answers to these questions before proceeding to the next section.

ESSAY QUESTIONS

1. You are a senior in college and are planning to continue your studies in graduate business school. You have a high grade point average and know that many good schools would be ready to accept you. List each phase of the consumer decision-making process and describe how you will go about deciding which school you'll go to.

2. Assume that you have been trying to become a member of a prestigious campus club and have finally been invited to join. The members of this club typically wear extremely expensive "pump" style athletic shoes. List and briefly describe four factors that could influence your level of involvement in the purchase of these shoes. How involved will you be in this purchase and why?

3. List and define each level of Maslow's hierarchy of needs.

4. What is a reference group? Name and describe four types of reference groups as they may relate to your purchase of a new wardrobe.

5. Apply the three methods of changing attitudes about brands to the marketing activities of the company that makes your brand of toothpaste.

CHAPTER NOTES

Use this space to record notes on the topics you are having the most trouble understanding.

ANSWERS TO THE END-OF-CHAPTER DISCUSSION AND WRITING QUESTIONS

Use this space to work on the questions at the end of the chapter.

ANSWERS TO THE END-OF-CHAPTER CASE

Use this space to work on the questions at the end of the chapter.

CHAPTER 6 Business Marketing

LEARNING OBJECTIVES

1 Describe business marketing

Business marketing is the marketing of goods and services to individuals and organizations for purposes other than personal consumption. Business products include those that are used to manufacture other products, that become part of other products, that aid in the normal operations of an organization, or that are acquired for resale without any substantial change in form. The key characteristic distinguishing business products from consumer products is intended use, not physical characteristics.

2 Describe the role of the Internet in business marketing

The Internet has made business markets more competitive than ever before. The number of business buyers and sellers using the Internet is rapidly increasing. Firms are seeking new and better ways to expand markets and sources of supply, increase sales and decrease costs, and better serve customers. With the Internet, every business in the world is potentially a local competitor.

3 Discuss the role of relationship marketing and strategic alliances in business marketing

Relationship marketing is the name of the strategy that entails seeking and establishing strategic alliances or partnerships with customers. Relationship marketing is driven by strong business forces: quality, cost, speed, cost-effectiveness, and new design techniques.

A strategic alliance is a cooperative agreement between business firms. It may take the form of a licensing or distribution agreement, joint venture, R&D consortium, or multinational partnership.

Companies form strategic alliances for a variety of reasons, and some fail miserably. The keys to success appear to be choosing partners carefully and creating conditions where both parties benefit.

Some benefits may be shared R&D costs, provision of capital, shared production expertise or capacity, ability to penetrate global markets, and reduction of the threat of competition.

Three common problems of strategic alliances are coordination problems which result when the partners are organized quite differently for making marketing and product design decisions; partners' inability to effectively use their skills in other countries; and the inability of one partner to continue meeting their partner's needs, which is often due to rapid technological change.

4 Identify the four major categories of business market customers

The business market consists of four major categories of customers: producers, resellers, governments, and institutions. The producer segment of the business market is quite large. It consists of individuals and organizations that buy goods and services to produce other products, to become part of other products, or to facilitate the organization's daily operations.

The reseller market consists of retail and wholesale businesses that buy finished goods and resell them for a profit.

Government organizations include thousands of federal, state, and local buying units and account for 20 percent of the U.S. gross domestic product, a substantial market. The fourth major segment of the business market consists of institutions seeking to achieve goals that differ from such ordinary business goals as profit, market share, and return on investment. This segment includes many schools, hospitals, churches, civic clubs, and private nonprofit organizations.

5 Explain the North American Industry Classification System

The North American Industry Classification System (NAICS) was developed jointly by the U.S., Canada, and Mexico to provide a common industry classification system for the NAFTA) Organizations can be identified and compared by a numeric code indicating business sector, subsector, industry and ndustry, and country industry. NAICS is a valuable tool for analyzing, segmenting, and targeting business markets.

6 Explain the major differences between business and consumer markets

Business demand is different from consumer demand in that business demand is derived from the demand for other products, is often price inelastic, may be jointly tied to other products, and is more volatile than consumer demand.

The business market has some other important characteristics: large purchase volume, large number of customers, geographically concentrated customers, direct distribution, professional buyers, negotiation of purchasing prices and terms, the practice of reciprocity, leasing, and the emphasis on personal selling.

7 Describe the seven types of business goods and services

Business products generally fall into one of the following seven categories, depending on how they are used.

Major equipment includes such capital goods as large or expensive machines, mainframe computers, blast furnaces, generators, airplanes, and buildings. These goods are often leased, custom-designed, sold direct, and depreciated over time. Accessory equipment is generally less expensive, standardized, and shorter-lived than major equipment and includes items such as power tools, word processors, and fax machines.

Raw materials are unprocessed extractive or agricultural products, such as mineral ore, logs, wheat, vegetables, or fish, that become part of the final product. Component parts are either finished items ready for assembly or products that need very little processing before they become part of some other product. Examples include spark plugs, tires, and electric motors. There is also a replacement market for component parts.

Processed materials are used directly in the production of other products, but unlike raw materials, they have had some processing. They do not retain their original identity in the final product. Examples include sheet metal, lumber, chemicals, and plastics.

Supplies are inexpensive, standardized, consumable items that do not become part of the product. Examples include lubricants, detergents, paper towels, pencils, and paper. Services are expense items that do not become part of a final product. Businesses retain outside providers to perform such tasks as advertising, janitorial, legal, maintenance, or other services.

8 Discuss the unique aspects of business buying behavior

A buying center includes all those persons in an organization who become involved in the purchase decisions of that organization. Membership and relative influence of the participants in the buying center vary widely from organization to organization. Buying centers do not appear on the formal organization chart, and many people may have informal yet important roles. In a lengthy decision-making process, people may move in and out of the buying center. Buying center roles include initiators, influencers/evaluators, gatekeepers, deciders, purchasers, and users.

There are three typical buying situations in business marketing. In a new-buy situation, a good or service is purchased when a new need arises. A modified rebuy is normally less critical and time-consuming than new-buy purchasing. In a modified-rebuy situation, the purchaser wants something new or something added to the original good or service. In a straight rebuy the order is placed on a routine basis, and the goods or services are provided just as in previous orders.

The business purchase process is a multiple-step procedure that begins with the recognition of a need, the subsequent definition of the type of product that will fill the need, and the development of product specifications. Next, various potential suppliers/vendors are contacted, and proposals are sought. Vendor analysis is the comparison of various alternative suppliers.

After analyzing alternative vendors, selecting a source of supply, and negotiating the terms of the purchase, the buying firm issues a purchase order. After the products are received, inspected, and checked into inventory, the payment process begins. After completion of the transaction, the supplier's performance is periodically monitored to determine whether future purchases should be made with this vendor.

PRETEST

Answer the following questions to see how well you understand the material. Re-take it after you review to check yourself.

1. What is business marketing?

2. What is a strategic alliance?

3. List and briefly describe the four major categories of business market customers.

4. List and briefly describe the seven types of business goods and services.

5. List and briefly describe five important aspects of business buying behavior.

CHAPTER OUTLINE

1 Describe business marketing

I. **Business marketing** is the marketing of goods and services to individuals and organizations for purposes oth than personal consumption.

A. Business products include those that:

1. Are used to manufacture other products
2. Become part of another product
3. Aid the normal operations of an organization
4. Are acquired for resale without substantial change in form

B. The key characteristic distinguishing business products is intended use, not physical characteristics.

2 Describe the role of the Internet in business marketing

II Business Marketing on the Internet

The innovations developed by the computing, telecommunications, and electronic media industries will affect every business. Over 95 percent of Fortune 1000 companies use the Internet in one way or another. Many business marketers now realize that the Internet is a valuable tool for expanding markets and better serving customers.

3 Discuss the role of relationship marketing and strategic alliances in business marketing

III. Relationship Marketing and Strategic Alliances

A. Relationship marketing entails seeking and establishing strategic alliances or partnerships with customers. Relationship marketing is driven by strong business forces: quality, cost, speed, cost-effectiveness, and new design techniques.

B. A **strategic alliance** is a cooperative agreement between firms. It may take the form of licensing or distribution agreements, joint ventures, R&D consortia, or multinational partnerships.

4 Identify the four major categories of business market customers

IV. Major Categories of Business Customers

The business market consists of four major categories of customers: producers, resellers, governments, and institutions.

A. Producers

The producer segment is quite large and consists of individuals and organizations that buy goods and services used in producing other products, incorporating into other products, or facilitating the organization's daily operations.

B. Resellers

1. The reseller market consists of retail and wholesale businesses that buy finished goods and resell them for a profit.

2. Many retailers and most wholesalers carry large numbers of items.

C. Governments

Government organizations include thousands of federal, state, and local buying units and represent what is considered to be the largest single market for goods and services in the world.

1. The Federal Government

The U.S. federal government is the world's largest customer. It purchases almost every imaginable good and service.

Many different agencies and departments handle federal purchasing, as if they were separate companies.

2. State, County, and City Government

A business marketer may find over 82,000 state, county, and city governmental units likely to buy its wares.

The paperwork and regulations involved in selling to these government agencies may be less complicated than selling to the federal government, but the sheer volume of potential clients may be frustrating.

D. Institutions

The fourth major segment of the business market consists of institutions that seek to achieve goals different from such ordinary business goals as profit, market share, and return on investment.

This segment includes many schools, hospitals, churches, civic clubs, and private nonprofit organizations.

5 Explain the North American Industry Classification System

V. Classifying Business and Government Markets

The **North American Industry Classification System (NAICS)** is an industry classification system introduced in 1997 to replace the standard industrial classification system (SIC).

A. Provides a common industry classification system for the NAFTA partners.

B. A valuable tool for business marketers in analyzing, segmenting, and targeting markets.

6 Explain the major differences between business and consumer markets

VI. Business Versus Consumer Markets

Many characteristics of business markets are different from those of consumer markets. Thus marketing strategies are dissimilar.

A. Demand

1. Derived Demand

Derived demand is the demand for business products that results from the demand for consumer products.

Because demand for business products is derived, business marketers must carefully monitor demand patterns and changing preferences in final consumer markets.

2. Inelastic Demand

Inelastic demand means that an increase or a decrease in the price of a product will not significantly affect demand or it.

The demand for many business products is inelastic because the price of many products used in the production of a final product has an insignificant effect on the total price of the final consumer product. The result is that demand for the final consumer product is not affected.

3. Joint Demand

Joint demand occurs when two or more items are used together in a final product. An increase in demand for the final product will affect all of the jointly demanded products.

4. Fluctuating Demand

The demand for business products tends to be more volatile than the demand for consumer products.

A small increase or decrease in consumer demand produces a much larger change in demand for the facilities and equipment needed to manufacture the consumer product. This is known as the **multiplier effect** or **accelerator principle**.

B. Purchase Volume

Business customers buy in much larger quantities than consumers.

C. Number of Customers

Business marketers typically have far fewer customers than consumer marketers.

1. Business marketers may have an advantage in identifying prospective customers, monitoring their needs, and providing personal attention.

2. The reduced number of customers can also be a disadvantage, because each customer is so overwhelmingly important to the business.

D. Location of Buyers

Business customers tend to be much more geographically concentrated than consumers.

More than half the nation's industrial purchasers are located in just seven states: New York, California, Pennsylvania, Illinois, Ohio, Michigan, and New Jersey.

E. Distribution Structure

Direct channels are much more common in business marketing than in consumer marketing.

F. Nature of Buying

Business buyers, who are often professionally trained purchasing agents or buyers, normally take a more formal approach to buying compared to consumers.

G. Nature of Buying Influence

More people are usually involved in a single business purchase decision than in a consumer purchase decision.
Some purchase decisions rest with a buying center, which is a panel of experts from a variety of fields within an organization.

H. Type of Negotiations

Negotiation of price, product specifications, delivery dates, payment terms, and a variety of other conditions of sale is common in business marketing.

I. Use of Reciprocity

Business purchasers often choose to buy from their own customers, a practice known as **reciprocity**. Reciprocity is neither illegal nor unethical unless one party coerces the other; it is generally considered a reasonable business practice.

J. Use of Leasing

Consumers normally buy products rather than lease them. But businesses commonly lease expensive equipment, such as computers, construction equipment and vehicles, and automobiles.

K. Primary Promotional Method

Personal selling tends to be emphasized by business marketers in their promotion efforts.

Many business products are expensive, require customization, are ordered in large volumes, or involve intricate negotiations. All these situations necessitate a great deal of personal contact.

7 Describe the seven types of business goods and services

VII. Types of Business Products

A. Major Equipment

Major equipment consists of capital goods, such as large or expensive machines, mainframe computers, blast furnaces, generators, airplanes, and buildings.

1. Major equipment is also called an **installation**.

2. Major equipment always depreciates over time.

3. Major equipment is often leased, custom-designed, and sold direct from the producer.

B. Accessory Equipment

Accessory equipment is generally less expensive and shorter-lived than major equipment.

1. Accessories include such items as power tools, word processors, and fax machines.
2. Accessories are often standardized.
3. Accessories are sold to a broad array of businesses.
4. Accessories use less direct sales channels.
5. Accessories use advertising as an important promotional tool.

C. Raw Materials

Raw materials are unprocessed extractive or agricultural products, such as mineral ore, lumber, wheat, vegetables, and fish, which become part of the final product.

1. Personal selling is very important.
2. Channels of distribution are usually direct from the producer to the business user.

D. Component Parts

Component parts are either finished items ready for assembly or products that need very little processing before becoming part of some other product; examples include spark plugs, tires, and electric motors.

1. Component parts often retain their identity after becoming part of some other product.
2. Component parts may wear out and need to be replaced during the life of a product.
3. The two markets for component parts are the original equipment manufacturer (OEM) market and the replacement market.

E. Processed Materials

Processed materials are used directly in manufacturing other products; unlike raw materials, they have had some processing. Examples include sheet metal, lumber, chemicals, corn syrup, and plastics.

1. Processed materials do not retain their original identity in the final product.
2. Processed materials are usually marketed to OEMs or to distributors servicing the OEM market.

F. Supplies

Supplies are consumable items that do not become part of the final product, such as lubricants, detergents, paper towels, pencils, and paper.

1. Supplies are normally standardized, inexpensive items sold through local distributors.
2. This category is often referred to as MRO items, because supplies generally fall into one of three categories: maintenance, repair, or operating supplies.

G. Business Services

Business services are expense items that do not become part of a final product. Businesses retain outside providers to perform such tasks as advertising, janitorial, payroll, legal, market research, maintenance, or other services.

8 Discuss the unique aspects of business buying behavior

VIII. Business Buying Behavior

A. Buying Centers

A **buying center** includes all those persons in an organization who become involved in the purchase decision.

Membership in the buying center and relative influence of the participants vary widely from organization to organization.

Buying centers do not appear on the formal organizational chart, and members informal yet important roles.

In a lengthy decision process people may move in and out of the buying center.

1. Roles in the Buying Center

a. The ***initiator*** is the person who first suggests making a purchase.

b. ***Influencers*** or evaluators often define specifications for the purchase or provide information for evaluating options.

c. ***Gatekeepers*** regulate the flow of information about the purchase to the deciders and others.

d. The ***decider*** is the person who possess formal or informal power to choose or approve the selection of the supplier or brand.

e. The ***purchaser*** is the person who actually negotiates the purchase.

f. ***Users*** are the members of the organization who will actually use the product.

2. Implications of Buying Centers for the Marketing Manager

Vendors need to identify and interact with the true decision makers. Other critical issues are each member's relative influence and the evaluative criteria used by each member.

B. Evaluative Criteria

The three most important and commonly used criteria are quality, service, and price--in that order.

1. ***Quality*** refers to technical suitability, the salesperson, and the salesperson's firm.

2. ***Service*** may range from prepurchase needs surveys to installation to dependability of supply.

3. ***Price*** is extremely important in most business purchases.

C. Buying Situations

Often business firms must decide whether to make a certain item or to buy it from an outside supplier. Essentially, this is an economic decision, concerning price and use of company resources.

If a firm does choose to purchase a product, it will do so under one of three basic conditions: new buy, modified rebuy, or straight rebuy.

1. New Buy

   A **new buy** situation requires the purchase of a product for the first time.

   a. A new buy situation is the greatest opportunity for a new vendor to sell to a business purchaser because no previous relationship with a vendor has been established.

   b. New buys often result from **value engineering**, a systematic search for less-expensive substitute goods or services.

2. Modified Rebuy

   A **modified rebuy** is normally less critical and time-consuming than a new buy.

   a. In a modified rebuy situation, the purchaser wants some change in the original good or service.

   b. In some cases the purchaser just works with the original vendor, but in other cases the modified rebuy is opened to outside bidders.

3. Straight Rebuy

   a. In a **straight rebuy**, the purchaser reorders the same goods or services without looking for new information or investigating other suppliers.

   b. One common technique in a straight rebuy is the use of a purchasing contract for items that are purchased often and in high volume, which further automates the purchase process.

D. Purchasing Ethics

The ethics of business buyer and seller relationships are often scrutinized and sometimes criticized by superiors, associates, other prospective suppliers, the general public, and the news media.

E. Customer Service

Business marketers are increasingly recognizing the benefits of developing a formal system to monitor customer opinions and perceptions of the quality of customer service.

VOCABULARY PRACTICE

Fill in the blank(s) with the appropriate term or phrase from the alphabetized list of chapter key terms.

accessory equipment
business marketing
business services
buying center
component parts
derived demand
joint demand
major equipment
modified rebuy
multiplier effect (or accelerator principle)
new buy
North American Industry Classification System (NAICS)
processed material
raw materials
reciprocity
straight rebuy
strategic alliance (strategic partnership)
supplies

1 Describe business marketing

1. The marketing of goods and services to individuals and organizations for purposes other than personal consumption is termed ___________________________.

3 Discuss the role of relationship marketing and strategic alliances in business marketing

2. A cooperative agreement between business firms is a(n) ___________________________.

5 Explain the North American Industry Classification System

3. The ___________________________ is used by the US, Canada and Mexico to provide a common system for grouping business organizations according to their primary economic activity.

6 Explain the major differences between business and consumer markets

4. Demand characteristics in business markets are different from those in consumer markets. Organizations purchase products to be used in producing consumer products. This is termed ___________________________. Additionally, the demand for many business products is not sensitive to price changes; it is inelastic. When two or more items are used in combination in a final product, ___________________________ occurs. Finally, the demand for business products tends to be unstable, which is called fluctuating demand. A small percentage change in consumer demand can produce a much larger change in demand for the facilities and equipment needed to manufacture the consumer product. Economists refer to this as the ___________________________.

5. Business purchasers often choose to buy from their own customers, a practice known as ______________________________. This is generally considered a reasonable business practice.

7 Describe the seven types of business goods and services

6. Business products fall into several categories, depending on their use. Capital goods, such as large and/or expensive machines, are called ______________________________. Items that are less expensive and shorter-lived, such as office equipment or hand tools, are called ______________________________. Unprocessed extractive or agricultural products are termed ______________________________ and become a part of finished products. If goods are finished and ready for assembly into the final product, they are called ______________________________ and needs very little or no processing. If products require some processing, they are termed ______________________________ and are used directly in the manufacturing of other products; they do not retain their identity in the final product. Consumable items that do not become part of the final product are called ______________________________. Finally, ______________________________ are expense items that do not become part of the final product and are tasks performed by outside providers.

8 Discuss the unique aspects of business buying behavior

7. All those persons in an organization who become involved in the purchase decision-making process make up the ______________________________.

8. If a firm decides to buy a product rather than make it, it will do so under three different types of buying situations. The first situation requires the purchase of a product or service for the first time and is called a(n) ______________________________. The second buying situation is one in which the purchase is normally less critical and time-consuming. In this case, the purchaser may want something new or added to the original goods or services. This is termed a(n) ______________________________. Finally, there are purchases in which the purchaser is not looking for new information or at other suppliers, and the order is consistently placed with the same provider on a routine basis. This is called a(n) ______________________________.

Check your answers to these questions before proceeding to the next section.

TRUE/FALSE QUESTIONS

Mark the statement **T** if it is true and **F** if it is false.

1 Describe business marketing

_____ 1. Tokia is a manufacturer of cellular phones. Tokia sells its phones through electronics retailers, which in turn sell the phones to households. Because Tokia does not sell directly to the consumer, Tokia is engaged in business marketing.

2 Describe the role of the Internet in business marketing

_____ 2. Because of the Internet, General Electric's consumer sales far exceed those of its business markets since consumers are becoming more savvy on the Internet.

3 Discuss the role of relationship marketing and strategic alliances in business marketing

_____ 3. A strategic alliance allows two companies to share resources and penetrate global markets.

4 Identify the four major categories of business-market customers

_____ 4. One way that business organizations differ from consumer organizations is in their primary goals; business firms typically do not have profit, market share, or return on investment as primary goals.

5 Explain the North American Industry Classification System

_____ 5. NAICS is an industry classification system used by most nations of the world.

6 Explain the major differences between business and consumer markets

_____ 6. If the American Sugar Association showed advertisements that encouraged people to bake more desserts, this would be an attempt to influence derived demand.

_____ 7. Roxanne's company manufactures ball bearings for the aerospace industry. Roxanne noticed that a dramatic increase in the price of ball bearings only slightly reduced demand. This is an example of fluctuating demand.

_____ 8. Channels of distribution for business products are often direct, unlike consumer product channels of distribution, which usually have several intermediaries.

_____ 9. In the business purchase process, unlike in consumer purchase processes, advertising plays one of the most important roles.

8 Discuss the unique aspects of business buying behavior

_____ 10. Kendall is the secretary for the buyer of an aircraft equipment firm. As secretary, Kendall determines which vendors are seen by the firm's buyer. Kendall is acting as a gatekeeper.

_____ 11. For the last five years, the Michigan Jams company has bought its birch bark gift baskets from Woodside, Inc. Michigan Jams wants a new kind of handle and decorative ribbons added to the largest type of basket. This is an example of a new buy.

Check your answers to these questions before proceeding to the next section.

AGREE/DISAGREE QUESTIONS

For the following statements, indicate reasons why you may agree and disagree with the statement.

1. Strategic alliances are more important in business marketing than in consumer marketing.

   Reason(s) to agree:

   Reason(s) to disagree:

2. Because business customers are fewer and geographically more concentrated, business marketing is easier than consumer marketing.

   Reason(s) to agree:

   Reason(s) to disagree:

3. Businesses use basically the same process for purchase decision making as consumers but on a more formal scale.

   Reason(s) to agree:

   Reason(s) to disagree:

4. Selling to business markets would be more rewarding than selling to consumer markets.

   Reason(s) to agree:

   Reason(s) to disagree:

MULTIPLE-CHOICE QUESTIONS

Select the response that best answers the question, and write the corresponding letter in the space provided.

1 Describe business marketing

_____ 1. You have decided that your firm will enter the business market arena. You have been a consumer-products producer selling directly to consumers previously and so you have little knowledge about this new marketing opportunity. Which of the following possible transactions might fit this new arena of business?
a. sales of products that are used to manufacture other products
b. sales of products that are to become parts of other products
c. sales of products that facilitate the normal operations of an organization
d. sales of products that are acquired for resale without any substantial change in form
e. all of the above are possible transactions that might fit this new arena of business

2 Describe the role of the Internet in business marketing

_____ 2. Which of the following is NOT likely to happen with the emergence of the Internet?
a. increase in competition
b. target markets become global
c. increase in the number of vendors evaluated
d. consumer markets become larger and more powerful than business markets
e. efficiency in data exchange

3 Discuss the role of relationship marketing and strategic alliances in business marketing

_____ 3. Benefits of strategic alliances could include all of the following EXCEPT:
a. allowing companies to share research and development ideas and costs
b. reducing the threat of competition, and providing capital
c. guaranteeing a technologically sophisticated partner in the long-term
d. allowing firms to share production expertise or capacity
e. increasing quality and service levels

4 Identify the four major categories of business-market customers

_____ 4. Aztec Industries owns and operates a metal ore mining operation, a trucking company, and an insurance firm. For classification purposes, Aztec Industries would be which type of business customer?
a. government
b. wholesaler
c. reseller
d. institution
e. producer

_____ 5. Sysco Foods purchases truckload quantities of consumer products such as frozen foods, cereals, and paper towels from manufacturers. Sysco then breaks down the items into case quantities and distributes them to restaurants and grocery stores. Sysco would be best classified as a(n):
a. inventory carrier
b. producer
c. institution
d. reseller
e. transportation company

_____ 6. A manager has decided that her company should sell to the government market. What is one common procedure for trying to get a government contract?
a. The government offers a publication of what it is willing to pay for certain goods, and companies offer to supply them for that price.
b. The company submits a sealed bid that states price and delivery terms.
c. The company advertises the availability of its product and has the government contact it.
d. It is difficult to get a government contract because once a business is a supplier for the government, it will have that contract for the life of the company.
e. The company submits a quotation, and if the quoted price is in the midrange of all quotes received, it will probably be selected as the fairest and most average supplier.

_____ 7. Phil's Paper Company specializes in serving institutions. Phil would probably have all of the following as customers EXCEPT:
a. restaurants
b. hospitals
c. public grade schools
d. state universities
e. Red Cross emergency shelter kitchens

5 Explain the North American Industry Classification System

_____ 8. Which of the following is false regarding the North American Industry Classification System (NAICS)?
a. In spite of its name, NAICS is also used to classify firms in other regions, such as Europe or Asia.
b. A code can identify whether a firm is based in the US, Canada, or Mexico.
c. The more digits in a NAICS code, the more homogenous the group is.
d. NAICS facilitates trade among firms in NAFTA countries.
e. NAICS can help firms identify the number, size, and geographic dispersion of potential business customers.

_____ 9. The North American Industry Classification System is currently used for ______ economic sectors.
a. 10
b. 13
c. 20
d. 34
e. 50

6 Explain the major differences between business and consumer markets

_____ 10. When the demand for soft drinks grew by 12 percent in one year, the demand for aluminum cans and glass bottles grew also. The aluminum and glass industries are enjoying the effects of:
a. joint demand
b. inelastic demand
c. derived demand
d. fluctuating demand
e. elastic demand

_____ 11. The Edison Electric Company increased its rates to business customers by 10 percent. Though customers complained, the company saw very little change in the volume of electricity used. The demand for electricity can be described as
a. inelastic
b. derived
c. joint
d. elastic
e. resistant

____ 12. 3M manufactures 3.5-inch computer diskettes. Each disk is made from four components: the magnetic media, the plastic casing, a metal hub, and a metal slide. The supplies purchaser at 3M has noticed that each time the case order for magnetic media increases, so does the case order for plastic casings, metal hubs, and metal slides. This is because the products in this situation have:
a. inelastic demand
b. derived demand
c. elastic demand
d. fluctuating demand
e. joint demand

____ 13. The Wisconsin Wild Rice Company has slowly changed its customer base. They used to sell small bags of rice to retail walk-in consumers, but now they deliver rice to grocery stores and restaurants. The company should expect:
a. more customers
b. a greater reliance on advertising to gain new customers
c. a smaller order size
d. a more formal purchasing process
e. less use of reciprocity

____ 14. Gary owns his own typewriter repair firm. He decided that he needed a photocopier for his business, and he checked his customer list first for copier dealers. He found three copier dealers that were regular customers, and he shopped only at those dealers. Gary's actions are:
a. a normal business practice called reciprocity
b. an example of illegal influence
c. an example of nested demand
d. unethical because of competitive discouragement
e. a business practice called circular buying

7 Describe the seven types of business goods and services

____ 15. Star-Kist purchases tuna fish directly from fishing fleets daily and then processes and cans the tuna. The fish are an example of:
a. OEM parts
b. supplies
c. component parts
d. processed materials
e. raw materials

____ 16. Flyair uses preassembled engines in its radio-controlled model airplanes. These engines are classified as:
a. raw materials
b. component parts
c. accessory equipment
d. processed materials
e. supplies

____ 17. Manufacturers of computers diskettes such as 3M or Sony sell their products to software developers, such as Microsoft, to be used to store software programs. Microsoft would be considered a(n):
a. replacement part market
b. processed materials market
c. finished goods market
d. accessory equipment market
e. OEM market

8 Discuss the unique aspects of business buying behavior

____ 18. When Mike visited a potential customer, he found that the manager was too busy to meet with him. The secretary offered to look through Mike's brochures and pass them on to the manager if it was a product that the firm could use. Which role in the buying center does the secretary have?
a. gatekeeper
b. influencer
c. purchaser
d. decider
e. user

____ 19. The Western Printing Press Company has been purchasing all of its small machined shafts and gears from a machine shop for $130,000 annually. If Western buy two lathes, a drill press, and a milling machine it could produce these parts itself. The equipment, materials and two machinists would cost the company $450,000 over five years, a savings of $200,000. Should Western pursue this option?
a. no, because the best solution would be to spend the money on research and development to find more advanced parts to go into the computers
b. yes, because it is cheaper to make rather than to buy
c. no, because it is not a large enough savings to justify the change
d. yes, but only if this is the best use of the company's resources and the money cannot be used to increase profit even more with some other project
e. no, because it is cheaper to buy these parts than make them

____ 20. Hitec Corp is interested in selecting a supplier for unusual parts used in its recently developed highly technical machine tooling equipment. Which of the following buying processes will most likely be employed:
a. extensive buying process
b. low-involvement buying process
c. new-buy process
d. modified-rebuy process
e. straight-rebuy process

____ 21. The local university buys many desk calculators each year and has an ongoing relationship with a local supplier. The accounting department secretary finds that he needs a calculator with more functions than his current one has, and the university buyer discusses the specifications with the supplier. This is an example of a:
a. modified rebuy
b. value engineering task
c. straight rebuy
d. new task
e. derived rebuy

____ 22. A new edition of a marketing textbook has just been released. The Business Department of Hudson University has always liked the textbook and authorizes the purchase of the new edition without even looking at other textbooks. This is an example of a:
a. modified rebuy
b. value engineering task
c. straight rebuy
d. new task
e. derived rebuy

Check your answers to these questions before proceeding to the next section.

ESSAY QUESTIONS

1. What is a strategic alliance? Why are strategic alliances becoming more important in business marketing?

2. Briefly describe the four major categories of customers in business marketing. Give examples of companies or organizations in each category.

3. Name and briefly describe eight major differences between business and consumer markets. Are there any ways that business purchasing behavior is similar to consumer buying behavior?

4. Name and describe the seven types of business goods and services.

5. Assume that you are the vice-president of marketing in a small firm that includes the following departments: marketing, finance, purchasing, and management. The sales force manager has mentioned to you that one of the salespersons thought that cellular telephones would help the sales force become more efficient. The manager requests that eight such phones be purchased. Illustrate the six buying decision roles that would take place for the purchase of these cellular phones.

6. Pike's Print Shop has decided to purchase some color copying machines. Describe the conditions under which each of the three business buying situations would take place.

CHAPTER NOTES

Use this space to record notes on the topics you are having the most trouble understanding.

ANSWERS TO THE END-OF-CHAPTER DISCUSSION AND WRITING QUESTIONS

Use this space to work on the questions at the end of the chapter.

ANSWERS TO THE END-OF-CHAPTER CASE

Use this space to work on the questions at the end of the chapter.

CHAPTER 7 Segmenting and Targeting Markets

LEARNING OBJECTIVES

1 Describe the characteristics of markets and market segments

A market is a group of people or organizations with wants and needs that can be satisfied by particular product categories. This group has the ability to purchase these products, is willing to exchange resources for the products, and has the authority to do so. If a group lacks any of these characteristics, it is not a market.

A market segment is a subgroup of people or organizations sharing one or more characteristics that cause them to have relatively similar product needs. The process of dividing a market into meaningful groups that are relatively similar and identifiable is called market segmentation.

2 Explain the importance of market segmentation

The purpose of segmentation is to enable the marketer to tailor marketing mixes to meet the needs of one or more specific segments. Market segmentation assists marketers in developing more precise definitions of customer needs and wants.

3 Discuss criteria for successful market segmentation

To be useful, a segmentation scheme must produce segments that meet four basic criteria: substantiality, measurability, accessibility, and responsiveness.

A selected segment must be substantial, or large enough, to justify the development and maintenance of a special marketing mix. The marketer must be able to profit by serving the specific needs of this segment, whatever its size. The segments must be identifiable and their size measurable. The target market must be accessible by the firm's customized marketing mixes. A market segment must respond differently to some aspect of the marketing mix than other segments do.

4 Describe bases commonly used to segment consumer markets

A segmentation base (or variable) is a characteristic of individuals, groups, or organizations used to divide a market into segments. Many different characteristics can be used. Some common bases are geographic, demographic, psychographic, benefits offered by product, and usage rate by consumers.

5 Describe the bases for segmenting business markets

Business markets can be segmented on two bases. First, macrosegmentation divides markets according to general characteristics, such as location and customer type. Second, microsegmentation focuses on the decision-making units within macrosegments. Microsegmentation variables include the key purchasing criteria, purchasing strategies, importance of purchase, and personal characteristics of buyers.

6 List the steps involved in segmenting markets

Six steps are involved when segmenting markets: (1) Select a market or product category for study; (2) choose a basis or bases for segmenting the market; (3) select segmentation descriptors; (4) profile and analyze segments; (5) select target markets; (6) design; implement; and maintain appropriate marketing mixes.

7 Discuss alternative strategies for selecting target markets

There are three general strategies for selecting target markets. A firm using an undifferentiated targeting strategy essentially adopts a mass-market philosophy, viewing the world as one big market with no individual segments. It formulates only one marketing mix and assumes that the individual customers in the market have relatively similar needs.

A concentrated targeting strategy entails focusing marketing efforts on one segment of a market. Because the firm is appealing to a single segment, it can concentrate on understanding the needs, motives, and satisfactions of the members of that segment and develop a highly specialized marketing mix. The term *niche targeting strategy* is sometimes used to describe this strategy.

When a firm chooses to serve two or more well-defined market segments and develops distinct marketing mixes for each, it is practicing multisegment targeting. A firm can use various methods to achieve this goal. Some companies have very specialized marketing mixes, including specialized products for each segment; other companies may only customize the promotional message that is directed to the various target markets.

8 Explain how and why firms implement positioning strategies and how product differentiation plays a role

Positioning is the development of a specific marketing mix to influence potential customers' overall perception of a brand, product line, or organization in general. The term *position* refers to the place a product, brand, or group of products occupies in consumers' minds relative to competing offerings. Firms use the elements of the marketing mix to clarify their position or to reposition the product in the consumers' minds.

The purpose of product differentiation is to distinguish one firm's products from another's. The differences can be real or perceived. The marketer attempts to convince customers that a brand is significantly different from the others and should therefore be demanded over competing brands.

9 Discuss global market segmentation and targeting issues

The key tasks in market segmentation, targeting, and positioning are the same whether the selected target market is local, regional, or multinational. The main differences are the variables used by marketers in analyzing markets and assessing opportunities and the resources needed to implement strategies.

PRETEST

Answer the following questions to see how well you understand the material. Re-take it after you review to check yourself.

1. What is a market? A market segment?

2. List four basic criteria for segmenting markets.

3. Name and briefly describe five bases for segmenting consumer markets.

4. Name and briefly describe two bases for segmenting business markets.

5. List the six steps in segmenting a market.

6. List and describe three strategies for selecting target markets.

7. What is positioning?

CHAPTER OUTLINE

1 Describe the characteristics of markets and market segments

I. Market Segmentation

A **market** is people or organizations with needs or wants and with the ability and the willingness to buy.

A **market segment** is a subgroup of people or organizations sharing one or more characteristics that cause them to have similar product needs.

Market segmentation is the process of dividing a market into meaningful groups that are relatively similar and identifiable.

The purpose of segmentation is to enable the marketer to tailor marketing mixes to meet the needs of one or more specific segments.

2 Explain the importance of market segmentation

II. The Importance of Market Segmentation

A. Market segmentation developed in the 1960s in response to growing markets with a variety of needs and in response to increasing competition.

B. Market segmentation assists marketers in developing more precise definitions of customer needs and wants. Segmentation helps decision makers more accurately define marketing objectives and better allocate resources.

3 Discuss criteria for successful market segmentation

III. Criteria for Successful Segmentation

A. Substantiality: A selected segment must be large enough to justify the development and maintenance of a special marketing mix. Serving the specific needs of this segment, whatever its size, must be profitable.

B. Identifiability and measurability: The segments must be identifiable and their size measurable.

C. Accessibility: The firm must be able to reach members of targeted segments with customized marketing mixes.

D. Responsiveness: A market segment must respond differently to some aspect of the marketing mix than do other segments.

4 Describe bases commonly used to segment consumer markets

IV. Bases for Segmenting Consumer Markets

A **segmentation base** (or **variable**) is a characteristic of individuals, groups, or organizations that is used to divide a total market into segments. Markets can be segmented using a single or multiple variables.

A. Geographic Segmentation

1. **Geographic segmentation** is a method of dividing markets based on the region of the country or world, market size, market density (number of people within a unit of land), or climate.

2. Climate is frequently used because of its dramatic impact on residents' needs and purchasing behavior.

3. Regional marketing, using specialized marketing mixes in different parts of the country, has become prevalent among consumer goods companies for four principal reasons:

   a. Many firms need to find new ways to generate sales because of sluggish or intensely competitive markets.

   b. Computerized checkout stations enable retailers to accurately assess which products and brands are selling best in each region.

   c. New regional brands have been introduced that are intended to appeal to local preferences.

   d. A regional approach allows consumer goods companies to react more quickly to competition.

B. Demographic Segmentation

Demographic segmentation is the method of dividing markets on the basis of demographic variables, such as age, gender, income, ethnic background, and family life cycle.

Marketers use demographic information to segment markets because it is widely available and often related to consumers' purchasing and consuming behavior.

1. Age segmentation: Specific age groups are tremendously attractive markets for a variety of product categories. For example, children 14 and younger spend an estimated $20 billion per year.

2. Gender segmentation: Marketers of many items, such as clothes, footwear, personal care items, magazines, and cosmetics, commonly segment by gender.

3. Income segmentation: Income level influences consumers' wants and determines their buying power.

4. Ethnic Segmentation: Many companies are segmenting their markets by ethnicity, such as African-American, Asian-American or Hispanic markets. These three groups are projected to make up one-third of the country's population by 2010.

   a. Researchers have found differences in consumption patterns between ethnic groups.

   b. Even packaging choices differ between ethnic groups.

   c. At least half of all Fortune 500 companies have launched some ethnic marketing activities.

5. Family life cycle segmentation: The **family life cycle (FLC)** is a series of life stages, which are defined by a combination of age, marital status, and the presence or absence of children.

C. Psychographic Segmentation

Psychographic segmentation is the method of dividing markets based on personality, motives, lifestyle, and geodemographics.

1. Personality: Individual characteristics reflect traits, attitudes, and habits.

2. Motives: Consumers have motives for purchasing products. Marketers often try to appeal to such motives as safety, rationality, or status.

3. Lifestyles: Lifestyle segmentation divides individuals into groups according to activities, interests, and opinions. Often certain socioeconomic characteristics, such as income and education, are included.

4. Geodemographics: **Geodemographic segmentation** is the method of dividing markets on the basis of neighborhood lifestyle categories and is a combination of geographic, demographic, and lifestyle segmentations.

5. Geodemographics allows marketers to engage in the development of marketing programs tailored to prospective buyers who live in small geographic regions, such as neighborhoods, or who have very specific lifestyle and demographic characteristics

5. VALS approach: The **Values and Lifestyles Program**, or **VALS** is a consumer psychographic segmentation tool developed by SRI International to categorize U.S. consumers is

    a. VALS categorizes U.S. consumers by their values, beliefs, and lifestyles rather than by traditional demographic segmentation variables.

    b. The VALS groups are described by the intersection of two dimensions: resources and self-orientation.

    c. The eight VALS segments are Actualizers, Fulfillers, Believers, Achievers, Strivers, Experiencers, Makers, and Strugglers.

D. Benefit Segmentation

Benefit segmentation is the method of dividing markets on the basis of benefits consumers seek from the product.

1. By matching demographic information to interest in particular types of benefits, typical customer profiles can be built.

2. Customer profiles can be matched to certain types or times of media usage to develop promotional strategies for reaching these target markets.

E. Usage Rate Segmentation

Usage rate segmentation is the method of dividing a market based on the amount of product purchased or consumed.

1. The most common usage-rate categories are former users, potential users, first-time users, light users, medium users, and heavy users.

2. The **80/20 principle** is a business heuristic (rule of thumb) stating that 20 percent of all customers generate 80 percent of the demand. Although the percentages are not always

exact, a close approximation of this rule is often true. This principle reinforces the concept that heavy users, usually the most important and profitable part of a business, are actually a small percentage of the total number of customers.

5 Describe the bases for segmenting business markets

V. Bases for Segmenting Business Markets

Business market segmentation variables can be classified into two major categories.

A. Macrosegmentation

The first category, **macrosegmentation**, entails dividing business markets into segments based on general characteristics:

1. Geographic location: The demand for some business products varies considerably from one region to another.

2. Customer type: Segmentation by customer type allows business marketers to tailor their marketing mixes to the unique needs of particular organization types or industries.

3. Customer size: An organization's size may affect its purchasing procedures, the types and quantities of products that it needs, and its responses to different marketing mixes.

4. Product use: How customers use a product, particularly raw materials, may have a great deal of influence on the amount they buy, their buying criteria, and their selection of vendors.

B. Microsegmentation

The second category in business market segmentation is **microsegmentation**, the process of dividing business markets into segments based on characteristics of decision-making units within a macrosegment.

Some of the more commonly used microsegmentation variables are the following:

1. Key purchasing criteria: Each business places various amounts of importance on factors such as quality, delivery, technical support, and price.

2. Purchasing strategies:

   a. **Satisficers** usually contact familiar suppliers and place an order with the first to satisfy product and delivery requirements.

   b. **Optimizers** consider numerous suppliers, both familiar and unfamiliar, and then solicit bids and analyze options.

3. Importance of purchase: Classifying business customers according to the significance they attach to the purchase of a product is appropriate when the purchase is considered routine by some customers but very important for others.

4. Personal characteristics: The personal characteristics of purchase decision makers (their demographics, job responsibilities, and psychological makeup) influence their buying behavior and offer a viable basis for segmenting some business markets.

6 List the steps involved in segmenting markets

VI. Steps in Segmenting a Market

The purpose of market segmentation is to identify marketing opportunities (and eventually target markets) for existing or potential markets.

A. Select a market or product category for study

B. Choose a basis or bases for segmenting the market

1. The decision is a combination of managerial insight, creativity, and market knowledge; no scientific procedure has been developed for selecting segmentation variables.

2. The number of segmentation bases is limited only by decision makers' imagination and creativity.

C. Select segmentation descriptors

Descriptors are the specific segmentation variables to be used. For example, if demographics are the chosen base, age could be a descriptor variable.

D. Profile and evaluate segments

A profile should include a segment's size, expected growth rates, purchase frequency, current brand usage, brand loyalty, and overall long-term sales and profit potential.

The analysis stage can rank potential market segments by profit opportunity, risk, and consistency with the organizational mission and objectives.

E. Select target markets

F. Design, implement, and maintain appropriate marketing mixes

7 Discuss alternative strategies for selecting target markets

VII. Strategies for Selecting Target Markets

A target market is a group of people or organizations for which an organization designs, implements, and maintains a marketing mix to fit the needs of that group or groups, resulting in mutually satisfying exchanges.

A. Undifferentiated Targeting

The **undifferentiated targeting strategy** is a marketing approach based on the assumption that the market has no individual segments and thus requires a single marketing mix. It is a mass market philosophy, viewing the world as one big market with no individual segments.

1. There is only one marketing mix with an undifferentiated strategy.

2. It assumes that individual customers have similar needs.

3. The first firm in an industry sometimes uses this targeting strategy, because no competition exists.

4. With this strategy, production and marketing costs are often at their lowest.

5. This strategy leaves opportunities for competitors to enter the market with more specialized products and appeals to specific parts of the market..

B. Concentrated Targeting

The **concentrated targeting strategy** is a marketing approach based on appealing to a single segment of a market.

1. It focuses a firm's marketing efforts on a single segment or market **niche.**

2. A firm can concentrate on understanding the needs, motives, and satisfactions of the members of one segment and on developing and maintaining a highly specialized marketing mix.

3. A concentrated strategy allows small firms to be competitive with very large firms because of a small firm's expertise in one area of the market.

4. The dangers associated with this type of strategy include changes in the competitive environment which could destroy the only segment targeted.

C. Multisegment Targeting

The **multisegment targeting strategy** is a marketing approach based on serving two or more well-defined market segments, with a distinct marketing mix for each.

1. Some companies have very specialized marketing mixes for each segment, including a specialized product; other companies may only customize the promotional message for each target market.

2. The multisegment strategy offers many benefits, including potentially greater sales volume, higher profits, larger market share, and economies of scale in marketing and manufacturing.

3. This strategy also includes many extra costs, such as:

   - Product design costs
   - Production costs
   - Promotion costs
   - Inventory costs
   - Marketing research costs
   - Management costs
   - **Cannibalization**, which occurs when sales of a new product cut into sales of a firm's existing products..

8 Explain how and why firms implement positioning strategies and how product differentiation plays a role

VIII. Positioning

Positioning is the development of a specific marketing mix to influence potential customers' overall perception of a brand, product line, or organization in general.

The term **position** refers to the place a product, brand, or group of products occupies in consumers' minds relative to competing offerings.

A. Product Differentiation

Product differentiation is a positioning strategy designed to distinguish one firm's products from another's.

1. The difference can be real or perceived

2. The aim of differentiation is to convince customers that a brand is significantly different from the others and should therefore be demanded over competing brands

B. Perceptual Mapping

Perceptual mapping is a means of displaying or graphing, in two or more dimensions, the location of products, brands, or groups of products in the minds of present or potential customers.

C. Positioning Bases

1. A product is associated with an attribute, product feature, or customer benefit.

2. The position may stress high price as a symbol of quality or low price as an indicator of value.

3. Stressing uses and applications can be an effective means of positioning a product.

4. Another base may focus on a personality or type of user.

5. One position is to associate the product with a particular category of products.

6. Positioning against competitors is often a part of any positioning strategy.

D. Repositioning

Repositioning is changing consumers' perceptions of a brand in relation to competing brands.

9 Discuss global market segmentation and targeting issues

IX. Global Issues in Market Segmentation and Targeting

A. Two divergent trends, toward global marketing standardization and toward micromarketing in global markets, are occurring at the same time.

B. The steps in global market segmentation and targeting are the same as in local marketing; the main difference is the segmentation variables commonly used.

VOCABULARY PRACTICE

Fill in the blank(s) with the appropriate term or phrase from the alphabetized list of chapter key terms.

benefit segmentation
cannibalization
concentrated targeting strategy
demographic segmentation
80/20 principle
family life cycle (FLC)
geodemographic segmentation
geographic segmentation
macrosegmentation
market
market segment
market segmentation
microsegmentation
multisegment targeting strategy
niche
optimizer
perceptual mapping
position
positioning
product differentiation
psychographic segmentation
repositioning
satisficer
segmentation bases (variables)
target market
undifferentiated targeting strategy
usage rate segmentation

1 Describe the characteristics of markets and market segments

1. People or organizations with needs or wants and with the ability, and the willingness, to buy are a(n) ______________________________.

2. A subgroup of people or organizations sharing one or more characteristics that cause them to have similar product needs is a(n) ______________________________. The process of dividing a market into meaningful, relatively similar, and identifiable groups is called ______________________________.

4 Describe bases commonly used to segment consumer markets

3. Characteristics of individuals, groups, or organizations that are used to divide a total market into segments are called ______________________________.

4. Dividing markets based on region of the country or world, market size, market density, or climate is referred to as ______________________________.

5. Marketers can segment markets according to age, gender, income, and ethnic background. This is called ______________________________. Another common basis for this type of segmentation is a series of stages that uses a combination of age, marital status, and the presence or absence of children. This is known as the ______________________________.

6. Market segmentation based on personality, motives, or lifestyles, is known as ______________________________. One type of this segmentation combines geographic, demographic, and lifestyle variables to cluster potential customers into neighborhood lifestyle categories. This is called ______________________________.

7. The process of dividing a market according to different benefits customers seek from the product is called ______________________________. Dividing the market based on the amount of product purchased or consumed is called ______________________________. Segmenting in this manner allows marketers to concentrate on heavy users. Heavy users account for a disproportionate share of the total consumption of many products as stated by the ______________________________.

5 Describe the bases for segmenting business markets

8. Business market segmentation variables can be classified in tow major categories. The first category entails dividing markets into segments according to such general characteristics as geographic location, type of organization, customer size, or product use. This is termed ______________________________. The second category focuses on the characteristics of the decision-making units within the segments identified in category one. This is termed ______________________________. One characteristic of the decision-making unit is purchasing strategies. Two purchasing profiles have been identified. The first type contacts familiar suppliers and places an order with the first to satisfy product and delivery requirements. This type is called a(n) ______________________________. The second type considers numerous suppliers, both familiar and unfamiliar, solicits bids, and examines all alternatives carefully. This type is called a(n) ______________________________.

7 Discuss alternative strategies for selecting target markets

9. A group of people or organizations for which a marketer designs, implements, and maintains a marketing mix intended to meet the needs of that group is called a(n) ______________________________.

10. There are three general strategies for selecting target markets. If a firm adopts a mass-marketing philosophy and views the market as one big market with no segments, the firm is using a(n) ______________________________. If a firm selects one segment of a market for targeting its marketing efforts, it is using a(n) ______________________________, and the one segment is called a(n) ______________________________. Finally, if a firm chooses to serve two or more well-defined market segments and develops a distinct marketing mix for each segment, it is practicing ______________________________. One of the dangers of this last segmenting strategy is that sales of a firm's new product directed at one segment may cause a decline in sales of the firm's other products targeted at existing segments. This is called ______________________________.

8 Explain how and why firms implement positioning strategies and how product differentiation plays a role

11. There are two important marketing strategies related to targeting. The first strategy develops a specific marketing mix to influence potential customers' overall perception of a brand, product line, or organization in general. This strategy is called ______________________________. The strategy assumes that competing products are arranged in consumers' minds along various dimensions. The place a product occupies in consumers' minds relative to competing offerings is the ______________________________. Another strategy is used to distinguish one firm's products from another's. These distinctions can be either real or perceived. This strategy is called

______________________________. A means of displaying these dimensions and locations is known as ____________________________. These can be also used for plans to change consumers' perceptions, a process called ______________________________.

Check your answers to these questions before proceeding to the next section.

TRUE/FALSE QUESTIONS

Mark the statement **T** if it is true and **F** if it is false.

1 Describe the characteristics of markets and market segments

_____ 1. Critter Chow, Inc., has developed a new animal food. The company has identified pet ferrets, skunks, otters, weasels, raccoons, and badgers as animals that would eat the food. These animals make up the market for Critter Chow's new food.

2 Explain the importance of market segmentation

_____ 2. Large companies like Coca-Cola do not practice segmentation because their products are targeted at the mass market.

3 Discuss criteria for successful market segmentation

_____ 3. It is important for a targeted market segment to be identifiable and measurable because it must sustain long-term sales and profit for the marketer.

4 Describe bases commonly used to segment consumer markets

_____ 4. Levi's produces Dockers clothing designed for the baby boomer generation. Levi's is using the family life cycle as a segmentation variable for Dockers.

_____ 5. The Sharper Image sells a variety of high-tech items through its catalogs and specialty shops. The items are designed to appeal to the wealthy "yuppie" market, especially those consumers who are adventurous and active and have a flair for self-expression. The Sharper Image should use demographic variables for market segmentation.

_____ 6. Oral-B makes a variety of toothbrushes. One type has hard nylon bristles for extra cleaning power, while another type has soft bristles for gentle care of sensitive teeth. Smaller brushes with cartoon characters on the handle are designed for children. Finally, a toothbrush with an angled handle is for taking care of hard-to-clean areas. Oral-B uses benefit segmentation.

_____ 7. Many beer companies target college-age males because this group consumes a disproportionate amount of the product. These beer companies practice usage rate segmentation.

7 Discuss alternative strategies for selecting target markets

_____ 8. The C&H sugar company does not single out any particular subgroup within the population of sugar users but instead directs one marketing mix at everyone. This is an undifferentiated targeting strategy, also known as mass marketing.

_____ 9. Johnson & Johnson makes bandages and other first-aid products. Some products are designed for adults, and others are designed specifically for children, with smaller sizes and cartoon figure decorations. Johnson & Johnson engages in multisegment targeting.

8 Explain how and why firms implement positioning strategies and how product differentiation plays a role

_____ 10. Perceptual maps indicate where firms are trying to position their products.

Check your answers to these questions before proceeding to the next section.

AGREE/DISAGREE QUESTIONS

For the following statements, indicate reasons why you may agree and disagree with the statement.

1. Marketers should always practice market segmentation.

   Reason(s) to agree:

   Reason(s) to disagree:

2. The most important criterion for market segmentation is substantiality.

   Reason(s) to agree:

   Reason(s) to disagree:

3. To a certain extent, all firms practice geographic segmentation.

   Reason(s) to agree:

   Reason(s) to disagree:

4. Psychographic segmentation should not be used, as other bases for segmentation (demographic and geographic) are based on more accurate data.

   Reason(s) to agree:

Reason(s) to disagree:

5. It is better to launch a new product than to reposition an old one.

   Reason(s) to agree:

   Reason(s) to disagree:

MULTIPLE-CHOICE QUESTIONS

Select the response that best answers the question, and write the corresponding letter in the space provided.

1 Describe the characteristics of markets and market segments

____ 1. Which of the following is NOT necessarily a characteristic of a market?
   a. Willingness to buy the product.
   b. Ability to buy the product.
   c. Are aware of the product.
   d. A group of people or organizations.
   e. Have needs or wants.

____ 2. Pepper Planes has decided that all of its customers are not similar enough to respond to one marketing mix. In particular, some of its accounts are large corporations that need planes for executive transportation, and others are farmers who need planes for crop dusting. The procedure of dividing this market into identifiable, similar groups is called:
   a. micromarketing
   b. positioning
   c. cannibalization
   d. market segmentation
   e. perceptual mapping

2 Explain the importance of market segmentation

____ 3. Shirley owns the Coachlight Travel Agency and would like to improve customer satisfaction and increase repeat business. You ask Shirley to describe a typical customer, and she says it is hard to find one kind of customer. With corporate travel, family vacations, retirement cruises, college spring breaks, and honeymoons, it is hard to know how to serve all these accounts. You suggest that it is time for market segmentation because:
   a. Shirley needs to learn how to group these markets together into one market to serve them adequately
   b. Shirley needs to reduce the size of her market served
   c. it will enable Shirley to build an accurate description of the customer needs by group and design a marketing mix to fit each segment
   d. it will help develop a generalized definition of the market as a whole and the optimal marketing mix for this market
   e. this will position Shirley's company in the minds of her consumers as compared to the competition

____ 4. The Poquet Company, a manufacturer of hand-held computers, completed a thorough examination and analysis of its business customers two years ago. They grouped the customers into four segments based on size, geographic region, and benefits sought. Would you recommend a new segmentation analysis this year?
a. Yes, I would recommend one even more regularly because of the rapidly changing nature of most markets.
b. No, once every five years is about average.
c. No, business customer markets are not rapidly changing or developing like consumer goods markets.
d. Yes, but use different bases to get some variety.
e. No, segmentation is rarely done by business-to-business marketers because it is not useful.

3 Discuss criteria for successful market segmentation

____ 5. Which of the following factors should a museum consider prior to engaging in segmentation?
a. ability to identify and measure segments
b. differentiated responses among segments
c. ability to reach targeted segments
d. sufficient size to warrant developing a unique marketing mix
e. all of the above

____ 6. A manufacturer of photographic equipment segments the camera market by use: family and personal, hobbyists, professional, and scientific. In order for this segmentation scheme to be successful, all of the following criteria must be met EXCEPT:
a. substantiality
b. accessibility
c. identifiability and measurability
d. responsiveness
e. complexity

____ 7. The manager of CritterEats noted that due to an increase in owners of nontraditional pets such as ferrets, the exotic animal feed segment could be classified as having substantiality. This means that it:
a. has enough special stores, magazines, and other outlets that it will be possible to direct advertisements at this group
b. is too large and needs to be reduced to a more easily identifiable and measurable size
c. exhibits a response rate to marketing variables different from the rates of other segments
d. is large enough to permit a viable market effort toward its members
e. will be difficult to develop a product to match this group of buyers

____ 8. Marketers sometimes ignore small market niches—even if it has unfulfilled needs—because these niches lack:
a. substantiality
b. identifiability and measurability
c. responsiveness
d. accessibility
e. causality

____ 9. The Help-U Program will be offering a course to help people become more assertive. It wants to segment the market and to slant its ads based on a consumer's amount of timidity and shyness. The first segmentation criterion problem that would greet this proposal is:
a. identifiability and measurability
b. responsiveness
c. accessibility
d. substantiality
e. responsibility

_____ 10. The city of Calsey has a literacy program for migrant farm workers but has had a difficult time reaching this group with information about the program, even though radio and television stations have provided public service announcements free of charge. This is a segmentation problem called:
a. substantiality
b. accessibility
c. responsiveness
d. identifiability and measurability
e. causality

_____ 11. The Good Doggie firm has a new training program especially for puppies. After placing fliers around town, the classes filled up within two days. Which segmentation success criterion is in force?
a. accessibility
b. identifiability and measurability
c. responsiveness
d. substantiality
e. causality

4 Describe bases commonly used to segment consumer markets

_____ 12. The Merry Maid Home Cleaning Service offers three levels of cleaning: luxury-deluxe service (which includes a pet shampoo and ceiling scrubbing), the moderately priced basic package, and budget quick-clean service (vacuuming and laundry only). The income characteristic that has been used to divide its customers into segments is called the:
a. accessibility quotient
b. perceptual map
c. segmentation base
d. differentiation rule
e. 80/20 principle

_____ 13. A new sport utility vehicle is targeting people who live in regions with rugged terrain, such as the desert or mountainous areas. This is an example of:
a. topographic segmentation
b. geographic segmentation
c. demographic segmentation
d. benefit segmentation
e. geodemographic segmentation

_____ 14. At an affordable price, Dodge Neon is targeting Generation X. This is an example of:
a. usage rate segmentation
b. benefit segmentation
c. psychographic segmentation
d. demographic segmentation
e. geodemographic segmentation

_____ 15. The director of marketing has just completed demographic research on the total market for home improvement supplies. Which of the following demographic data would be his best choice for developing his market segmentation strategy?
a. age data about home improvement buyers
b. lifestyles of home improvement buyers
c. home improvement usage patterns among buyers
d. the benefits sought by home improvement buyers
e. values of home improvement buyers

_____ 16. A new minivan is targeting households with a married couple who have three or more young children. This kind of demographic targeting is based on:
a. micromarkets
b. the family life cycle
c. the VALS program
d. geodemographics
e. psychographics

____ 17. Early to Rise hot air balloon company targets people who are adventurous and fun-loving. This is an example of:
a. demographic segmentation
b. usage rate segmentation
c. benefit segmentation
d. psychographic segmentation
e. family life cycle segmentation

____ 18. Wrigley's Chewing Gum ran an advertising campaign that targeted heavy smokers. The advertising indicated that "when you can't smoke, chew Wrigley's." This is an example of what type of segmentation?
a. geodemographic
b. benefit
c. demographic
d. psychographic
e. usage rate

____ 19. The Parker Pen Company manufactures several different categories of writing instruments: inexpensive disposable ballpoints, pens with erasable ink, pens specifically designed for artwork, and expensive executive pens. The Parker Pen Company is using which type of segmentation?
a. benefit
b. usage rate
c. demographic
d. psychographic
e. geodemographic

____ 20. When Jon opened his music store he noticed that he sold most of his cassette tapes to the same small group of customers (about eighty people), even though his sales records showed that he had over 400 regular cassette tape purchasers. This is a marketing phenomenon called the:
a. optimizer principle
b. music loyalty rule
c. 80/20 principle
d. cannibalization rule
e. majority fallacy

5 Describe the bases for segmenting business markets

____ 21. The Teletech Telephone Supply Company has categorized its business customers by their purchasing strategy used. Teletech has found that it is much easier to serve the customer that prefers to recontact familiar suppliers and place an order immediately if product and delivery requirements are acceptable. These customers can be described as:
a. experiencers
b. satisficers
c. optimizers
d. strugglers
e. actualizers

____ 22. The Teletech Telephone Supply Company finds that the most difficult firms to service in the purchasing-strategy-used segmentation scheme are the ___________. This group considers numerous, even unfamiliar suppliers, solicits bids, and carefully analyzes options; therefore, it requires a higher level of customer service and follow-up on quotations.
a. actualizers
b. strivers
c. satisficers
d. achievers
e. optimizers

7 Discuss alternative strategies for selecting target markets

_____ 23. Industry-Quip has decided to enter the industrial heating and air-conditioning market, and all the current competitors in that market use a differentiated product market strategy. Which strategy would make the LEAST sense for Industry-Quip?
a. product differentiation strategy
b. specialized product strategy
c. concentrated or niche targeting strategy
d. undifferentiated targeting strategy
e. multisegment targeting strategy

_____ 24. Which of the following marketers is most likely to be able to practice undifferentiated marketing?
a. the owner of the only hardware store in a small rural town located in Texas
b. the owner of a three-store chain of budget outlets on the lower eastern shore of Maryland
c. the owner of a print-shop in Chicago
d. the administration of a small private college in upstate New York
e. the operator of a year-round resort in Wisconsin

_____ 25. Left-Out, Inc., manufactures knives, scissors, and other utensils specifically designed for left-handed consumers. Left-Out is using which type of strategy?
a. concentrated or niche targeting
b. undifferentiated targeting
c. multisegment marketing
d. universal product coding
e. specialized marketing

_____ 26. When Procter & Gamble introduced Liquid Tide to a new segment, consumers in the traditional powdered detergent segment switched to the liquid product. Rather than real sales growth, P&G simply experienced the shifting of existing customers to a new product. This drawback of a multisegment targeting strategy is called:
a. shift fallacy
b. perceptual confusion
c. undifferentiation
d. repositioning
e. cannibalization

8 Explain how and why firms implement positioning strategies and how product differentiation plays a role

_____ 27. A producer of teas commissioned a large research project on different brands of tea. The result was a graphical display of how consumers viewed these brands. Some brands were viewed as being "for iced tea," while others were considered as "hot teas." Also, some brands were viewed as traditional, while others were considered to have a variety of flavors. This is an example of:
a. a positioning map
b. cannibalization
c. a segmentation study
d. a perceptual map
e. target marketing

_____ 28. In the late 1980s, Oldsmobile tried to change its image from being a car for older people to being a car for younger people by using an advertising campaign, "This is Not Your Father's Oldsmobile." This effort is an example of:
a. reimaging
b. perceptual mapping
c. resegmenting
d. market segmentation
e. repositioning

9 Discuss global market segmentation and targeting issues

_____ 29. You are a marketing manager wanting to segment European and Asian food markets. You might use all of the following segmentation variables EXCEPT:
  a. per capita GNP
  b. VALS
  c. religion
  d. culture
  e. political system

Check your answers to these questions before proceeding to the next section.

ESSAY QUESTIONS

1. You are given the following limited information about a market consisting of ten people. Describe all the possible ways to segment this market.

Age Group	Geographic Region
Adult	Northeast
Child	South
Child	South
Adult	West
Child	South
Adult	South
Adult	Northeast
Adult	West
Child	West
Child	West

2. Head and Shoulders Shampoo has targeted a new segment of people with dry scalps. To be useful, a segmentation scheme must produce segments that meet four basic criteria. Name and briefly describe each of these four criteria, and assess whether a dry scalp segment meets these criteria.

3. You would like to market a line of aromatherapy candles that not only provide light but also help people relax. Using all five variables for segmenting consumer markets, describe a target market for your products.

4. For toothpaste, list eight benefits that might be sought by consumers. For each benefit, give an existing brand name that best exemplifies segmentation according to that benefit.

5. You would like to start a new food store concept that sells only wholesome and organic foods. Describe three possible target markets for your new concept, using the three strategies for targeting markets.

6. You are the marketing manager for an automobile firm. Your firm would like to introduce a new electric car to the marketplace. You have determined that two important dimensions for positioning a product are (1) styling (sporty or traditional) and (2) price (high or low). Place the competing cars listed below on the perceptual map provided. Assuming the listed cars comprise all competitive cars, what position should your firm's new car have?

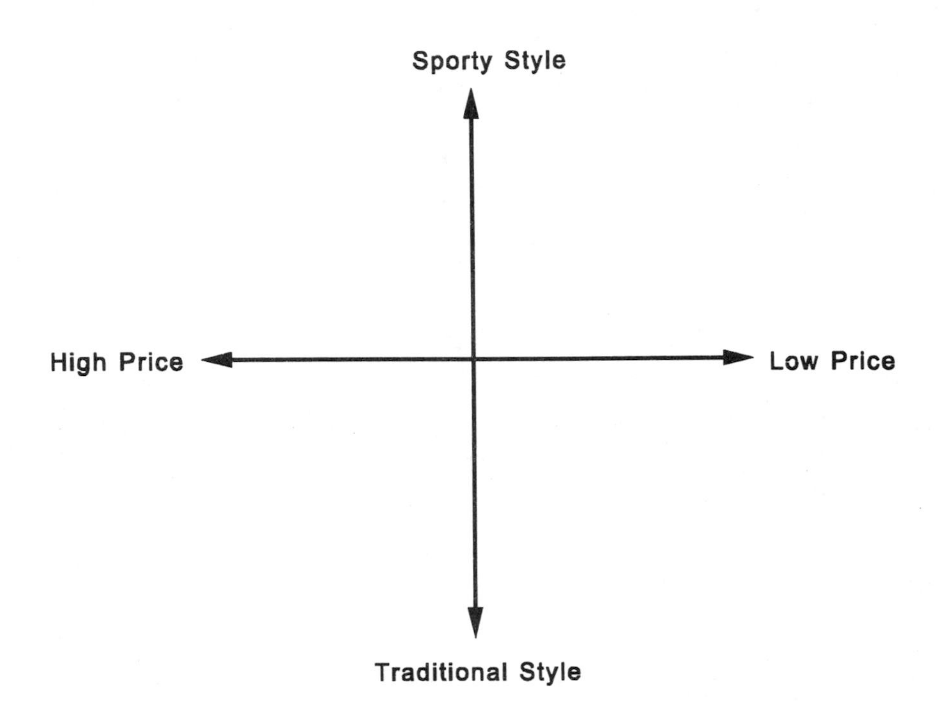

Porsche, Cadillac sedan, Jaguar, Lincoln Continental, Ferrari, Oldsmobile sedan, Honda Prelude, Buick sedan, Corvette, Ford F-150 utility truck, Acura 4-door

CHAPTER NOTES

Use this space to record notes on the topics you are having the most trouble understanding.

ANSWERS TO THE END-OF-CHAPTER DISCUSSION AND WRITING QUESTIONS

Use this space to work on the questions at the end of the chapter.

ANSWERS TO THE END-OF-CHAPTER CASE

Use this space to work on the questions at the end of the chapter.

CHAPTER 8 Decision Support Systems and Marketing Research

LEARNING OBJECTIVES

1 Explain the concept and purpose of a marketing decision support system

A decision support system (DSS) is an interactive, flexible information system that enables managers to obtain and manipulate information as they are making decisions. A DSS bypasses the information-processing specialist and lets managers instantly access data from their terminals. Successful use of a DSS allows the firm to respond quickly and creatively to marketing trends and opportunities.

2 Define marketing research and explain its importance to marketing decision making

Marketing research is the planning, collection, and analysis of data relevant to marketing decision making. The results of this analysis are then communicated to management. Marketing research provides managers with data on the effectiveness of the current marketing mix and provides insights for changes. It is an important tool for reducing uncertainty in decision making.

3 Describe the steps involved in conducting a marketing research project

The first step is to define the problem or questions that this research needs to examine. The next step, planning the research design, specifies method that will be used to collect data. Then the sampling procedures that best fit this situation are selected. Next the data are collected, often by an outside firm. Then data analysis takes place, and the results are interpreted. Subsequently, a report is drawn up and presented to management. A follow-up on the usefulness of the data and the report is the final step.

4 Discuss the growing importance of scanner-based research

A prominent goal of marketing research is to develop an accurate, objective picture of the direct causal relationship between different kinds of sales and marketing efforts and actual sales. Single-source research is bringing marketers closer to that goal. Scanner-based research gathers information from a single panel of respondents by continuously monitoring the advertising, promotion, and pricing the panel is exposed to and what it subsequently buys.

Two electronic tools help to manage the scanner system: television meters and laser scanners that read the UPC codes on products. These tools allow researchers to observe what advertisements reach a home and what products are purchased by the home.

5 Explain when marketing research should and should not be conducted

Marketing research is not always the correct solution to a problem. In several situations it may be best not to conduct marketing research. Sometimes a company already possesses extensive consumer information from years of research and interaction with the market. It would be redundant and a waste of money to conduct more research in this situation. Another point to consider is that information is frequently very costly to acquire and may actually cost more than the earnings provided by the improvement in the decision. Research should only be undertaken when the expected value of the information is greater than the cost of obtaining the data.

PRETEST

Answer the following questions. After you finish, check your answers with the solutions at the back of this Study Guide.

1. List four characteristics that a DSS system should have.

2. What is marketing research? List and explain the three roles of marketing research.

3. List the seven steps of the marketing research process.

4. List nine types of survey research.

5. What is scanner-based research?

CHAPTER OUTLINE

1 Explain the concept and purpose of a marketing decision support system

I. Marketing Decision Support Systems

A. Accurate and timely information is the lifeblood of marketing decision making. **Marketing intelligence** is everyday information about developments in the marketing environment that managers use to prepare and adjust marketing plans. The system for gathering marketing intelligence is called a marketing decision support system (DSS).

B. A **marketing decision support system (DSS)** is an interactive, flexible information system that enables managers to obtain and manipulate information as they are making decisions.

A DSS bypasses the information-processing specialist and gives managers access to the data from their terminals.

The four characteristics of a DSS are that it is interactive, flexible, discovery-oriented, and accessible.:

2 Define marketing research and explain its importance in marketing decision making

II. The Role of Marketing Research

A. **Marketing research** is the planning, collection, and analysis of data relevant to marketing decision making. It plays a key role in the marketing system.

1. provides managers with data on the effectiveness of the current marketing mix
2. provides insights for changes
3. is the primary data source for and DSS

B. Roles of Marketing Research

1. Descriptive: gathering and presenting statements of fact
2. Diagnostic: explaining data
3. Predictive: attempting to estimate the results of a planned marketing decision

C. Differences between Marketing Research and DSS

Marketing research is oriented to specific problems

DSS continually collects a variety of information and channels it into the organization. Marketing research is part of the DSS system.

D. Management Uses of Marketing Research

1. Marketing research improves the quality of marketing decision making.
2. Marketing research is also used to discover what went wrong with a marketing plan.

3. Marketing research is used to retain customers by having an intimate understanding of their needs.

4. Marketing research helps managers understand what is going on in the marketplace and take advantage of opportunities.

3 Describe the steps involved in conducting a marketing research project

III. Steps in a Marketing Research Project

Virtually all firms that have adopted the marketing concept engage in marketing research, which can range from a very formal study to an informal interview with a few customers.

The marketing research process is a scientific approach to decision making that maximizes the chance for getting accurate and meaningful results.

A. Problem/Opportunity Identification and Formulation

1. The research process begins with the recognition of a marketing problem or opportunity. It may be used to evaluate product, promotion, distribution or pricing alternatives, and/or find and evaluate new market opportunities.

2. The **marketing research problem** involves determining what information is needed and how that information can be obtained efficiently and effectively.

3. The **marketing research objective** is to provide insightful decision-making information. This requires specific pieces of information needed to answer the marketing research problem.

4. The **management decision problem** is broader in scope and far more general. It is action oriented.

5. **Secondary data** are data previously collected for any purpose other than the one at hand.

a. Data may already exist that can quickly and inexpensively meet present needs.

b. Advantages of using secondary data are the savings in time and money, the aid in formulating the problem statement, and the basis for comparison they provide.

c. Disadvantages include the lack of fit between this unique problem and the data that have already been collected.

d. Another disadvantage is that it assessing the quality and accuracy of secondary data is hard.

6. The **Internet** is a world-wide telecommunications network. The **World Wide Web (or Web)** is one component of the Internet which was designed to simplify transmission of text and images.

a. Search Engines

- A web address or URL (**Uniform Reference Locator**) is similar to a street address in that it identifies a particular location.

- Search engines contain collections of links to documents throughout the world. You can use the search engine to locate sites that include information that meets your requirements.

b. Discussion Groups include newsgroups that function much like bulletin boards for a particular topic or interest.

c. Databases on CD ROM include segmentation and demographic studies and mapping.

- The Department of Commerce makes census data available and TIGER files which contain street maps of the United States.

B. Planning the Research Design and Gathering Primary Data

1. The **research design** specifies which research questions must be answered using primary data, how and when the data will be gathered, and how the data will be analyzed.

2. **Primary data** are pieces of information collected for the first time and used for solving the particular problem under investigation.

   a. Must be collected if the specific research question cannot be answered by available secondary data

   b. Answer specific questions

   c. Are current

   d. Are gathered using methodology specified by the researcher

   e. Can be gathered in such a way as to maintain accuracy and secrecy.

3.. A major disadvantage of primary data collection is the expense.

   a. To limit the cost of data collection, researchers may cut back on the number of interviews to be performed.

   b. They may also economize by piggybacking, or collecting data on two different projects using one questionnaire. But piggybacking may confuse participants and lead to long, tiring interviews.

4. The most popular technique for gathering primary data is **survey research**, in which a researcher interacts with people to obtain facts, opinions, and attitudes.

   a. In-home interviews often provide high-quality information, offer the opportunity to gain in-depth responses, but are extremely expensive because of the interviewer's time and mileage

   b. **Mall intercept interviews** are a survey research method that involves interviewing people in the common areas of shopping malls. They must be briefer than in-home interviews, often do not provide a representative sample of consumers, provide an overall quality of information similar to that of telephone interviews, and are less costly than in-home interviews.

   c. Another technique is **computer-assisted personal interviewing**, with the interviewer reading the questions from a computer screen and entering the respondents comments. A second approach is **computer-assisted self interviewing.**

d. Telephone interviews offer the advantage of lower cost than personal interviews, and Have the potential for the best sample among interview types

1) Sometimes interviews are conducted from the interviewer's home, but most often are executed from a **central-location telephone (CLT) facility**, which is a specially designed room for conducting telephone interviews for survey research. These facilities have a number of phone lines and monitoring equipment in one location. Many CLT facilities offer **computer-assisted interviewing**, in which interviewers read questions from a screen and input responses directly into the computer.

2) A new trend is in-bound telephone interviewing, where consumers call in after using samples sent to them as part of a survey. The customer calls an interactive voice mail system.

e. Mail surveys, have relatively low costs, eliminate interviewers and field supervisors, centralize control, provide (or promise) anonymity for respondents, and produce low response rates.

1). Computer-disc by mail survey is another new technique. It has the Advantage of building in skip patterns and can display graphics.

f. **Mail panels** consist of a sample of households recruited to participate by mail for a given period, responding repeatedly over time.

g. Internet surveys provide the advantages of speed, low cost, creation of longtidinal studies, cost effectiveness of short surveys, the ability to reach large audiences, and eye appeal.

Disadvantages encountered when using the Internet to conduct surveys include the unrepresentativeness of the sample, security, the respondents are self-selecting and may complete the survey more than once.

Internet samples may be classified as unrestricted, screened, and recruited.

Screened Internet samples adjust for the unrepresentativeness of the self-selected respondents by imposing quotas based on some deserved sample characteristics.

Recruited Internet samples are used for targeted populations in surveys that require more control over the make-up of the sample.

h. **Focus groups**: are groups of seven to ten people with desired characteristics who participate in a group discussion about a subject of interest to a marketing organization. The **group dynamics** and interaction are essential to the success of this method.

1) The newest development in qualitative research is the on-line or cyber focus group.

2) The benefits of online focus groups are-speed, cost-effectiveness, broad

geographic scope, accessibility
and honesty

3) The disadvantages of online focus groups are an unrepresentative sample, uncertainty as to how group dynamics work, the honesty of behavior, opinion and feelings, and the inability to observe appearances and body language of the participants

4. All forms of survey research require a questionnaire to ensure that all participants are asked the same series of questions.

   a. **Open-ended questions** are those worded to encourage unlimited answers phrased in the respondent's own words.

   b. **Closed-ended questions** are those in which the respondent is provided with a list of responses and is requested to make a selection from that list. These questions may be dichotomous (only two possible answers) or multiple choice.

   c. A **scaled-response question** is a closed-ended question designed to measure the intensity of a respondent's answer.

   d. Qualities of good questionnaires include clearness, conciseness, objectiveness, and reasonable terminology.

   e. Care should be taken to ask only one question at a time.

   f. The purpose of the survey and expectations of the participants' behavior should be stated up front.

5. **Observation research** is a research method that relies on three types of observation: people watching people, people watching physical phenomena, and machines watching people.

   a. People watching people is one form of observation research.

      1) Mystery shoppers are researchers who pose as customers to observe the quality of service being offered.

      2) One-way mirror observations are used to see how consumers use and react to products.

   b. Another form of observation research is people watching physical phenomena.

      An **audit** is the examination and verification of the sale of a product.

      Because of the availability of scanner-based data, physical audits at the retail level may someday all but disappear.

   c. Machines watching people has become a more prevalent mode of research in the last few years, primarily used to measure response to advertisements. Four types of machines are used:

      1) Traffic counters measure the flow of vehicles or people over a stretch of roadway.

2) Passive people meters provide information on which television shows are being watched, the number of households watching, and which family members are watching. The passive system will someday be programmed to recognize faces and record electronically who is watching what.

d. Advantages of observation research include the elimination of interviewer bias and the information gathered does not have to rely on the willingness of the respondent.

e. Disadvantages of observation research include data-collection costs may run high unless observed behavior patterns occur frequently, briefly, or somewhat predictably, and motivations, attitudes, and feelings are not measured, limiting the richness of the results.

6. In an **experiment**, the researcher changes one or more variables (such as price or package design) while measuring the effects of those changes on another variable (usually sales).

C. When specifying the sampling procedures, the **sample** or subset of the larger population to be drawn for interviewing must be determined.

- First, the population from which a sample is drawn, or **universe** of interest, must be defined.
- Next, the researcher must determine if the sample for this study should be representative of the population as a whole.
- If yes, then a probability sample is called for.

1. With a **probability sample**, every element in the population has a known nonzero probability of being selected. Scientific rules are used to ensure that the sample represents the population.

   a. A **random sample** is set up so that every element of the population has an equal chance of being selected as part of the sample.

   b. Often a random sample is selected by using random numbers from a table found in statistics books.

2. A **nonprobability sample** is any sample in which little or no attempt is made to get a representative cross section of the population.

   a. A **convenience sample** uses respondents who are readily accessible to the researcher.

   b. Nonprobability samples are frequently used in market research, because they cost less than probability samples.

3. Several types of errors are associated with sampling.

   a. **Measurement error** occurs when the information desired by the researcher differs from the information provided by the measurement process.

   b. **Sampling error** occurs when a sample is not representative of the target population in certain respects.

1) Nonresponse error, occurs when certain members of the sample refuse to respond and their responses may have differed from the sample as a whole.

2) **Frame error** arises if the sample drawn from a population differs from the target population.

3) **Random error** occurs when the selected sample does not represent the overall population.

D. Collecting the Data

Most data collection is done by marketing research field services. **Field service firms** specialize in interviewing respondents on a subcontract basis.

Field service firms also conduct focus groups, mall intercepts, retail audits, and other data-collection services.

E. Analyzing the Data

1. Three types of analysis are common in marketing research.

a. One-way frequency counts are the simplest, noting how many respondents answered a question a certain way. This method provides a general picture of the study's results.

b. **Cross tabulations** relate the responses to one question to the responses to one or more other questions.

c. Statistical analysis, the most sophisticated type, offers a variety of techniques for examining the data.

F. Prepare and Present the Report

The research must communicate the conclusions and recommendations to management. Usually both oral and written reports are required.

G. Follow Up

The researcher should determine if management followed the recommendations and why or why not.

4 Discuss the growing importance of scanner-based research

IV. Scanner-based Research

Scanner-based research is a system for gathering information from asingle group of respondents by continuously monitoring the advertising, promotion, and pricing panel members are exposed to and the things they buy.

A. Two electronic tools help to manage the single-source system: television meters and laser scanners, which read the bar codes on products.

B. The two major suppliers of single-source research are Information Resources Incorporated (IRI) and the A.C. Nielsen Company.

1. IRI offers information collected through its **BehaviorScan** program, a single-source research program that tracks the purchases of 3,000 households through store scanners. Household purchasing is monitored, as well as exposure to such marketing variables as TV advertising and consumer promotions.

2. Another IRI product, **InfoScan**, is a scanner-based national and local sales-tracking service for the consumer packaged-goods industry. Retail sales, detailed consumer purchasing information, and promotional activity are monitored and evaluated for bar-coded products.

3. The next generation of scanners, Scanner Plus, will be able to communicate with personal computers in homes and analyze the consumption patterns of individual households. With this information, retailers can offer menu or product-use suggestions with an associated shopping list.

5 Explain when marketing research should and should not be conducted

V. When Should Marketing Research Be Conducted?

A. Market research is not always the correct solution to a problem. In several situations it may be best not to conduct market research:

1. When decision-making information already exists

2. When the costs of conducting research are greater than the benefits

VOCABULARY PRACTICE

Fill in the blank(s) with the appropriate term or phrase from the alphabetized list of chapter key terms.

audit
BehaviorScan
central-location telephone (CLT) facility
closed-ended question
computer-assisted personal interviewing
computer-assisted self interviewing
computer disk by mail survey
convenience sample
cross-tabulation
decision support system (DSS)
experiment
field service firms
focus group
frame error
group dynamics
in-bound telephone surveys
InfoScan
integrated interviewing
Internet
mall intercept interview
management decision problem
marketing intelligence
marketing research
marketing research objectives
marketing research problem
measurement error
nonprobability sample
observation research
open-ended question
primary data
probability sample
random error
random sample
recruited Internet sample
research design
sample
sampling error
scaled-response question
scanner-based research
screened Internet sample
secondary data
survey research
uniform reference locator (URL)
universe
unrestricted Internet sample
World Wide Web (web)

1 Explain the concept and purpose of a marketing decision support system

1. Everyday information about developments in the marketing environment that managers use to prepare and adjust marketing plans is called ______________________________. The system used for gathering this information is an interactive, flexible system that enables managers to obtain and manipulate information as they are making decisions. This ______________________________ bypasses the information-processing specialist and lets managers access data from their own terminals.

2 Define marketing research and explain its importance to marketing decision making

2. The process of planning, collecting, and analyzing of data relevant to marketing decision making defines ______________________________.

3 Describe the steps involved in conducting a marketing research project

3. The first step in the marketing research process is defining the ______________________________. Once this step has been completed, the researcher should determine what specific pieces of information are needed and will formulate ______________________________. Marketing research may provide managers with enough information to determine the ______________________________, which tends to be broader in scope and more action oriented.

4. One important type of information used for decision making in the marketing research process is ______________________________, which are data previously collected for any purpose other than the one at hand.

Gathering this type of data has become easier with the advent of the ______________________ that allows researchers to access data, pictures, sound, and files without regard to their physical location or the type of computer in which they reside. One component of this world-wide network, known as the ______________________, simplifies transmission of text and images. A researcher can navigate his or her way around the network by typing in an address, known as a ______________________.

5. Upon completing the situation analysis, researchers may conclude that secondary data does not answer the research question. This means that ______________________ must be collected, which is information gathered for the first time. The ______________________ specifies the research questions to be answered, where and when these data will be gathered, and how the data will be analyzed.

6. The most popular technique for gathering primary data is through ______________________, in which a researcher interacts with people to obtain information. One technique is using in-home interviews. Another technique involves interviewing people in the common areas of shopping malls. This is called a(n) ______________________.

7. Marketing researchers are applying new technology in mall interviewing. One technique allows researchers to conduct in-person interviews by reading questions from a computer and keying responses into the computer. This method is known as ______________________. A second approach allows respondents to read questions off the computer screen and respond themselves. This technique is known as ______________________.

8. Most telephone interviewing is conducted from a specially designed phone room called a(n) ______________________. A new trend in telephone interviewing is the ______________________, which allows users to call a toll-free number and respond to an interactive voice-mail interviewer. Even newer technologies will allow respondents to be interviewed via this Internet. This technique is ______________________.

9. Mail surveys are one of the most common types of surveys but require transfer of data from the paper survey to a database. One way to eliminate this is to send surveys by disk, also known as a ______________________.

10. Though there are many advantages to conducting surveys through the Internet, there are also some disadvantages. One disadvantage occurs when anyone who desires completes a survey. This ______________________ may not be representative of targeted respondents. One solution is to select respondents by specific characteristics, such as gender or geographic region, resulting in a ______________________. Another solution maintains even more control over the sample. Researchers call, e-mail, or write to potential respondents and ask them to participate in a survey. This sample is called a ______________________.

11. One type of personal interviewing involves a small group of qualified consumers who participate in a group discussion led by a moderator. This type of survey research is called a(n) ______________________. The interactions provided by the groups of respondents is called ______________________ and is crucial to the success of this research.

12. There are a variety of question types that can be used on a survey questionnaire. To encourage unlimited answer choices phrased in the respondent's own words requires a(n) ______________________. Alternatively, a respondent may be asked to make a selection from a limited list of responses, which is called a(n) ______________________. A type of this second kind of question that measures the intensity of a respondent's answer is called a(n) ______________________.

13. In contrast to survey research, an alternative type of research does not rely on direct interaction with people. This technique is called ______________________ and takes three forms: people watching people, people watching physical phenomena, and machines watching people. One way to watch physical phenomena is with a(n) ______________________, which examines and verifies the sale of a product.

14. In another method that researchers use to gather primary data, the researcher alters one or more variables (such as price, advertising, or packaging) while observing the effects of those alterations on another variable (usually sales). This is called a(n) ______________________.

15. Once the primary data collection technique has been determined, the sampling procedures must be selected. First, a population, or ______________________ of interest must be defined. A(n) ______________________ is a subset of that population.

16. After the universe is defined, the researcher must then decide if the sample must be representative of the population. If the sample must be representative, a(n) ______________________ is needed. One type of this kind of sample is arranged in such a way that every element of the population has an equal chance of being selected as part of the sample. This is called a(n) ______________________. If population representativeness is not required, a(n) ______________________ can be used. A common form of this type of sample is based on using respondents who are readily accessible to the researcher. This is called a(n) ______________________.

17. Whenever a sample is used in marketing research, two major types of error occur. When there is a difference between the information desired by a researcher and the information provided by the measurement process, ______________________ has occurred. The other major type of error occurs when the sample is not representative of the target population. This is known as ______________________. This second type of error can be categorized into several components. If the sample drawn from a population differs from the target population, ______________________ has occurred. Finally, if the sample is an imperfect representation of the overall population, ______________________ has occurred.

18. After specifying sampling procedures, the data must be collected. Most data collection is performed by ______________________ that specialize in interviewing respondents on a subcontract basis. After data collection, data analysis takes place. One method of analysis allows the analyst to look at responses to one question in relation to responses to one or more other questions. This is called a(n) ______________________. Other types of analysis include one-way frequency counts and statistical analysis.

4 Discuss the growing importance of scanner-based research

19. One type of marketing research gathers its information from a single panel of respondents by continuously monitoring the advertising, promotion, and pricing the panel is exposed to and its subsequent purchases. This is called ____________________________. IRI offers single-source data of two types. The first type uses panel members who shop with an ID card so that purchases can be tracked. This is called ____________________________. The second product tracks sales of consumer packaged goods by scanning UPC codes. This is known as ____________________________.

Check your answers to these questions before proceeding to the next section.

TRUE/FALSE QUESTIONS

Mark the statement **T** if it is true and **F** if it is false.

1 Explain the concept and purpose of a marketing decision support system

_____ 1. Marketing intelligence is a measure of the skills and education levels of employees in a marketing department.

2 Define marketing research and explain its importance to marketing decision making

_____ 2. Marketing research is but one component in a decision support system (DSS).

_____ 3. Marketing research focuses on gathering information about environmental trends and communicating that information to marketing managers.

3 Describe the steps involved in conducting a marketing research project

_____ 4. Cosmic Bowling has experienced a decline in sales over the past five years. Management at the bowling alley has defined the problem and would like to get started with some marketing research. The next step in the research process is to determine which methodology to use to gather information.

_____ 5. Jim and Kim operate a day-care center for children. They want to learn more about the parents of the children but don't have money available to conduct research. To save money, Jim suggests they use the results of a survey they conducted last year because they were asking some similar questions then. Jim is suggesting they use primary research data.

_____ 6. John is gathering information in the form of primary data on the drinking habits of underage, undergraduate college students who are business majors and live in apartments without roommates. The major advantage of using primary data is that this is probably the only way he could get questions about such a specific topic answered.

_____ 7. The Internet can be used for both primary and secondary research.

_____ 8. When Anheuser-Busch split the United States into three regions, decreased advertising in region one, left advertising levels the same in region two, and doubled advertising in region three, they were most likely conducting an experiment.

_____ 9. Rebecca needs twenty people to fill out a survey for her marketing class project. She stands outside the library and hands the survey to the first twenty people who come out. Rebecca is using a random sample.

5 Explain when marketing research should and should not be conducted

_____ 10. Gregg's Frozen Custard has operated in the same county for the last ten years, has developed a loyal customer base, and is pleased with its marketing research to date. Management plans to introduce hot soups to three of its stores, with the knowledge that this addition has been successful in the other four outlets. Gregg's probably does not need to conduct more marketing research before introducing the soups to the remaining stores.

Check your answers to these questions before proceeding to the next section.

AGREE/DISAGREE QUESTIONS

For the following statements, indicate reasons why you may agree and disagree with the statement.

1. Secondary data rarely provide enough information for decision-making.

   Reason(s) to agree:

   Reason(s) to disagree:

2. Focus groups should not be used as a sole tool for making marketing decisions.

   Reason(s) to agree:

   Reason(s) to disagree:

3. There is no such thing as perfect information in marketing research.

   Reason(s) to agree:

   Reason(s) to disagree:

MULTIPLE-CHOICE QUESTIONS

Select the response that best answers the question, and write the corresponding letter in the space provided.

1 Explain the concept and purpose of a marketing decision support system

____ 1. Decision support systems have all the following characteristics EXCEPT:
a. flexibility
b. accessibility
c. interactive features
d. difficult to learn
e. discovery orientation

____ 2. Which of the following is the best example of the interactivity of a decision support system?
a. Managers who are not skilled with computers can use the system.
b. Managers give simple instructions to the system and see immediate results.
c. Managers can create "what if" scenarios.
d. The system can manipulate data in various ways.
e. The system is not costly.

____ 3. Marketing decision support systems have all the following characteristics EXCEPT:
a. flexibility
b. discovery-oriented
c. interactive
d. accessibility
e. difficult to learn

2 Define marketing research and explain its importance to marketing decision making

____ 4. The manager for Slurp Soda has requested marketing research to find out why a recent rebate offer failed to generate much response or any increase in sales. This type of research is:
a. descriptive
b. predictive
c. historical
d. normative
e. diagnostic

____ 5. An advertising agency is researching consumer attitudes toward a TV commercial that promotes a cellular phone service. This type of research is:
a. descriptive
b. predictive
c. historical
d. normative
e. diagnostic

____ 6. Which of the following best illustrates the predictive role of marketing research?
a. Exit poll research has found that 60 percent of voters voted for the incumbent senator.
b. A survey concludes that nine out of ten dentists recommend Brite toothpaste.
c. Consumers in a focus group explain why they like the new product.
d. Research has found that consumers do not like the new product improvement.
e. Research has found that increasing a product's price 20 percent will result in a sales decrease of 35 percent.

3 Describe the steps involved in conducting a marketing research project

____ 7. The manager of a video rental store is worried about his store's performance. Sales have declined steadily over the past two years, and she cannot explain why. She would like to do some research. What should her first step in the market research process be?
   a. collect the data
   b. plan the research design
   c. analyze the marketplace
   d. recognize the marketing problem
   e. specify the sampling plan

____ 8. The President of Clone Clothing stores has seen her chain of stores experience a decline in sales. She disliked the recent advertising campaign and feels the campaign caused the sales decline. She conducted research showing potential customers were not enthusiastic toward the ads for Clone Clothing. Which of the following actions should be taken?
   a. conduct an experiment to show which of several alternative advertising campaigns are most appealing to potential Clone customers
   b. perform a survey to see how knowledgeable potential customers are with Clone Clothing
   c. investigate to reveal the exact problem with the advertising and why it caused sales to decline
   d. do a survey first to see if an experiment would be cost effective
   e. find out what exactly caused the sales decline

____ 9. Which of the following is the best example of a marketing research objective?
   a. To determine why sales have decreased over the past three years.
   b. To create a print advertisement emphasizing customer service.
   c. To determine what role children have in influencing family decisions about vacation destinations.
   d. To increase sales by 10 percent.
   e. To develop a method to track consumer attitudes.

____ 10. The Play-It-Again used sporting equipment outlet cannot afford to collect primary data and has decided to use secondary data to answer some research questions. Play-It-Again should realize the potential disadvantages of secondary data, which include:
   a. the difficulty of holding certain variables constant while varying one factor of interest
   b. the high cost of collecting secondary data
   c. the information may not fit the needs of the research problem
   d. the measurement error that can occur during the collection process
   e. the length of time it takes to collect secondary data

____ 11. When assessing the quality of secondary data, it is:
   a. not necessary to know why the data were collected in the first place
   b. important to be able to have easy access to the data
   c. not important to know when the data were collected
   d. important to know the purpose for which the data were originally collected
   e. imperative to use the same methods and procedures when primary data are collected

____ 12. You would like to find some annual sales figures of a major competitor as part of your secondary data research. You have decided to use the World Wide Web as your source. Which of the following is the most direct way of finding out this information?
   a. Type the words "sales" and the competitor's name in a search engine.
   b. Type the URL (Uniform Reference Locator) of the competitor and search through annual reports.
   c. Type the competitor's name in a search engine and scan for sales information.
   d. Type the URL (Uniform Reference Locator) of a trade association to which you both belong.
   e. Type the word "sales" in a search engine and scan for the competitor's name.

____ 13. Which of the following World Wide Web domain name extensions would be used by a for-profit company?
a. .org
b. .net
c. .gov
d. .com
e. .pro

____ 14. Which of the following is NOT a search engine on the World Wide Web?
a. Infoscan
b. Alta Vista
c. Yahoo
d. Excite
e. Infoseek

____ 15. The Porter Finance Company has decided to collect primary data but knows that this has the disadvantage of:
a. high cost compared to secondary data
b. accessibility only through complex computerized data bases
c. being nonsecure data because of its availability to any interested party for use
d. potential irrelevance to the problem at hand
e. not providing a realistic picture of the marketplace

____ 16. The Acme Marketing Department is collecting data on consumers' spending habits and their attitudes toward environmental legislation, using one questionnaire. This is an example of a(n):
a. piggyback study
b. single-source study
c. dual-probe survey
d. dichotomous question
e. experiment

____ 17. The Robo-Vac company wants to get consumer feedback about the style, features, and a demonstration of the operation of its prototype automatic mini-vacuum cleaner. What form of survey would allow them to do this?
a. mall intercept
b. mail questionnaire
c. telephone interview
d. observation study
e. laboratory experiment

____ 18. A local political party has decided to use telephone interviews to help predict the next election because this technique offers:
a. speed in gathering data
b. few nonresponses
c. a potential for census rather than sample data
d. the ability to collect complex and large amounts of data
e. the least amount of sampling error

____ 19. The biggest disadvantage of Internet surveys is that:
a. the cost is prohibitive.
b. it has a lower response rate than mail or telephone surveys.
c. the respondents are not representative of the population as a whole.
d. research firms can only communicate with respondents during business hours.
e. the respondents may not provide accurate answers to questions.

____ 20. Behavior Research Company is calling several people to determine who would be appropriate respondents to a long Internet survey it would like to administer. The people who agree to take the survey are then directed to a Web site that contains a link to the survey. The respondents taking the Internet survey are an example of:
a. a recruited Internet sample.
b. a screened Internet sample.
c. an Internet focus group.
d. an unrestricted Internet sample.
e. a mass market sample.

____ 21. You have been hired to view store customers through a video camera and record the number of times a shopper pauses beside the special display case. You are engaging in:
a. survey sampling research
b. experimental research
c. observation research
d. focus group research
e. panel research

____ 22. Cox's Department Stores wanted to learn more about its customers and their shopping habits in Cox's stores. After considering several different methods, Cox's decided on observation research because:
a. biases from the interviewing process are eliminated
b. it relies on the respondent's willingness to provide the desired data
c. it has a relatively high response rate
d. attitudes and motivations are clearly demonstrated in observation research
e. it is actually an experiment using natural environments

____ 23. A research manager decides to have his telephone interviewers dial every tenth number in the telephone directory. This is a:
a. convenience sample
b. probability sample
c. stratified sample
d. quota sample
e. nonprobability sample

____ 24. Caryn will be sending a survey to 25 percent of her carpet-cleaning service customers. She used a random number table to select customers from her master customer list. This is an example of:
a. field service sample
b. random sample
c. nonprobability sample
d. regular sample
e. convenience sample

____ 25. Marty has decided that it is not absolutely necessary to have a sample representative of the population. However, a problem associated with nonprobability samples is:
a. setting up the method
b. that they create measurement errors
c. that there is no way of knowing how much sample error has accumulated
d. finding respondents
e. that so few researchers understand them

____ 26. Jerome has to interview eighty people for his marketing research class project and decides to use his fellow dorm residents. This is a:
a. convenience sample
b. probability sample
c. simple random sample
d. field service sample
e. multirespondent sample

_____ 27. After Yellow-Belly Jelly Beans, Inc., received the results of a survey that measured parents' intentions to buy jelly bean packages for Easter, the firm worried that many parents overstated their intentions and would not actually buy. This is what kind of sampling error?
a. nonsampling
b. frame
c. random
d. measurement
e. nonresponse

4 Discuss the growing importance of scanner-based research

_____ 28. Pietro's Pasta Products gathers weekly information from the same consumers. Pietro's continuously monitors the advertising the consumers are exposed to and records their purchases. Pietro's uses cable TV to send different commercials to different areas, and the consumers keep a record of all grocery purchases in return for a fee. This is a:
a. television meter investigation
b. single-source research system
c. CLT interview
d. laser scanner study
e. one-way TV observation

5 Explain when marketing research should and should not be conducted

_____ 29. Marketing research should not be undertaken when:
a. the perceived costs are greater than the projected benefits
b. it will take a long time to complete
c. there is no secondary data in existence to guide the project definition
d. the actual costs are less than the forecasted benefits
e. the perceived costs are the same as the forecasted benefits

Check your answers to these questions before proceeding to the next section.

ESSAY QUESTIONS

1. What is marketing research? Name and briefly describe the three functional roles of marketing research. Then discuss two possible benefits of marketing research to managers.

2. What are the steps of the marketing research process?

3. Name six advantages and three disadvantages to conducting a survey on the Internet. What can be done to avoid getting an unrestricted sample on the Internet?

4. A marketing researcher has many options when designing a questionnaire. However, some questions are better than others for obtaining responses. Match each survey question example with its definition by placing the matching letter in the blank to the left of the example. Each letter is used only once.

A. Open-ended question
B. Closed-ended dichotomous question
C. Closed-ended multiple-choice question
D. Scaled-response question
E. Ambiguous question
F. Two questions in one
G. Biased/leading question

_____ What is your favorite brand of athletic shoe?
_____ Nike _____ L.A. Gear _____ Reebok _____ Other

_____ Do you currently own a pair of athletic shoes?
_____ YES _____ NO

_____ What are your opinions about owning a pair of Nike shoes?

_____ Do you think Nikes and Reeboks are good quality shoes?

_____ Why do you think Nikes are such excellent shoes?

_____ Now that you know about Nike athletic shoes, would you ...
_____ Definitely buy
_____ Probably buy
_____ Might or might not buy
_____ Probably not buy
_____ Definitely not buy

_____ If you are thinking of buying a pair of Nikes, will you be buying soon?
_____ YES _____ NO

CHAPTER NOTES

Use this space to record notes on the topics you are having the most trouble understanding.

ANSWERS TO THE END-OF-CHAPTER DISCUSSION AND WRITING QUESTIONS

Use this space to work on the questions at the end of the chapter.

ANSWERS TO THE END-OF-CHAPTER CASE

Use this space to work on the questions at the end of the chapter.

CHAPTER 9 Creating a Competitive Advantage Using Competitive Intelligence

LEARNING OBJECTIVES

1 Describe the concept of competitive advantage and the three types of competitive advantages

A competitive advantage is a set of unique features of a company and its products that are perceived by the target market as significant and superior to the competition. Three types of competitive advantages are cost, differentiation, and niche strategies.

2 Discuss the concept of competitive intelligence

Competitive intelligence is the creation of an intelligence system that helps managers assess their competition and their vendors in order to become more efficient and effective competitors. Intelligence is information that is analyzed and that can help an organization make future decisions for long-term growth.

3 Identify the sources of internal competitive intelligence data

The best source of internal competitive intelligence data is the company sales force. Since sales people are "in the trenches" and closest to customers, they are in a good position to gather data on competitors. They can find this data through conversations with buyers or in-store or other on-site promotions.

4 Identify the non-computer-based external sources of competitive intelligence

There are many non-computer-based external sources of competitive intelligence: experts, CI consultants, government agencies, UCC filings, suppliers, photographs, newspapers and other publications, yellow pages, trade shows, speeches by competitors, neighbors of competitors, and advisory boards.

5 Explain how the Internet and databases can be used to gather competitive intelligence

The Internet and databases accessed from the Internet offer excellent and quick sources of data. CD ROM databases, such as Global Data Manager, can also prove valuable to CI researchers.

6 Discuss the problem of industrial espionage

Industrial espionage is an attempt to learn a competitor's trade secrets by illegal or unethical means. Many cases of espionage are conducted against high-tech firms.

PRETEST

Answer the following questions to see how well you understand the material. Re-take it after you review to check yourself.

1. What is a competitive advantage? List and describe three types of competitive advantages.

2. What is competitive intelligence? What benefits do firms gain from having a competitive intelligence system?

3. Name at least five sources of internal competitive intelligence.

4. Name twelve sources of non-computer-based competitive intelligence.

5. What is industrial espionage?

CHAPTER OUTLINE

1 Describe the concept of competitive advantage and the three types of competitive advantages

I. Competitive advantage

Discuss the importance of competitive advantage in a company's efforts to sustain profitability and maintain a loyal customer base. Competitive advantage is also referred to as differential advantage.

A. **Competitive advantage** consists of a set of unique features of a company and its products that are perceived by the target market as significant and superior to the competition.

B. Factor or factors which cause customers to patronize a firm and not the competition.

C. There are three types of competitive advantages.

1. A **cost-competitive advantage** results from being the low cost competitor in an industry while maintaining satisfactory profits.

a. Sources of cost-competitive advantages include:

1) **Experience curves** tell us that costs decline at a predictable rate as experience with a product increases.

2) Efficient labor resulting from pools of cheap labor.

3) Removing frills and options from a product or service.

4) Government subsides which effectively lower the cost of production by the amount of the subsidy.

5) Designing products for ease of production or using reverse engineering to cut research and design costs.

6) Reengineering through downsizing, deleting unprofitable product lines, closing obsolete factories, or renegotiating supplier contracts.

7) Use of production innovations such as new technology and simplified production techniques.

8) Developing new, more efficient, methods of service delivery.

b. Cost competitive advantages are rarely sustainable as they can be adopted by competitors.

2. A **differential competitive advantage** exists when a firm provides something unique that is valuable to buyers.

a. Sources of differential advantage include:

1) Brand names offer enduring competitive advantage.

2) **Value impressions** are features that signal value to the customer.

3) **Augmented products** where features that are not expected by the customer.

b. Sources of differential advantage are difficult for competitors to match.

3. A **niche competitive advantage** seeks to target and effectively serve a single segment of the market.

D. Building Tomorrow's Competitive Advantage

1. The source of future competitive advantages are the skills and assets of the organization.

2. A sustainable competitive advantage is a function of the speed with which competitors can imitate a leading company's strategy plans.

2 Discuss the concept of competitive intelligence

II. Competitive intelligence

A. **Competitive intelligence** is the creation of a system that helps managers assess their competitors and their vendors in order to become a more efficient and effective competitor.

B. Competitive intelligence allows managers to predict changes in business relationships, identify opportunities, guard against threats, forecast a competitor's strategy, and develop a successful marketing plan.

3 Identify the sources of internal competitive intelligence data

III. Sources of internal competitive intelligence.

A. Sales personnel have the greatest ability to gather CI on a day in and day out basis.

B. Companies can uncover internal sources of CI by conducting a **CI audit** which includes information from:

1. Employees

2. Independent databases maintained on non-network computers

3. Collections of marketing research studies

C. A CI directory consists of CI audit data entered into a database management file.

4 Identify non-computer-based external sources of competitive intelligence

IV. Non-computer-based external sources of competitive intelligence.

A. An **expert** has an in-depth knowledge of a subject or activity.

B. CI consultants gather needed information quickly and efficiently.

C. U.S. government agencies and departments are useful sources of information, e.g. OSHA, EPA, FCC, and FDA.

D. **Uniform Commercial Code filings** identify goods that are leased or pledged as collateral.

E. Suppliers will often provide information regarding products shipped to competitors.

F. Photographs can reveal competitor activities.

G. Newspaper clipping services will gather current information about specified topics that is printed in newspapers and other publications.

H. Yellow pages are one of the cheapest and easiest ways to obtain competitive intelligence.

I. Trade shows are a useful source of competitive intelligence.

J. Speeches by company executives often reveal new strategic directions, new products, plant openings, etc.

K. Conversing with neighbors to competitors often reveals what is happening inside the facility.

L. Advisory boards supply expert advice and act as a sounding board to particular areas of management.

5 Explain how the Internet and databases can be used to gather competitive intelligence

V. Using the Internet and databases for competitive intelligence.

A. Information available using Internet access to databases includes such topics as:

1. Media sources and specific articles

2. Industry participants and trends

3. Specific company activities

B. CD ROM databases are also valuable sources of information.

6 Discuss the problem of industrial espionage

VI. Industrial espionage.

A. **Industrial espionage** is an attempt to learn a competitor's trade secrets by illegal and/or unethical means.

1. Governments as well as companies are guilty of engaging in industrial espionage.

2. CI is an area of business that has had more than its share of ethics abuses.

B. It is not just foreign companies but governments as well who are conducting industrial espionage.

1. Companies or governments of 23 countries are currently involved in the illicit acquisition of U.S. trade secrets.

2. The FBI reports about 800 pending probes of thefts by foreign companies or governments per year.

C. The Economic Espionage Act of 1996 makes it a federal criminal act for any person to convert a trade secret to his own benefit or the benefit of others.

D. Industrial espionage is also conducted by domestic companies.

VOCABULARY PRACTICE

Fill in the blank(s) with the appropriate term or phrase from the alphabetized list of chapter key terms.

Augmented product
CI audit
CI directory
Competitive advantage
Competitive intelligence
Cost competitive advantage
Differential competitive advantage
Experience curve
Expert
Industrial espionage
Niche competitive advantage
Uniform Commercial Code (UCC)
Value impressions

1 Describe the concept of competitive advantage and the three types of competitive advantages

1. The set of unique features of a company and its products that are perceived by the target market as significant and superior to the competition is called a ______________________________ . A firm can follow one of three different strategies to maintain this. The first strategy is called a ______________________________, in when a firm provides something that is unique that is valuable to buyers beyond simply offering a lower price. Features of a product or service that signal value to the customer are called ______________________________. A firm can also maintain this strategy with a(n) ______________________________ when it takes the product or service and adds features not expected by the customer.

2. A second strategy that a firm can follow is ______________________________, in which a firm is the low cost competitor in an industry while maintaining satisfactory profit margins. A firm's ______________________________ shows production costs declining at a predictable rate as it becomes efficient at producing a particular product.

3. A third strategy that a firm can follow is ______________________________, in which the firm seeks to target and effectively serve a small segment of the market.

2 Discuss the concept of competitive intelligence

4. The creation of a system that helps manager assess their competition and their vendors in order to become more efficient and effective competitors is called ____________________________.

3 Identify the sources of internal competitive intelligence data

5. In order to gather competitive intelligence, a firm might require that its employees file reports about competitors' activities. These reports may be collected in a(n) ____________________________, which documents employee expertise on competitive activity, collects information from independent in-house databases, or summarizes marketing research studies internally. When all this information is entered into a database management file specifically for competitive intelligence, it becomes a(n) ____________________________.

4 Identify the non-computer-based external sources of competitive intelligence

6. There are many non-computer-based external sources of competitive intelligence. One such source is the use of ____________________________ who or someone with in-depth knowledge of a subject or activity, such as developments in an industry. Another external source of intelligence is ____________________________ filings, which are filings by banks with government agencies that identify goods that are leased or pledged as collateral.

6 Discuss the problem of industrial espionage

7. One problem that may occur in the gathering of competitive intelligence is ____________________________, or an attempt to learn competitors' trade secrets by illegal or unethical means.

Check your answers to these questions before proceeding to the next section.

TRUE/FALSE QUESTIONS

Mark the statement T if it is true and F if it is false.

1 Describe the concept of competitive advantage and the three types of competitive advantages

_____ 1. A cost competitive advantage is relatively easy to sustain because a firm can usually controls its production costs by tracking its experience curve.

_____ 2. With the launch of the Macintosh, Apple Computer was able to provide consumers with a computer that was user friendly compared to IBM and other personal computers in the 1980s. This is an example of a differential competitive advantage.

_____ 3. A small company with limited resources can use a niche competitive strategy in order to circumvent its larger, richer competitors.

2 Discuss the concept of competitive intelligence

_____ 4. Competitive intelligence is basically the same as marketing intelligence: they both help managers in firms make good decisions about the future.

3 Identify the sources of internal competitive intelligence data

_____ 5. Probably the best source of internal competitive intelligence data is a company's sales people, since they are the closest to the customers and to competitors.

_____ 6. A database management file that contains employees' knowledge of competitive activities, independent databases maintained on nonnetwork computers, and collections of marketing research studies is called a CI audit.

4 Identify the non-computer-based external sources of competitive intelligence

_____ 7. Advisory boards are a good source of internal competitive intelligence since they have access to information at outside firms.

_____ 8. Sarah, a CI consultant, checks UCC filings to identify which goods have been leased by competitors. Sarah is engaging in unethical industrial espionage.

6 Discuss the problem of industrial espionage

_____ 9. The largest threat of industrial espionage by another country against U.S. companies is France, which has stolen many software trade secrets.

_____ 10. The Economic Espionage Act of 1996 makes it a federal crime to either sell a trade secret to competitors or to receive trade secrets from a competitor.

Check your answers to these questions before proceeding to the next section.

AGREE/DISAGREE QUESTIONS

For the following statements, indicate reasons why you may agree and disagree with the statement.

1. In the free enterprise system, it is difficult to maintain any kind of competitive advantage over the long-term.

   Reason(s) to agree:

Reason(s) to disagree:

2. Competitive intelligence efforts will eventually lead to industrial espionage, even without intent.

   Reason(s) to agree:

   Reason(s) to disagree:

3. Using the sales force as a source of competitive intelligence is not prudent.

   Reason(s) to agree:

   Reason(s) to disagree:

4. External sources of competitive intelligence are more accurate than internal sources.

   Reason(s) to agree:

   Reason(s) to disagree:

5. Industrial espionage is necessary in order to compete effectively in today's global markets.

   Reason(s) to agree:

   Reason(s) to disagree:

MULTIPLE CHOICE QUESTIONS

Select the response that best answers the question, and write the corresponding letter in the space provided.

1 Describe the concept of competitive advantage and the three types of competitive advantages

_____ 1. Nordstrom's is a department store chain that has a reputation for high quality customer service. The store has a "no questions asked" return policy and treats customers with great respect. Nordstrom's customer service would be considered:
a. a company distinction
b. a competitive advantage
c. competitive intelligence
d. an augmented product
e. quality assurance

_____ 2. Southwest Airlines offers low airfares because of its no frills service, its general efficiency in operations, and its short haul flights. Southwest is an example of a company that has a(n):
a. cost competitive advantage
b. distinctive competence
c. differential competitive advantage
d. augmented product
e. niche competitive advantage

_____ 3. Some firms can maintain a low cost competitive advantage by producing mass quantities and by gaining learning from production technologies. Management is able to predict that per unit costs will decline over time by analyzing:
a. production curves
b. economies of scale
c. efficient labor
d. value impressions
e. experience curves

_____ 4. All of the following explain how a firm can maintain a cost competitive advantage EXCEPT:
a. no frills goods and services
b. government subsidies
c. augmented products
d. efficient labor
e. experience curves

_____ 5. Federal Express is a delivery company that competes with the U.S. Postal Service and UPS. Though Federal Express charges somewhat higher prices than its competitors, the company offers high quality customer service, including guarantees on on-time delivery and a sophisticated tracking system. Federal Express enjoys a(n):
a. cost competitive advantage
b. experience curve
c. differential competitive advantage
d. efficient labor force
e. niche competitive advantage

_____ 6. High-end cosmetics brands, such as Lancome, probably spend more on packaging than on the products themselves. The beautiful designs of the packaging are what appeal to consumers. These cosmetics companies enjoy a differential competitive advantage due to:

a. an augmented product
b. an efficient labor force
c. extraordinary customer service
d. a value impression
e. an experience curve

_____ 7. The Hudson School is an expensive private school serving children who come from affluent families around the world but who need remedial education. Hudson is the only school in the metropolitan area that targets this group of people and "owns" this positioning. Hudson has a(n):

a. cost competitive advantage
b. experience curve advantage
c. differential competitive advantage
d. augmented product
e. niche competitive advantage

2 Discuss the concept of competitive intelligence

_____ 8. In order for managers to assess competition to become a more efficient and effective competitor, a company should engage in:

a. a competitive advantage
b. competitive intelligence
c. industrial espionage
d. an audit
e. a differential competitive advantage

_____ 9. Which of the following is NOT an advantage of competitive intelligence?

a. It guarantees the long-term success of a company.
b. It allows managers to predict changes in business relationships.
c. It identifies marketplace opportunities.
d. It helps companies guard against threats, such as new product launches.
e. It helps companies learn from the successes and failures of others.

3 Identify the sources of internal competitive intelligence data

_____ 10. High-Tech, Inc, is in a very competitive industry and would like to create a world class ompetitive intelligence system. Some of the company's best sources are its own employees. In order to create the system, High-Tech should conduct a(n):

a. CI directory
b. comprehensive marketing research project
c. employee skills inventory
d. industrial espionage
e. CI audit

_____ 11. CI audit data entered into a database management file is known as a(n):

a. decision support system (DSS)
b. management information system (MIS)
c. competitive intelligence directory
d. marketing intelligence system
e. computer information system

_____ 12. Which of the following is NOT an internal source for competitive intelligence?
a. Company salespeople.
b. Company engineers.
c. Marketing research studies commissioned by the company.
d. Company advisory board.
e. Company engineers.

4 Identify the non-computer-based external sources of competitive intelligence

_____ 13. Geena Maven has over twenty years of experience at marketing financial services. She recently resigned from her post of Vice-President of marketing at a major financial institution and started her own business to help other organizations gather competitive intelligence in the same industry. Geena would be considered a(n):
a. expert
b. financial services consultant
c. marketing consultant
d. management consultant
e. CI consultant

_____ 14. Arnold would like to set up a tool and die business in his hometown but has no idea who his competitors might be. Arnold should first use which of the following external sources of competitive intelligence?
a. yellow pages of his hometown phone book
b. the Internet
c. regional trade shows
d. UCC filings
e. advertisements in the local newspaper

_____ 15. Which of the following would NOT be considered a government agency from which competitive intelligence data can be gathered?
a. OSHA
b. FCC
c. SEC
d. FOIA
e. EPA

_____ 16. Fred Jones is a marketing manager at a large computer manufacturer. He is trying to find some information about a competitor's financial position, including detailed reports about the company's financial performance. Fred should check with:
a. the Securities and Exchange Commission (SEC)
b. the Occupational Safety and Health Administration (OSHA)
c. the Federal Communications Commission (FCC)
d. the Food and Drug Administration (FDA)
e. the Federal Trade Commission (FTC)

_____ 17. Managers at a car manufacturer would like to find out if their major competitors have leased goods or have pledged certain goods as collateral. This will help them understand if the competitors are planning major expansion of production capacity. These managers should search:
a. competitors' banks
b. UCC filings
c. the Internet
d. FDA reports
e. SEC filings

_____ 18. The yellow pages can be used for competitive intelligence in all of the following ways EXCEPT:

a. to define a realistic trading area
b. to locate suppliers and distributors
c. to gather information about competitors' manufacturing costs
d. to retrieve marketing and product line information
e. to determine how many competitors exist in a specified area

_____ 19. The best way of picking up competitors' sales literature quickly is to:

a. have the entire sales force go to competitors' headquarters.
b. ask your advisory board to get the information for you.
c. call all competitors and ask them to send you the literature.
d. find out who prints the literature and get copies from their files.
e. attend an industry trade show.

5 Explain how the Internet and databases can be used to gather competitive intelligence

_____ 20. One of the best first places to search for general information about a competitor, as well as any press releases recently issued by the company, is:

a. newspaper databases.
b. periodicals on-line databases.
c. Lexis-Nexis.
d. the competitor's Web site.
e. Standard & Poor's Register.

_____ 21. Greta is a paralegal that works in a large corporate law office. She has been asked by a law partner to research any lawsuits that have been brought up against her client's major competitor. Greta should use:

a. Standard & Poor's Register.
b. Lexis-Nexis.
c. the competitors' web sites.
d. the Legal Registry.
e. Martindale Hubbel's directory.

6 Discuss the problem of industrial espionage

_____ 22. A woman posing as a janitor is cleaning the offices of a cosmetics manufacturer. When all the managers leave for the evening, the woman turns on computers and searches for any information that would help her client, a competitor of the cosmetics company. The woman posing as a janitor is engaging in:

a. competitive intelligence
b. corporate espionage
c. industrial espionage
d. infringement of intellectual property
e. contract violation

_____ 23. The country that poses the biggest threat to U.S. companies in industrial espionage is:

a. France
b. Japan
c. Taiwan
d. Russia
e. China

_____ 24. The act that makes it a federal crime for any person to convert a trade secret to his own benefit or to the benefit of others intending or knowing that the offense will injure any owner of the trade secret is:

a. the Industrial Espionage Act of 1996
b. the Economic Espionage Act of 1996
c. the Corporate Espionage Act of 1996
d. the Trade Violation Act of 1996
e. the FTC Act

_____ 25. Greg is a software engineer who works in a highly competitive environment. He recently received a phone call from Keith, another engineer and a college friend, who is working at a competitor's company. Keith offered to pay Greg $100,000 for any information that Greg could give about his company's new software line. If Greg accepts the offer, and if his company suffers because of it, who would be guilty under U.S. law?

a. Greg would be the only guilty one; he didn't have to accept the offer.
b. Keith would be the only guilty one; he was the one who gave the offer.
c. Both Greg and Keith would be guilty since they both engaged on either side of the offer.
d. Keith's company would be the only guilty party, since Keith acts as an agent for his company.
e. Both Greg's company and Keith's company would be guilty.

Check your answers to these questions before proceeding to the next section.

ESSAY QUESTIONS

1. You would like to open a new child care center in the city in which you live. You believe that there are market gaps in current offerings of child care and would like to position yourself to fill those gaps. Name three competitive advantages that you could use, and give an example of a new child care concept that might fit these three strategies.

2. You are a new Product Manager who has just taken over the management of a pharmaceutical brand just acquired by your employer. There is very little that your company knows about the market, and you decide to put together a comprehensive competitive intelligence report on the brand. The problem is that you have no budget with which to conduct this research. Name five sources of internal competitive intelligence and five sources of external competitive intelligence that you could use to complete your report.

3. What is industrial espionage? List at least five actions that constitute industrial espionage.

CHAPTER NOTES

Use this space to record notes on the topics you are having the most trouble understanding.

ANSWERS TO THE END-OF-CHAPTER DISCUSSION AND WRITING QUESTIONS

Use this space to work on the questions at the end of the chapter.

ANSWERS TO THE END-OF-CHAPTER CASE

Use this space to work on the questions at the end of the chapter.

CHAPTER GUIDES and QUESTIONS

PART THREE

PRODUCT DECISIONS

CHAPTER 10 Product Concepts

LEARNING OBJECTIVES

1 Define the term "product"

A product may be defined as everything, both favorable and unfavorable, that one receives in an exchange. It can be a tangible good, a service, an idea, or a combination of these things.

2 Classify consumer products

Consumer products can be differentiated from business products according to intended use. Consumer products are purchased to satisfy an individual's personal wants. The most widely used classification approach has four categories: convenience products, shopping products, specialty products, and unsought products. This approach classifies products on the basis of the amount of effort that is normally expended in the shopping process.

3 Define the terms "product item," "product line," and "product mix"

A product item is a specific version of a product that can be designated as a distinct offering among an organization's products. A product line is a group of closely related products offered by the organization. An organization's product mix includes all the products it sells.

Marketing strategies and mixes may be built around individual product items, product lines, or the entire product mix. Product line and mix decisions can affect advertising economies, packaging strategies, standardization of components, and the efficiency of the sales and distribution functions.

4 Describe marketing uses of branding

Branding is the major tool marketers have to distinguish their products from those of the competition. A brand is a name, term, symbol, design, or combination thereof that identifies a seller's products and differentiates them from competitors' products. Branding has three main objectives: identification, repeat sales, and new product sales.

5 Describe marketing uses of packaging and labeling

Packaging is an important strategic part of the marketing mix. The product and its package are often inseparable in the consumer's mind. Subtle changes in packaging can dramatically alter the consumer's perception of the product. A package should communicate an image to the consumer that will help achieve the positioning objectives of the firm.

An integral part of any package is its label. Labeling strategy generally takes on one of two forms. Persuasive labeling focuses on a promotional theme or logo, with information for the consumer taking secondary importance. Informational labeling is designed to help consumers in making proper product selections and to lower cognitive dissonance after the purchase.

6 Discuss global issues in branding and packaging

In addition to brand piracy, global marketers must address a variety of concerns regarding branding and packaging. These include selecting a brand name policy, translating labels, meeting host-country labeling requirements, making packages aesthetically compatible with host-country cultures, and offering the sizes of packages preferred in host countries.

7 Describe how and why product warranties are important marketing tools

One part of the product is its warranty, a protection and information device for consumers. A warranty guarantees the quality or performance of a good or service. An express warranty is made in writing; an implied warranty is an unwritten guarantee that a good or service is fit for the purpose for which it was sold. All sales have an implied warranty under the Uniform Commercial Code, and regulations covering express warranties have been passed by Congress.

PRETEST

Answer the following questions to see how well you understand the material. Re-take it after you review to check yourself.

1. What is a product? Name and briefly describe four types of consumer products.

2. Define the terms "product item," "product line," and "product mix." Give an example of each.

3. What is a brand? Name and briefly describe three branding strategies.

4. What are the three most important functions of packaging?

5. List and briefly describe three different branding strategies when entering international markets.

6. Why are warranties important as marketing tools?

CHAPTER OUTLINE

1 Define the term "product"

I. What Is a Product?

A. A **product** may be defined as everything, both favorable and unfavorable, that one receives in an exchange. It can be a tangible good, a service, an idea, or a combination of these things.

B. A product is the heart of the organization's marketing program, the starting point in creating a marketing mix.

2 Classify consumer products

II. Types of Consumer Products

Products are classified as either business or consumer products depending on the buyer's intentions for the product's use.

A. A **business product** is purchased for

1. Use in the manufacture of other goods or services

2. Use in an organization's operations

3. Resale to other customers

B. A **consumer product**:

1. Is purchased to satisfy an individual's personal wants

2. Can be classified on the basis of the amount of effort that is normally expended in the shopping process

a. A **convenience product** is an inexpensive item that requires little shopping effort. These products are purchased regularly, usually with little planning, and require wide distribution.

b. A **shopping product** requires comparison shopping, because it is usually more expensive than convenience products and is found in fewer stores. Consumers usually compare items across brands or stores.

1) *Homogeneous* shopping products are products that consumers see as being basically the same, and consumers shop for the lowest price.

2) *Heterogeneous* shopping products are seen by consumers to differ in quality, style, suitability, and lifestyle compatibility. Comparisons between heterogeneous shopping products are often quite difficult because they may have unique features and different levels of quality and price.

c. A **specialty product** is searched for extensively, and substitutes are not acceptable. These products may be quite expensive, and distribution is limited.

d. An **unsought product** is a product that is not known about or not actively sought by consumers. Unsought products require aggressive personal selling and highly persuasive advertising.

3 Define the terms "product item," "product line," and "product mix"

III. Product Items, Lines, and Mixes

A **product item** is a specific version of a product that can be designated as a distinct offering among an organization's products.

A **product line** is a group of closely related products offered by the organization.

An organization's **product mix** includes all the products it sells.

A. Why Form Product Lines?

1. *Advertising economies* occur when several products can be advertised under the umbrella of the line.

2. *Package uniformity* increases customer familiarity, and the different items actually help to advertise one another.

3. Product lines provide an opportunity for *standardizing components*, thus reducing manufacturing and inventory costs.

4. Product lines *facilitate sales and distribution*, leading to economies of scale for managing the sales force, warehousing, and transportation.

5. Product lines based on a brand name help consumers *evaluate quality*. Consumers usually believe that all products in a line will be of similar quality.

B. Width and Depth

Product mix width refers to the number of product lines that an organization offers. Firms increase product mix width to

1. Spread risk across multiple lines

2. Capitalize on established reputations

Product line depth is the number of product items in a product line. Firms increase product line depth to

1. Attract buyers with widely different preferences

2. Capitalize on economies of scale

3. Increase sales and profits by further segmenting the market

4. Even out seasonal sales patterns

C. Adjustments to Product Items, Lines, and Mixes

1. Modifying Existing Products

A product modification is a change in one or more of a product's characteristics.

a. A *quality modification* entails changing a product's dependability or durability.

b. A *functional modification* is a change in a product's versatility, effectiveness, convenience, or safety.

c. A *style modification* is an aesthetic product change rather than a quality or functional modification.

2. **Repositioning** involves changing customers' perceptions of a product.

3. **Line extension** is the practice of adding products to a product line, by

a. Targeting new market segments

b. Offering new features

4. **Contraction** of a product line may be undertaken if the line is overextended. These are the symptoms of overextension

a. Some products are not contributing to profit because of low sales or cannibalization.

b. Manufacturing or marketing resources are being disproportionately allocated to slow-moving products.

c. Items in the line have become obsolete because of new product entries in the line.

4 Describe marketing uses of branding

IV. Branding

Branding is the major tool marketers have to distinguish their products from those of the competition.

A **brand** is a name, term, symbol, design, or combination thereof that identifies a seller's products and differentiates them from competitors' products

A **brand name** is that part of the brand that can be spoken.

The **brand mark** is the element of the brand that cannot be spoken, such as symbols.

B. Benefits of Branding

1. Identification is the most important objective. The brand allows the product to be differentiated from others and serves as an indicator of quality to consumers.

a. The term **brand equity** refers to the value of company and brand names.

b. **Brand loyalty**, a consistent preference for one brand over all others, is quite high in some product categories.

1) Brand identity is quite important to developing brand loyalty.

2) Brand loyalty leads to repeat purchasing, a marketer's traditional aspiration.

c. The term **master brand** refers to a brand so dominant in consumers' minds that they think of it immediately when a product category, use situation, product attribute, or customer benefit is mentioned.

2. An established brand name can facilitate the introduction of new products because a familiar brand is more quickly accepted by consumers.

3. Multilingual suitability is an important factor as companies go global.

4. The Internet provides a new alternative for generating brand awareness.

C. Branding Strategies

1. **Generic products** vs branded products

a. A **generic product** is typically a no-frills, no-brand-name, low-cost product that is identified by its product category

b. Are sold for 30 to 40 percent less than manufacturers' brands and 20 to 25 percent less than retailer-owned brands

c. Are obtaining substantial market shares in some product categories, such as pharmaceuticals

2. Manufacturer's brands vs private brands

a. A **manufacturer's brand** strategy is used when manufacturers use their own name on their products.

b. A **private brand** is a brand name that a wholesaler or retailer uses for the products it sells.

3. Individual brands vs family brands

a. **Individual branding** is the practice of using a different brand name for each product.

b. Individual brands are used when products differ greatly in use or in performance, quality, or targeted segment.

c. A company that markets several different products under the same brand name is using a **family brand.**

Consumer familiarity with the brand name facilitates the introduction of new products.

4. Co-branding

a. **Co-branding** is the use of a combination of brand names to enhances the perceived value of a product or when it produces incremental benefits to brand owners and users.

b. Co-branding may be used to identify product ingredients or components, when two organizations wish to collaborate to offer a product, or to add value to products that are generally perceived to be homogeneous shopping goods

D. Trademarks

1. A **trademark** is a legal term indicating the owner's exclusive right to use the brand or part of the brand. Others are prohibited from using the brand without permission. A **service mark** performs the same functions for service businesses.

a. Many parts of a brand (such as phrases and abbreviations) and symbols associated with product identification (such as shapes and color combinations) qualify for trademark protection.

b. The mark has to be used continuously to be protected.

c. Rights to a trademark continue for as long as it is used.

2. Companies must guard against the unauthorized use of their brands, slight alterations to the brand by mimics, and counterfeit merchandise that is labeled with the brand.

a. Under the Lanham Act of 1946, severe penalties may be imposed for trademark infringement.

b. The injured party can sue for triple damages and recovery of any profit made from the products and can destroy all materials with the infringing mark.

c. The Trademark Revision Act of 1988 allows organizations to register trademarks based on the intention to use them for 10 years.

d. Companies that fail to protect their trademarks face the problem of their product names becoming generic. A **generic product name** identifies a product by class or type and cannot be trademarked.

5 Describe marketing uses of packaging and labeling

V. Packaging

A. Packaging is a container for protecting and promoting a product. Packaging has traditionally been viewed as a means of holding contents together and as a way of protecting the physical good as it moves through the distribution channel.

B. Packaging Functions

1. Containing and Protecting Products

These are the most obvious functions of packaging.

Some packaging has to be quite sophisticated to protect the product from spoilage, tampering, or children.

2. Promoting Products

a. Packaging can differentiate a product from the competition by its convenience and utility.

b. Packages are the last opportunity marketers have to influence buyers before they make purchase decisions.

c. Packages are very important in establishing the brand image.

3. Facilitating Storage, Use, and Convenience

a. Wholesalers and retailers prefer packages that are easy to ship, store, and stock on shelves. They also like packages that protect the product, prevent spoilage or breakage, and extend shelf life.

b. Customers seek items that are easy to handle, open, and reclose.

c. Packaging is often used to segment markets, particularly by offering different sizes for different segments.

4. Facilitating Recycling and Reducing Environmental Damage

1. An important recent issue is the compatibility of the package and environmental concerns.

2. Many consumers demand recyclable, biodegradable, and reusable packages.

C. Labeling

1. **Persuasive labeling** focuses on a promotional theme or logo, with information for the consumer taking secondary importance.

2. **Informational labeling** is designed to help consumers in making proper product selections and to lower cognitive dissonance after the purchase. Informational labeling

a. often includes care and use information.

b. also may help explain construction features.

3. The Nutrition Labeling and Education Act of 1990 directed the Food and Drug Administration to require detailed nutritional information on most food packages and to establish standards for health claims on food packaging.

4. The **Universal Product Codes (UPC)** that appear on most items found in supermarkets and other high-volume outlets were first introduced in 1974. The bar codes, series of thick and thin vertical lines, are read by computerized optical scanners.

6 Discuss global issues in branding and packaging

VI. Global Issues in Branding and Packaging

A. Branding

1. Counterfeiting and brand imitations are a major problem in some countries.

2. When planning to enter a foreign market, a firm has three major alternatives for brand name choices:

a. One Brand Name Everywhere strategy uses one name with no adaptations for local markets. The strategy is useful when the company markets primarily one product and the brand name does not have negative connotations in any local market.

b. Adaptations and Modifications strategy uses different names in different markets for the same product. It is necessary if the original brand name is not pronounceable in the local language, is owned by someone else in the local market, or has a negative connotation in the local language.

c. Different Brand Names in Different Markets strategy uses different names in different markets for the same products. Local brand names are used to make the brand appear to be local to overcome translation and pronunciation problems.

B. Packaging

1. Three issues are particulary important in global markets

a. Meeting local labeling requirements and translating correctly

b. Making the package aesthetically pleasing under local cultural standards

c. Dealing with extreme climate conditions and long-distance shipping

7 Describe how and why product warranties are important marketing tools

VII. Product Warranties

A. Another part of the product is its warranty, a protection and information device for consumers.

1. A **warranty** guarantees the quality or performance of a good or service.

2. An **express warranty** is made in writing; an **implied warranty** is an unwritten guarantee that a good or service is fit for the purpose for which it was sold.

3. All sales have an implied warranty under the Uniform Commercial Code.

4. Under the Magnuson-Moss Warranty-Federal Trade Commission Improvement Act, a manufacturer that promises a full warranty must meet certain minimum standards. A limited warranty must be conspicuously promoted by the manufacturer; otherwise, a full warranty is assumed.

VOCABULARY PRACTICE

Fill in the blank(s) with the appropriate term or phrase from the alphabetized list of chapter key terms.

brand
brand equity
brand loyalty
brand mark
brand name
business (industrial) product
co-branding
consumer product
convenience product
express warranty
family brand
generic product
generic product name
implied warranty
individual branding
informational labeling
manufacturer's brand
master brand
persuasive labeling
planned obsolescence
private brand
product
product item
product line
product line depth
product line extension
product mix
product mix width
product modification
service mark
shopping product
specialty product
trademark
universal product code (UPC)
unsought product
warranty

1 Define the term "product"

1. Everything, both favorable and unfavorable, that a person receives in an exchange, is a(n) __________________________. This may take the form of a tangible good, a service, an idea, or any combination of these three.

2 Classify consumer products

2. Products can be classified in two ways, depending on the intended use of the product. If a product is used to manufacture other goods or services, to facilitate an organization's operations, or is resold to other customers, it is called a(n) __________________________, or industrial product. If the product is purchased to satisfy an individual's personal wants, it is a(n) __________________________.

3. Consumer products can be classified into four categories. A product like gum or gasoline that is relatively inexpensive and requires little shopping effort is called a(n) __________________________. A product that is more expensive and found in fewer stores is called a(n) __________________________; these products can be homogeneous or heterogeneous. A product that consumers search for extensively and that is usually expensive is called a(n) __________________________. Finally, a product unknown to the potential buyer or a known product that the buyer does not actively seek is referred to as a(n) __________________________.

3 Define the terms "product item," "product line," and "product mix"

4. A specific version of a product that can be designated as a distinct offering among an organization's products is called a(n) ______________________________. A group of closely related products offered by the organization is called a(n) ______________________________. Finally, all the products that an organization sells is called the ______________________________.

5. The number of product lines that an organization offers is referred to as ______________________________. The number of product items in one product line is ______________________________.

6. One job of the marketing manager is to decide when and if to alter existing products by changing one or more of a product's characteristics. This alteration procedure is ______________________________, and can change functional, style, or quality attributes. Frequent product modifications can cause products to become outdated before they actually need replacement. This practice describes ______________________________.

7. Adding additional products to an existing product line to compete more broadly in the industry is called ______________________________.

4 Describe marketing uses of branding

8. A name, term, symbol, or design that identifies a seller's products and differentiates them from competitors' products is a(n) ______________________________. The elements that can be spoken make up the ______________________________, and the elements that cannot be spoken are called the ______________________________.

9. When a company owns a brand name that is extremely familiar to consumers, the dollar value of the company is raised. This value of brand names is ______________________________. If the brand name is so familiar it is dominant in consumers' minds and they think of it immediately when a product category, usage situation, product attribute, or customer benefit is mentioned, the brand is termed a(n) ______________________________. If a customer consistently prefers one brand over all others and is a repeat purchaser, the consumer has ______________________________.

10. Firms face a variety of branding decisions, the first of which is whether to brand at all. A no-brand name, low-cost product is a(n) ______________________________. If a firm chooses to brand, it has several options. The manufacturer's name used as a brand name is called a(n) ______________________________, sometimes called a national brand. If the brand name is owned by a wholesaler or retailer, the brand is a(n) ______________________________. If a firm has products that vary substantially in use or performance, it should

use different brand names for different products, a strategy called ______________________. Alternatively, if the company markets several different products under the same brand name, it is using a(n) ______________________. If two or more brands names are placed on a product or its package, ______________________ is being used.

11. A legal term indicating the owner's exclusive right to use a product brand name or other identifying mark brand is a(n) ______________________. A(n) ______________________ performs the same legal function for services. If companies do not sufficiently protect their brand names, the name could become a(n) ______________________, which identifies a product by class or type and cannot be trademarked.

5 Describe marketing uses of packaging and labeling

12. The container for protecting and promoting a product is its package. An integral part of this is the label. Labeling strategy can take one of two forms. The first form focuses on a promotional theme or logo, and is known as ______________________. The second form is designed to help consumers make proper product selections and lower cognitive dissonance after the purchase; this is known as ______________________. A numerical code that appears on many product labels is the ______________________ for reading by computerized optical scanners.

7 Describe how and why product warranties are important marketing tools

13. When a company wants to guarantee the quality or performance of a good or service, it uses a(n) ______________________, which protects the buyer and gives essential information about the product. If it is in the form of a written guarantee, it is a(n) ______________________. If it is unwritten, a(n) ______________________ is being used; all sales have this.

Check your answers to these questions before proceeding to the next section.

TRUE/FALSE QUESTIONS

Mark the statement **T** if it is true and **F** if it is false.

1 Define the term "product"

_____ 1. When you buy a toothpaste, you are most interested in the chemical composition of the toothpaste.

2 Classify consumer products

_____ 2. Tammy considers her eye makeup to be very important. She spends considerable time comparing the prices and color options available at various cosmetic counters in department stores. For Tammy, makeup is a convenience product.

_____ 3. Joe Foxhunter makes unique silver jewelry. Loyal customers drive from miles around to buy the jewelry, which is sold only in his workshop. This jewelry is a specialty product.

3 Define the terms "product item," "product line," and "product mix"

_____ 4. Sony makes a variety of products, all under the Sony brand name. These products include CD players, televisions, and computer diskettes. These products represent Sony's product line.

_____ 5. Sugar Frosted Flakes have been marketed as a presweetened cereal for kids. Recently, however, advertisements have shown adults admitting that they too eat the cereal. The ads use the slogan, "You loved it as a kid, so try it again as an adult." The marketing term for the efforts to change perceptions of the cereal is "positioning."

_____ 6. Ultra-Comp produces a high-quality, expensive line of executive laptop computers. Recently, it introduced another line of computers with fewer features and a lower price, but which still carried the Ultra-Comp brand name. This is an example of a line extension.

4 Describe marketing uses of branding

_____ 7. The Nike "swoosh" is very well known and is considered to be the company's brand name.

_____ 8. Colgate-Palmolive makes Colgate toothpaste and has recently introduced Colgate mouthwash and Colgate toothbrushes. This is an example of family branding.

5 Describe marketing uses of packaging and labeling

_____ 9. The most important functions of packaging are to contain and protect products; promote products; facilitate product storage, use, and convenience; and facilitate recycling.

Check your answers to these questions before proceeding to the next section.

AGREE/DISAGREE QUESTIONS

For the following statements, indicate reasons why you may agree and disagree with the statement.

1. Branding is dead. Consumers make purchase decisions on price more than any other feature.

   Reason(s) to agree:

   Reason(s) to disagree:

2. Because they are typically less expensive, private brands are purchased by lower income consumers.

   Reason(s) to agree:

   Reason(s) to disagree:

3. Global marketers should adapt their products (including brand) to local markets instead of using global strategies.

   Reason(s) to agree:

   Reason(s) to disagree:

4. It is better for marketers to use family branding than to use individual branding.

   Reason(s) to agree:

   Reason(s) to disagree:

MULTIPLE-CHOICE QUESTIONS

Select the response that best answers the question, and write the corresponding letter in the space provided.

1 Define the term "product"

____ 1. The creation of a product is the starting point for the marketing mix because:
- a. the production department must know what to produce first
- b. the product is the first of the four P's in the marketing mix
- c. the product does not have to be the starting point--distribution or promotional strategies could also be the starting point
- d. product development takes the longest amount of time to complete
- e. determination of the price, promotional campaign, and distribution network cannot begin until the product has been specified

2 Classify consumer products

____ 2. David has wanted to purchase a Mercedes most of his adult life. He has done extensive research into different Mercedes models. Finally, at age 50, he has achieved the income level to be able to purchase one. The Mercedes is an example of a(n):
- a. specialty product
- b. consumer product
- c. convenience product
- d. business product
- e. unsought product

____ 3. Zack decides to purchase a filing cabinet. He watches television and newspaper ads until he sees one at a low price. For Zack, the filing cabinet is a(n):
- a. convenience product
- b. shopping product
- c. component product
- d. unsought product
- e. specialty product

____ 4. Joe jumps out of his truck and runs into a 7-Eleven store to grab a drink. While he is paying for his drink, he notices the candy bars at the counter and grabs one to buy. The candy bar in this case is considered to be a(n):
- a. specialty product
- b. consumer product
- c. convenience product
- d. business product
- e. unsought product

3 Define the terms "product item," "product line," and "product mix"

____ 5. In the Helidyne Industries' product portfolio, there is a list of all the product items that the company manufactures and markets. This list includes several types of helicopter engines, nuts, bolts, washers, rotor blades, and instrument systems. This is a description of its:
- a. product mix
- b. product line
- c. marketing mix
- d. line depth
- e. mix consistency

____ 6. Procter & Gamble makes four different brands of toilet tissue (Charmin, White Cloud, Banner, and Summit). This is an example of:

a. product line width
b. product mix
c. marketing mix
d. product mix consistency
e. product line depth

____ 7. A laptop manufacturer recently added a metal case and shock mountings to its laptop computers to make them more durable. It has not changed its prices. This is a ___________ modification.

a. quality
b. style
c. functional
d. repositioning
e. minor

____ 8. A manufacturer of car batteries has added a handle and an emergency reserve switch to its batteries. The manufacturer has not changed its prices. This is a(n) ___________ modification.

a. style
b. functional
c. quality
d. repositioning
e. upward extension

____ 9. Frequent style modifications by product manufacturers can also be called planned obsolescence. All the following statements are true about planned obsolescence EXCEPT:

a. planned obsolescence describes the practice of causing products to become obsolete before they actually need replacement
b. opponents of planned obsolescence argue that the practice is wasteful and unethical
c. marketers contend that consumers decide when styles are obsolete
d. marketers ignore quality, safety, and functional modifications to incorporate planned obsolescence into their marketing plans
e. consumers favor style modifications because they like changes in the appearance of goods such as clothing and cars

____ 10. Kentucky Fried Chicken (KFC) has added several items to its menu, including Hot Wings, skinless chicken, chicken nuggets, and chicken salad. This is an example of:

a. product line width
b. portfolio expansion
c. product line extension
d. contraction
e. repositioning

4 Describe marketing uses of branding

____ 11. Coca-Cola is the best known brand name in the world. The name has a high perceived quality and high brand loyalty among soft drink users. The company has developed the brand name for over 100 years. Coca-Cola has a valuable:

a. line extension
b. logo
c. private brand
d. package
e. brand equity

____ 12. You want to develop a brand that is so dominant in your customers' minds that they think of it immediately when they have a have need for the benefits derived by this product category. Which of the following brand issues would you concentrate on creating in your customers' minds?
a. trademarked brand
b. brand equity
c. master brand
d. generic brand
e. family brand

____ 13. The Safeway supermarket chain sells a brand of cookies, pasta, and other products under the name "Safeway Select." This is an example of a:
a. master brand
b. family brand
c. brand grouping
d. generic brand
e. private brand

____ 14. Fastop convenience stores sell only manufacturer's brands of snack foods. You explain to the manager that there are disadvantages to selling only manufacturer's brands such as:
a. a well-known manufacturer's brand will not enhance store image
b. manufacturers typically offer a lower gross margin than a dealer could earn on a private label
c. manufacturers rarely spend money advertising the brand name to consumers
d. manufacturers force stores to carry large inventories
e. relying on the manufacturer to deliver a national brand quickly is optimistic at best

____ 15. General Mills offers brands such as Bisquick pancake mix, Gold Medal flour, Betty Crocker cake mixes, and Yoplait yogurt. General Mills appears to be using a(n) ___________ strategy.
a. equity brand
b. individual brand
c. private brand
d. family brand
e. dealer brand

____ 16. Many personal computer manufacturers, such as Compaq and Dell, sell computers with the words "Intel inside" written on the central processing unit. This is an example of:
a. dealer branding
b. brand grouping
c. private branding
d. generic branding
e. co-branding

____ 17. Your boss has just told you to work on getting trademark protection for a new product brand the firm has developed, the "Zipper," which has a unique logo. You tell your boss that this protection will be difficult because:
a. brand logo designs, even as unique as yours, cannot be trademarked
b. of the catchy phrase used to promote your new brand
c. you also use abbreviated versions of your brand name, "Zip" and "Zipp"
d. your brand is considered a generic product name
e. the shape of your new and unusual product cannot be legally protected

5 Describe marketing uses of packaging and labeling

_____ 18. Which of the following is NOT one of the major functions of packaging?
a. guarantees product quality
b. contains the product
c. protects the product
d. promotes the product
e. facilitates storage and use of the product

6 Discuss global issues in branding and packaging

_____ 19. Companies considering global marketing should consider all of the following global aspects of branding and packaging EXCEPT:
a. whether to use one brand name with no adaptation to local markets, or whether to use one name but adapt and modify it for each local market
b. problems with brand imitations, brand piracy, and product counterfeits
c. whether to use different brand names in different markets for the same products
d. different currencies in each country, exchange rates, and final retail prices
e. product labeling, package aesthetics, and climate considerations

7 Describe how and why product warranties are important marketing tools

_____ 20. The label on a can of Elmo's glue that states "100 percent satisfaction guaranteed" is __________ warranty.
a. a descriptive
b. an express
c. an implied
d. a limited
e. not a

_____ 21. Realizing that their product needed a warranty to gain rapid market acceptance, the managers of Risktakers, Inc. produced:
a. a statement for salespeople to read to prospective buyers
b. an acknowledgement of company responsibilities for salespeople to build into presentations
c. a label stating that the product is the highest quality and backed by years of experience
d. an advertisement stating that buyers would not perceive the purchase of this product as risky
e. a written guarantee that the product would work as promised and that it is fit for the purpose it was sold

_____ 22. Greta purchased an electric can opener last week and attempted to open a can of soup with it this weekend for the first time. However, the blade on the can opener was too dull and succeeded only in denting the can, rather than cutting it. Under the __________, she has a right to demand that the machine perform the job for which it was purchased.
a. Uniform Commercial Code
b. Product Liability Act
c. Trademark Protection Act
d. Federal Communications Commission Code
e. Package Labeling Act

Check your answers to these questions before proceeding to the next section

ESSAY QUESTIONS

1. Name and briefly define four categories of consumer products. For each category, list three specific examples of products that would most likely be classified in that category.

2. The Kauphy Coffee Company's product portfolio is shown below:

Coffees	Appliances	Desserts
Fresh roast beans	Coffeemaker	Coffee ice cream
Fresh ground	Espresso maker	Coffee cakes
Instant	Coffee grinder	
Decaffeinated		
Gourmet		

What is the product mix width? What is the product line depth? What options does this company have to adjust its product lines or its mix as a whole?

3. You are a marketing manager for Gillette, a manufacturer of razors, blades, and various personal care items. You have been asked by the Vice-President of marketing to some up with some branding strategies for a new line of men's toiletries to be launched by the company. These toiletries will include shaving cream, deodorant and anti-perspirant, shampoo and conditioner, and disposable razors. List all the branding strategies you would use and list one advantage and one disadvantage to each strategy. Which one would you choose?

4. You are the brand manager of the family of Q-T-Pie custom-shaped pies that are distributed in the United States. The products include a wide variety of pie flavors, ingredients, and shapes. The brand is a favorite among gourmet gift givers and dessert aficionados. Your firm would like to enter several foreign markets. Name and describe the three major alternative brand name choices for this global strategy. What is the best alternative?

CHAPTER NOTES

Use this space to record notes on the topics you are having the most trouble understanding.

ANSWERS TO THE END-OF-CHAPTER DISCUSSION AND WRITING QUESTIONS

Use this space to work on the questions at the end of the chapter.

ANSWERS TO THE END-OF-CHAPTER CASE

Use this space to work on the questions at the end of the chapter.

CHAPTER 11 Developing and Managing Products

LEARNING OBJECTIVES

1 Explain the importance of developing new products and describe the six categories of new products

The product life cycle concept reminds us that developing and introducing new products is vital to business growth and profitability. Major manufacturers expect new products to account for a substantial portion of their total sales and profits.

New products can be classified as new-to-the-world products (discontinuous innovations), new product lines, additions to existing product lines, improvements to or revisions of existing products, repositioned products, or lower-cost products.

2 Explain the steps in the new-product development process

Most companies use a formal new product development process, which usually begins with identifying the firm's new product strategy. Each stage in the process acts as a screen, filtering out ideas that should not be considered further. After setting a new product strategy, the steps are as follows: idea generation, screening, business analysis, development, testing, and commercialization.

3 Explain why some products succeed and others fail

The most important factor in new product success is a good match between the product and market needs. Most new product failures can be linked to an inappropriate marketing strategy or to poor implementation of an appropriate marketing strategy.

4 Discuss global issues in new-product development

Multinational firms should consider new product development from a worldwide perspective. The main goal of the global product development process is to build adaptable products that can achieve worldwide appeal.

5 Describe the organizational groups or structures used to facilitate new-product development

Several organizational forms can facilitate a stream of new products. These include new product committees, new product departments, and venture teams.

A new product committee is a group of individuals, usually representing such functional interests as manufacturing, research and development, finance, and marketing, who manage the new product development process.

Some firms have established new product departments, which are separate departments that perform the same functions as a new product committee but on a full-time basis. Ideally, people in the product development department communicate with their peers in the operating departments. A formal department should have well-defined authority and responsibilities.

A venture team is an entrepreneurial, market-oriented, multidisciplinary group, comprising a small number of representatives from marketing, research and development, finance, and other areas, and focusing on a single objective: planning the company's profitable entry into a new business. Venture teams are frequently used to handle tasks that do not fit easily into the existing organization structure, that demand large amounts of time and money, and that require imaginative entrepreneurship. These teams are often full-time but temporary structures.

6 Explain the diffusion process through which new products are adopted

The diffusion process is the spread of a new product idea from its producer to the ultimate adopters. An adopter is a consumer who was sufficiently satisfied with his or her trial experience with a product to use it again. Adopters in the diffusion process belong to five categories: innovators, early adopters, early majority, late majority, and laggards.

Several product characteristics influence the rate of adoption: product complexity, compatibility with existing social values, relative advantage over substitutes, observability, and trialability. The process is facilitated by word-of-mouth promotion among consumers and communication from marketers to consumers.

One particular challenge to marketers is the fact that individual consumers are always in different stages of the adoption process. Marketing messages must deal with all these stages and different informational needs.

7 Explain the concept of product life cycles

The product life cycle is one of the most familiar concepts in marketing and a prevalent marketing management tool. The product life cycle has been adopted as a way to trace the stages of a product's acceptance from its introduction to its demise. The stages are introduction, growth, maturity, and decline.

The length of time a product spends in any one stage of the product life cycle may vary dramatically, from a few weeks to decades. The life cycle concept does not predict how long a product will remain in any one stage, rather, it is an analytical tool to help marketers understand where their product is now, what may happen, and which strategies are normally appropriate.

PRETEST

Answer the following questions to see how well you understand the material. Re-take it after you review to check yourself.

1. What are the six categories of new products?

2. List the seven stages of new product development.

3. List at least five reasons why new products may fail.

4. What is diffusion? List and briefly describe five categories of product adopters.

5. List and briefly describe the four stages of the product life cycle.

CHAPTER OUTLINE

1 Explain the importance of developing new products and describe the six categories of new products

I. The Importance of New Products

New products are important to sustain growth and profits, and to replace obsolete items.

A. Categories of New Products

1. **New-to-the-world products** are also called discontinuous innovations. The product category itself is new.

2. **New product lines** are products the firm has not offered in the past that allow it to enter an established market.

3. **Additions to existing product lines** are new products that supplement a firm's established line.

4. **Improvements or revisions** of existing products result in new products.

5. **Repositioned products** are existing products and targeted at new markets or market segments.

6. **Lower-priced products** are products that provide similar performance to competing brands at a lower cost.

2 Explain the steps in the new-product development process

II. The New Product Development Process

A. New Product Strategy

A **new product strategy** links the new product development process with the objectives of the marketing department, business unit, and corporation. All these objectives must be consistent.

1. New product strategy is a subset of the organization's overall marketing strategy.

2. New product strategy specifies the roles that new products must play in the organization's overall plan and describes the characteristics of the products the organization wishes to offer and the markets that it wishes to serve.

B. Idea Generation

New product ideas can come from a many sources:

1. The marketing concept suggests that customers' needs and wants should be the springboard for developing new products.

2. Because of their involvement in and analysis of the marketplace, employees who are not in the research and development department often come up with new product ideas.

3. A distributor is often more aware of customer needs than the manufacturer because the distributor or dealer is closer to end users.

4. Competitors may have new products that can or should be a source of new product ideas.

5. Research and development (R&D) may be a source of new product ideas and innovation. R&D is carried out in four ways:

    a. Basic research is scientific research aimed at discovering new technologies.

    b. Applied research attempts to find useful applications for new technologies.

    c. **Product development** goes one step further by converting the applications into marketable products.

    d. Sometimes research and development involves *product modification*, cosmetic changes in products or functional product improvements.

6. Consultant groups are available to examine a business and recommend product ideas.

7. A variety of approaches and techniques have been used to stimulate creative thinking and generate product ideas:

    a. **Brainstorming** is a process in which group members propose, without criticism or limitation, ways to vary a product or solve a problem.

    b. Excellent new product ideas are often generated by focus groups. These interviews usually consist of seven to ten consumers interacting in a structured discussion.

C. Idea Screening

Screening is the initial filter that serves to eliminate new product ideas that are inconsistent with the organization's new product strategy or are obviously inappropriate for some other reason.

1. Most new product ideas are rejected at this stage.

2. **Concept tests** are often used to rate concept (product) alternatives. A concept test is the evaluation of a new product idea, usually before any prototype has been created.

D. Business Analysis

In the **business analysis** stage, preliminary demand, cost, sales, and profitability estimates are made.

At the end of this stage, management should have a good idea of the market potential for the product.

E. Development

In the development stage, a prototype is developed and a marketing strategy is outlined.

Costs increase dramatically as the new product idea moves into the development stage.

1. In the early stages of development, the research and development or engineering department may develop a prototype or working model.

2. Decisions such as packaging, branding, and labeling will be made. Preliminary promotion, price, and distribution strategies should be established.

3. During the development stage, the technical feasibility of manufacturing the product at a reasonable cost is thoroughly examined, and the product may be modified.

4. Laboratory tests subject the product to much more severe or critical treatment than is anticipated by end users.

5. Many products that test well in the laboratory are next subjected to use tests, in which they are placed in consumers' homes or businesses for trial.

6. The development process works best when all the involved areas (departments) work together rather than sequentially. This process is called parallel engineering, simultaneous engineering, or concurrent engineering.

F. Test Marketing

After products and marketing programs have been developed, they are usually tested in the marketplace.

1. **Test marketing** is a limited introduction of a product and a marketing program to determine the reactions of potential customers in a market situation.

2. Selection of test market cities should ensure that they reflect market conditions in the new product's projected market area. There is no one "perfect" city that reflects the market as a whole, and selecting a test market city is a very difficult task.

3. Costs of test marketing are very high. Some companies choose to forgo this procedure altogether, especially for line extensions of well-known brands. Test marketing can take twelve to eighteen months and cost in excess of $1 million.

4. Many firms are looking for cheaper or faster alternatives to test marketing.

    a. Supermarket scanner testing (single-source data) keeps track of the sales response to marketing mix alternatives.

    b. Another alternative is **simulated (laboratory) market tests**, which usually entail showing members of the target market advertising for a variety of products. Purchase behavior, in a mock or real store, is then monitored.

5. One drawback of test marketing is that the product and its marketing mix are exposed to competitors before its introduction. Competitors may sabotage the test or rush an imitation of the new product to market.

6. Despite the problems associated with test marketing, most firms still consider it essential for new products. The high price of failure prohibits the widespread introduction of a product that might fail.

G. Commercialization

The final stage in the new product development process is **commercialization**, consisting of tasks necessary to begin marketing the product.

1. Several tasks are set in motion: ordering production materials and equipment, starting production, building inventories, shipping the product to field distribution points, training the sales force, announcing to the trade, and advertising to potential customers.

2. The total cost of development and introduction can be staggering, and the time period for commercialization can range from weeks to years.

3 Explain why some products succeed and others fail

III. Why Some New Products Succeed and Others Fail

A. Success Factors

1. The most important factor in successful new product introduction is a good match between the product and market needs.

Other factors that increase the chances of success for both consumer and business product introductions:

a. Unique but superior product

b. Meaningful and perceivable benefit to a sizable number of people or organizations

B. Failure Factors

The new product failure rate is estimated to be in the range of 80 to 90 percent. Many products fail because they do not meet consumers' needs but instead are built around what the company does best.

Failure can be absolute, in which case a company cannot regain its development, marketing, and production costs; or it can be relative, in which case the product makes a profit but does not meet sales or market share objectives.

4 Discuss global issues in new-product development

IV. Global Issues in New Product Development

Most multinational firms have to consider new product development from a worldwide perspective.

A. Many firms develop each product for potential worldwide use and then modify it as needed for each market.

B. Often products have to be designed to meet the regulations and requirements of various countries.

5 Describe the organizational groups or structures used to facilitate new-product development

V. Organizing for New Product Development

Many firms use an organized structure for generating new product ideas. A commitment to new product development implies a commitment by top management.

A. New Product Committees

A **new product committee** is an ad hoc group whose members represent various functional interests, such as manufacturing, research and development, finance, and marketing. It manages the new product development process and screens ideas.

B. New Product Departments

Some firms have established **new product departments**, separate departments that perform the same functions as a new product committee but on a full-time basis.

1. Ideally, people in the product development department communicate with their peers in the operating departments.

2. The new product department can be situated in one of several places within the organization. It can be a separate function, a high-level staff activity, a subfunction of marketing, or a subfunction of research and development.

C. New Product Venture Teams

A **venture team** is an entrepreneurial, market-oriented group, staffed by a small number of representatives from marketing, research and development, finance, and other areas, and focused on a single objective: planning the company's profitable entry into a new business.

1. Venture teams are frequently used to handle tasks that do not fit easily into the existing organization structure, that demand large amounts of time and money, and that require imaginative entrepreneurship.

2. Venture teams involve a full-time commitment but are formed and disbanded as needed.

3. The term *intrapreneur* (an entrepreneur working within a large organization) is often used to describe members of a venture team.

D. Simultaneous Product Development

The objective for many organizations is to shorten new product development time and be the first to market new products. A new team-oriented organizational form called **simultaneous engineering** has emerged to shorten the new product development process and to reduce its cost. With this process, all relevant functional areas and outside suppliers participate in all stages of the development process.

6 Explain the diffusion process through which new products are adopted

VI. The Spread of New Products

An **adopter** is a consumer who was happy enough with his or her trial experience with a product to use it again.

A. Diffusion of Innovation

1. An **innovation** is a product perceived as new by a potential adopter.

2. **Diffusion** is the process by which the adoption of an innovation spreads.

3. The **diffusion process** is the spread of a new idea from its source of invention or creation to its ultimate users or adopters.

B. Five categories of adopters participate in the diffusion process:

1. **Innovators** are eager to try new ideas and products, have higher incomes, and are better educated than noninnovators, and represent the first 2.5 percent of all those who will adopt.

2. **Early adopters** represent the next 13.5 percent to adopt the product. They are much more reliant on group norms, are oriented to the local community, and tend to be opinion leaders.

3. The **early majority**, the next 34 percent to adopt, collect more information and evaluate more brands than do early adopters. They rely on friends, neighbors, and opinion leaders for information and norms.

4. The **late majority**, the next 34 percent to adopt, do so because most of their friends have already done so. For them, adoption is the result of pressure to conform. This group is older than the others and tends to be below average in income and education.

5. **Laggards**, the final 16 percent to adopt, are similar to innovators in that they do not rely on the norms of the group. They are independent, however, because they are tradition-bound. Laggards tend to have the lowest socioeconomic status, are suspicious of new products, and are alienated from a rapidly advancing society.

C. Product Characteristics and the Rate of Adoption

Five product characteristics to predict and explain the rate of acceptance and diffusion of new products.

1. **Complexity** refers to the degree of difficulty involved in understanding and using a new product. The more complex the product, the slower its diffusion.

2. **Compatibility** refers to the degree to which the new product is consistent with existing values and product knowledge, past experiences, and current needs. Incompatible products diffuse more slowly than compatible products.

3. **Relative advantage** is the degree to which a product is perceived to be superior to existing substitutes.

4. **Observability** refers to the degree to which the benefits and other results of using a new product can be observed by others and communicated to target customers.

5. **Trialability** is the degree to which a product can be tried on a limited basis.

D. Marketing Implications of the Adoption Process

1. Two types of communication, *word-of-mouth communication* among consumers and *communication directly from the marketer to potential adopters*, aid the diffusion process.

2. The effectiveness of different messages and appeals depends on the type of adopter targeted.

7 Explain the concept of product life cycles

VII. Product Life Cycles

The **product life cycle** is as a way to trace the stages of a product's acceptance from its introduction to its demise.

1 The product life cycle refers to the life of the **product category**, which includes all brands that satisfy a particular type of need

2. The length of time a product category spends in any one stage of the product life cycle may vary dramatically, from a few weeks to decades.

The life cycle concept does not predict how long a product category will remain in any one stage, rather, it is an analytical tool to help marketers understand where their product is now, what may happen, and which strategies are normally appropriate.

> ➢ *The length of the product life cycle is usually related to the type of benefit the product offers its consumers. A fad or fashion offers limited value in contrast to a consumer durable like a washing machine.*

A. Stages in the Product Life Cycle

1. Introductory Stage

a. The **introductory stage** of the product life cycle represents the full-scale launch of a new product into the marketplace.

b. The introductory stage is typified by a high failure rate, little competition, frequent product modification, and limited distribution

c. The introductory stage usually has high marketing and production costs and negative profits as sales increase slowly

d. Promotion strategy in this stage focuses on developing product awareness and informing customers of product benefits. The aim is to stimulate primary demand for the product category.

e. Intensive personal selling to retailers and wholesalers is required.

2. Growth Stage

a. The **growth stage** of the life cycle is characterized by sales growing at an increasing rate, the entrance of competitors into the market, market consolidation, and healthy profits.

b. Promotion in the growth stage emphasizes heavy brand advertising and the differences between brands.

c. Gaining wider distribution is a key goal in this stage.

d. Toward the end of the growth stage, prices normally fall and profits reach their peak.

e. Development costs have been recovered by the end of the growth stage, and sales volume has created economies of scale.

3. Maturity Stage

a. The **maturity stage** of the life cycle is characterized by declining sales growth rates, markets approaching saturation, annual product changes that are more cosmetic that substantial, and a move toward the widening or extension of the product line .

b. During the maturity stage, marginal competitors begin dropping out of the market. Heavy promotions to both the dealers and consumers are required. Prices and profits begin to fall.

c. Emergence of "niche marketers" that target narrow, well-defined, under-served segments of a market.

4. Decline Stage

a. The **decline stage** is signaled by a long-run drop in sales.

b. The rate of decline is governed by how rapidly consumer tastes change or how rapidly substitute products are adopted.

c. Falling demand forces many competitors out of the market, often leaving a few small specialty firms manufacturing the product.

C. Implications for Marketing Management

1. Strategies used by marketing managers to prevent products from slipping into the decline stage include

a. Promoting more frequent use of the product by current customers

b. Finding new target markets for the product

c. Finding new uses for the product

d. Pricing the product below the market

e. Developing new distribution channels

f. Adding new ingredients or deleting old ingredients

g. Making a dramatic new guarantee

VOCABULARY PRACTICE

Fill in the blank(s) with the appropriate term or phrase from the alphabetized list of chapter key terms.

adopter
brainstorming
business analysis
commercialization
concept test
decline stage
development
diffusion
growth stage
innovation
innovators
introductory stage
maturity stage
new product
new product committee
new product department
new product strategy
product category
product development
product life cycle
screening
simulated (laboratory) test marketing
simultaneous product development
test marketing
venture team

1 Explain the importance of developing new products and describe the six categories of new products

1. Discontinuous innovations, new product lines, additions to existing product lines, and improvements/revisions all describe categories of a(n) ______________________________.

2 Explain the steps in the new-product development process

2. The strategy that links the new product development process with marketing, business unit, and corporate objectives is the ______________________________.

3. If managers create new products for present markets by converting applications for new technologies into marketable products, ______________________________ has taken place. If cosmetic or functional changes are made to products to improve them, product modification has taken place.

4. One option for generating new product ideas is to have a group think of unlimited ways to vary a product or solve a problem. This is called ______________________________.

5. After new ideas are generated, they pass through the first filter in the product development process. This stage, called ______________________________, eliminates new product ideas that are inappropriate. One tool that may be used in this stage evaluates a new product idea, usually before any prototype has been created. This tool is a(n) ______________________________ and is considered a relatively good predictor of product success.

6. New product ideas that survive the screening process progress to the ______________________________ stage, when preliminary demand, costs, sales, and profitability estimates are made. If financial estimates are favorable, then a prototype product may be developed and a marketing strategy sketched out. These activities occur during the ______________________________ stage.

7. A limited introduction of a product and a marketing program to determine the reactions of potential customers in a market situation is ____________________. An alternative to this method is a test that exposes consumers to the marketing mix and records subsequent purchasing behavior. This is called ____________________. If these tests are successful, the product goes to the final stage in the new product development process, ____________________.

5 Describe the organizational groups or structures used to facilitate new-product development

8. A firm has several options when organizing for new product development. One alternative is a group whose members represent various functional interests and manage the new product development process on a part-time basis. This is a(n) ____________________. If these functions are performed on a full-time basis, it is a(n) ____________________. Finally, an entrepreneurial, market-oriented group staffed by a small number of representatives from different disciplines is called a(n) ____________________.

9. A new organizational form has emerged to shorten the new product development process and to reduce cost. In this form, all relevant functional areas and outside suppliers participate in all stages of the development process. This organizational form is called ____________________.

6 Explain the diffusion process through which new products are adopted

10. If a person buys a new product, is satisfied, and becomes a repeat purchaser, that person is a(n) ____________________. A product that is perceived as new by this person is a(n) ____________________. The process by which the adoption of innovation spreads is ____________________.

7 Explain the concept of product life cycles

11. A marketing management tool that provides a way to trace the stages of a product's acceptance from its birth to its death is the ____________________. This concept refers to the life cycle of a class of products that satisfy a particular need, which is a(n) ____________________. The four stages, listed in the order from birth to death, begin with the ____________________, continue with the ____________________, along to the ____________________, and end with the ____________________.

Check your answers to these questions before proceeding to the next section.

TRUE/FALSE QUESTIONS

Mark the statement **T** if it is true and **F** if it is false.

1 Explain the importance of developing new products and describe the six categories of new products

____ 1. Ben & Jerry's addition of new ice cream flavors would not be considered new products since the company was already making ice cream. They would simply be called line extensions.

2 Explain the steps in the new-product development process

____ 2. If a firm adopts the marketing concept, then during the idea generation stage of the new product development process, the logical starting place should be the customers of the firm.

____ 3. The employees at Genco are all in a room together, shouting out potential ideas. The ideas are written down and evaluated as they are generated, with everyone voting on which ideas to accept or discard. This is an example of brainstorming.

____ 4. During the screening stage of the new product development process, it is appropriate to use concept tests.

____ 5. During the development stage of the new product process, the product may undergo laboratory tests.

____ 6. One of the great advantages to test markets is the low cost of conducting one.

____ 7. The Wheelsport Company has developed a new type of inline skateboard. The company has already gone through several stages of the new product development process, including screening, development, and testing. Assuming the results are consistent with profit and cost expectations, the most likely next step is commercialization.

5 Describe the organizational groups or structures used to facilitate new-product development

____ 8. Zapco, Inc., would like to reorganize the structure of its company to encourage constant development of new products. Three such organization forms are new product departments, venture capitalists, and new product committees.

____ 9. The president at the Newpro Corporation has given each department the same idea to develop a new product. Each department completes its version of the idea and submits it at the same time as the other departments. The president will compare the alternatives and pick the best one. NewPro is practicing parallel engineering.

6 Explain the concept of product life cycles

____ 10. The Kitchen-Pro Company is about to introduce a new kitchen gadget that combines a food processor with a toaster oven. As with other products of this type, Kitchen-Pro can expect sales of this new product to follow a bell-shaped curve over the next ten to fifteen years as it follows the product life cycle.

7 Explain the diffusion process through which new products are adopted

____ 11. Carol loves romantic comedy movies and is among the first 2 ½ percent to see a new movie when it is released. She either tries to see it in a preview showing or else goes during the first weekend launch. Carol is an example of an early adopter.

_____ 12. The Chipster Computer Company has two new products: One is a ready-to-go computer that is already assembled and preloaded with business software; the other is a kit of parts that the consumer must assemble to make his or her own computer. Of these two, the ready-to-go computer will probably be diffused more quickly because it is less complex.

_____ 13. The Robo-Vac company will be introducing its automated vacuum cleaner robot. The company plans to offer a one-week, in-home examination period to encourage people to buy its product. This plan would probably increase the product's rate of diffusion because it increases the product's trialability.

Check your answers to these questions before proceeding to the next section.

AGREE/DISAGREE QUESTIONS

For the following statements, indicate reasons why you may agree and disagree with the statement.

1. Following each step of the new product process is arduous and only delays a product launch.

   Reason(s) to agree:

   Reason(s) to disagree:

2. New product ideas that are based on a sound positioning will not fail.

   Reason(s) to agree:

   Reason(s) to disagree:

3. New product ideas are not the sole responsibility of a new products committee or department; they should be the responsibility of the entire organization.

   Reason(s) to agree:

   Reason(s) to disagree:

4. The product life cycle concept is old and not applicable to all products.

   Reason(s) to agree:

Reason(s) to disagree:

MULTIPLE-CHOICE QUESTIONS

Select the response that best answers the question, and write the corresponding letter in the space provided.

1 Explain the importance of developing new products and describe the six categories of new products

____ 1. Nabisco has come up with a lunchbox-sized snack pack of Teddy Grahams cookies to add to its popular line of Teddy Grahams products. Is this a new product?
- a. yes, additions to the product line are new products
- b. no, graham crackers cannot be considered an innovation
- c. no, it is only a product addition
- d. yes, this is a discontinuous innovation
- e. no, this is not a product improvement

____ 2. The artificial sweetener NutraSweet (aspartame), when first introduced, was a new-to-the-world product--a method of sweetening products with few calories and no bad aftertaste. This type of new product was a:
- a. diffusion
- b. discontinuous innovation
- c. moderate innovation
- d. venture product
- e. specialty product

____ 3. Recently, the Cheez Whiz cracker spread product has been featured in promotions as a cheese sauce for the microwave oven. This is an example of:
- a. revising existing products
- b. lowering costs
- c. repositioning
- d. segmenting existing product lines
- e. reformulation

____ 4. Klean Laundry detergent was just replaced with a "new and improved" Klean that makes clothes smell good as well as gets them clean. This new product is an example of:
- a. revising an existing product
- b. lowering costs
- c. repositioning
- d. a new-to-the-world product
- e. a line extension

2 Explain the steps in the new-product development process

____ 5. Veronica is the new products manager at a packaged foods company. She would like to put together a brainstorming team to generate ideas about new food products. She should consider inviting people from the following areas EXCEPT:
- a. account executives from her advertising agency
- b. a couple of big customers
- c. lawyers from the company's law firm
- d. a couple of large distributors
- e. sales people who have access to competitive intelligence

_____ 6. The first step in the new products process is:
a. concept testing
b. generating ideas ("brainstorming")
c. developing the new product
d. developing a new products strategy
e. test marketing

_____ 7. Fabrique, Inc. invented a fabric that was fireproof, tearproof, but also edible. The week after the product was invented, a group of the firm's employees got together and listed ways the product might be used. This is an example of:
a. an unfocus group
b. concept testing
c. brainstorming
d. venture group activities
e. screening

_____ 8. After the research team at Epsilon Corporation had generated a dozen new product ideas for gasoline additives, a committee of company executives met to analyze whether the product ideas were consistent with the organization's new product strategy. This is called:
a. screening
b. commercialization
c. business analysis
d. test marketing
e. idea generation

_____ 9. After Revvin' Cosmetics evaluated many new product ideas, the company selected three ideas to present to consumers. Before any prototype had been created, researchers presented the product ideas to groups of consumers for evaluation. This stage of new product development is called:
a. screening
b. market testing
c. concept testing
d. simulated market testing
e. use testing

_____ 10. As the Slimmy Company continued the development of its new product idea, diet fat, a group was assigned the task of estimating preliminary demand for the product, costs, sales, and future profitability. This is the ___________ stage in product development.
a. test marketing
b. idea generation
c. concept testing
d. screening
e. business analysis

_____ 11. When Philips developed a new extra-long life halogen light bulb, the company installed 100 bulbs at no charge in the homes of fifty consumers and monitored the bulbs for one year. This is a:
a. laboratory test
b. use test
c. concept test
d. market test
e. diffusion test

_____ 12. Sigma-Sunco has decided to introduce its new sunscreen eye drops in a limited market consisting of only two cities and closely monitor the reactions of potential customers to the product and marketing program. This is called:
a. diffusion analysis
b. use test
c. concept test
d. test marketing
e. laboratory test

____ 13. All of the following factors should be considered when choosing a test market EXCEPT:
a. the city's large population
b. the city's relative isolation from other cities
c. the city's good record as a test city, but not overly used
d. availability of retailers that will cooperate in the test market
e. similarity to planned distribution outlets

____ 14. Gramma's Treats tested consumers' reactions to a new dessert by getting consumers to look through a newspaper with grocery store ads, make out a grocery list, and then "shop" in a mock store filled with real products, including the new dessert. This is a:
a. real test market
b. simulated (laboratory) market test
c. concept test
d. use test
e. cable/scanner test

____ 15. SunGlass has ordered production materials and equipment for its new polarized auto windshields for all of its manufacturing plants. SunGlass is entering the ____________ stage of new product development.
a. business analysis
b. commercialization
c. product testing
d. product prototypes
e. market testing

3 Explain why some products succeed and others fail

____ 16. In the long run, products fail because of a poor match between:
a. advertising and personal selling
b. limited resources and unlimited consumer wants
c. the marketing mix and physical distribution
d. prices and consumer demand
e. product characteristics and consumer needs

5 Describe the organizational groups or structures used to facilitate new-product development

____ 17. Haytco, Inc., is considering several organization forms that can facilitate new product development. Haytco's options include all of the following EXCEPT:
a. new product department
b. intrapreneurship
c. new product committee
d. venture team
e. product-review committee

____ 18. The Blanko corporation is a highly consumer-oriented firm and has recently set up a new product department. Logically, Blanko should locate the department as a:
a. subfunction of the marketing department
b. subfunction of research and development
c. separate function
d. high-level staff activity
e. division of human resources

_____ 19. The Gallant Winery created a separate group of employees working outside the normal corporate structure to develop one special project, a nonalcoholic wine. What type of product development organization did Gallant use?
a. new product department
b. new product committee
c. venture team
d. new product matrix management
e. product manager

6 Explain the diffusion process through which new products are adopted

_____ 20. In an attempt to understand its target markets better, Carrot Personal Computers has been studying the product adoption stages and strategies. Carrot can expect to find all of the following EXCEPT that:
a. early adopters may also be opinion leaders
b. there are three categories of consumers who will adopt computer products
c. laggards are the last consumers to adopt a new product
d. early triers of a product are generally heavy users of that product
e. innovators were the first to purchase home computers

_____ 21. In a meeting with Sony corporate officials, you (a leading marketing consultant) are asked if the new digital tape technology will be quickly accepted and sought by the U.S. consumer. You tell them that this depends on many product characteristics, including all of the following EXCEPT:
a. how similar it is to existing tape players in use and fulfilling needs
b. the degree of difficulty involved in understanding and using it
c. how much of a relative advantage it has over current tape players
d. the "buy American" movement in the United States
e. its degree of trialability

7 Explain the concept of product life cycles

_____ 22. Wilderness, Inc., sells a tent that collapses down into a roll the size of one roll of toilet paper. The tent can be easily carried in a backpack. The tent has been in the market for two years, and the company is spending a lot on advertising. The tent is in which stage of the product life cycle?
a. maturity stage
b. introduction stage
c. innovation stage
d. growth stage
e. early majority stage

_____ 23. Pepsi and Coca-Cola have been battling in the "cola war" for years. Both products have been in the market for nearly 100 years and spend much of their marketing budgets on short-term promotions to steal market share from each other. Pepsi and Coca-Cola are in which stage of the product life cycle?
a. maturity stage
b. decline stage
c. promotional stage
d. growth stage
e. late majority stage

_____ 24. The Acme Slide Rule Company has seen most of its competition slip away, and it now holds 90 percent of the market share for slide rules and other mechanical calculation devices. Acme's own sales totals are slowly decaying, and it spends little on promotion. Acme's products are in which stage of the product life cycle?
a. maturity stage
b. decline stage
c. downfall stage
d. decay stage
e. laggard stage

_____ 25. When reports were published that aspirin consumption could reduce the chance of heart attacks, aspirin sales boomed. This is an example of product:
a. diffusion
b. adaptation
c. life cycle extension
d. modification
e. mix widening

Check your answers to these questions before proceeding to the next section.

ESSAY QUESTIONS

1. There are several correct definitions of the term "new product." Name and describe six categories of new products. For each category, give a specific real-life example of a product that fits into that category.

2. The multiple-step new product development process is an essential ingredient in new product development. List the steps of this process.

3. You are in charge of test marketing your company's new brand of nonalcoholic wine. List ten criteria you should consider when choosing a good test market city. What is an advantage for your company of using test marketing? What is a disadvantage of using test marketing?

4. Draw the sales line and the profit line of the product life cycle in the following diagram, and label each line. Indicate the names of the four stages of the product life cycle in the blanks provided. Then list the four stages of the product life cycle. For each stage, list five typical characteristics of that stage.

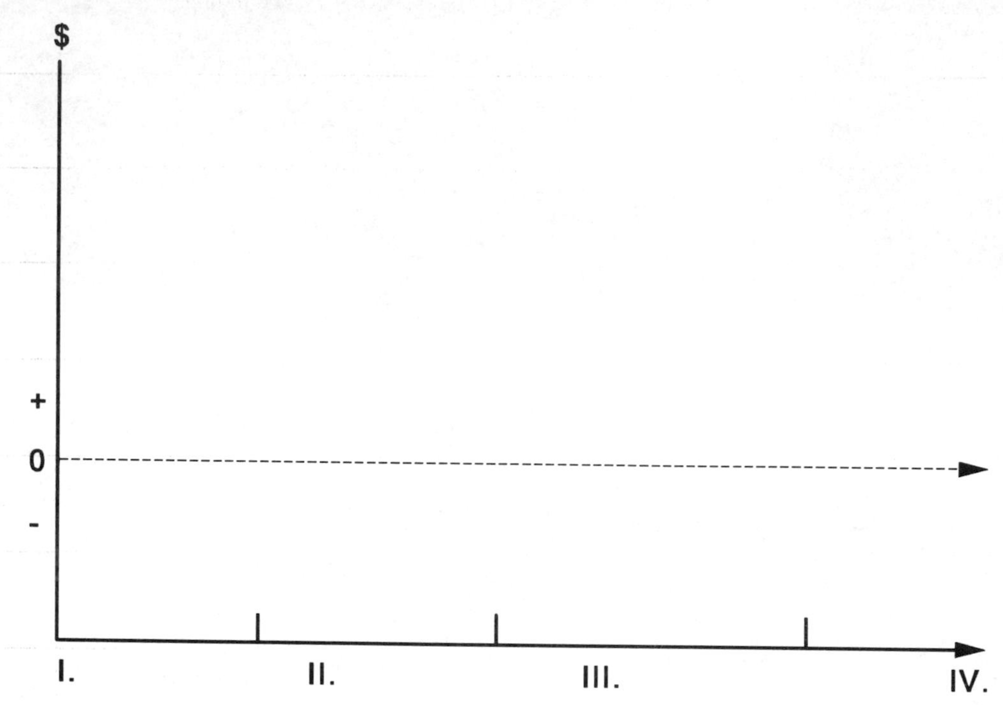

5. Name and briefly describe each of these five categories in the product adoption process. Using the launch of a new movie as an example, tell what each adopter might do.

CHAPTER NOTES

Use this space to record notes on the topics you are having the most trouble understanding.

ANSWERS TO THE END-OF-CHAPTER DISCUSSION AND WRITING QUESTIONS

Use this space to work on the questions at the end of the chapter.

ANSWERS TO THE END-OF-CHAPTER CASE

Use this space to work on the questions at the end of the chapter.

CHAPTER 12 Services and Nonprofit Organization Marketing

LEARNING OBJECTIVES

1 Discuss the importance of services to the economy

The service sector plays a crucial role in the U.S. economy, employing roughly 70 percent of the work force and accounting for about 60 percent of the gross domestic product.

2 Discuss the differences between services and goods

Services are distinguished from goods by four characteristics: intangibility, inseparability, heterogeneity, and perishability. Services are intangible in that they lack clearly identifiable physical characteristics, making it difficult for marketers to communicate their specific benefits to potential customers. The production and consumption of services are typically inseparable. Services are heterogeneous because their quality depends on variables such as the service provider, the individual consumer, location, and so on. Finally, services are perishable in the sense that they cannot be stored or saved.

3 Describe the components of service quality, and the gap model of service quality

There are five components of service quality: reliability (ability to perform the service dependably, accurately, and consistently), responsiveness (providing prompt service), assurance (knowledge and courtesy of employees and their ability to convey trust), empathy (caring, individualized attention), and tangibles (physical evidence of the service).

4 Explain why services marketing is important to manufacturers

Although manufacturers market mainly goods, the related services they provide often give them a competitive advantage, especially when competing goods are quite similar.

5 Develop marketing mixes for services

In contrast to positioning goods, positioning services favorably in the eyes and minds of potential customers is more difficult. This because of services' unique characteristics of intangibility, inseparability, heterogeneity, and perishability. Positive strategies to deal with these characteristics include carefully designed and implemented market analysis and customer targeting, along with careful decision making about product, price, promotion, and distribution.

6 Discuss relationship marketing in services

Relationship marketing involves attracting, developing, and retaining customer relationships. There are three levels of relationship marketing: Level 1 focuses on pricing incentives; level 2 uses pricing incentive and social bonds with customers; and level 3 uses pricing, social bonds, and structural bonds to build long-term relationships.

7 Explain internal marketing in services

Internal marketing means treating employees as customers and developing systems and benefits that satisfy their needs. Employees who like their jobs and are happy with the firm they work for are more likely to deliver good service. Internal marketing activities include competing for talent, offering a vision, training employees, stressing teamwork, giving employees freedom to make decisions, measuring and rewarding good service performance, and knowing employees' needs.

8 Discuss global issues in services marketing

Competition in the global market for services is increasing rapidly. Many U.S. service industries (such as financial institutions, construction and engineering, insurance, and leisure) can easily enter the global marketplace because of existing competitive advantages.

9 Describe nonprofit organization marketing

Nonprofit marketing is conducted by individuals and organizations to achieve goals other than profit, market share, and return on investment. Nonprofit organization marketing facilitates mutually satisfying exchanges between nonprofit organizations and their target markets.

10 Explain the unique aspects of nonprofit organization marketing

Several unique characteristics distinguish nonprofit marketing strategy, including a concern with services and social behaviors rather than manufactured goods and profit; a difficult, undifferentiated, and in some ways marginal target market; a complex product that may have only indirect benefits and elicit very low involvement; a short, direct, immediate distribution channel; a relative lack of resources for promotion; and prices only indirectly related to the exchange between the producer and the consumer of services.

PRETEST

Answer the following questions to see how well you understand the material. Re-take it after you review to check yourself.

1. List and briefly describe four characteristics of service products that make them different from goods.

2. What are five components of service quality?

3. List five "gaps" that can occur in service quality.

4. What is internal marketing?

5. Briefly describe the unique aspects of objectives, target markets, product, distribution, and promotion for nonprofit organization marketing.

CHAPTER OUTLINE

1 Discuss the importance of services to the economy

I. The Importance of Services

The basic marketing process is the same for all products and organizations. However, services and nonprofit businesses have many unique characteristics that affect marketing strategy.

A. The service sector substantially influences the U.S. economy

1. More than 8 in 10 workers in the U.S. work force are employed in services.

2. The service sector accounts for more than 74 percent of the U.S. gross domestic product.

3. The service sector has generated a $55.7 billion balance-of-trade surplus.

4. 73% of workers in Great Britain, 57% of workers in Germany, and 62% of workers in Japan work in service occupations.

2 Discuss the differences between services and goods

II. How Services Differ from Goods

Services are products in the form of an activity or a benefit provided to consumers, with four unique characteristics that distinguish it from a good: intangibility, inseparability, heterogeneity, and perishability.

A. Intangibility

The fundamental difference between services and goods is that services are intangible. **Intangibility** means that services cannot be touched, seen, tasted, heard, or felt in the same manner that goods can be sensed.

1. Evaluating the quality of services is more difficult than evaluating the quality of goods.

a. Services exhibit fewer **search qualities**, characteristics that can be easily assessed before purchase.

b. Services have more **experience qualities**, characteristics that can only be assessed after use.

c. Services also have **credence qualities**, characteristics that cannot be easily assessed even after purchase and experience.

2. Marketers often rely on tangible cues, such as the environment and atmosphere at facilities, to communicate a service's nature and quality.

B. Inseparability

1. Services are often sold and then produced and consumed at the same time. **Inseparability** refers to the characteristic of services that allows them to be produced and consumed simultaneously.

2. The consumer of services is usually involved at the time of production. Services cannot be produced in a centralized location and consumed in decentralized locations.

C. Heterogeneity

Heterogeneity means that services tend to be less standardized and uniform than goods.

1. Consistency and quality control are often hard to achieve, because services are labor-intensive and vary by employee.

2. Consistency and reliability are strengths that many service firms actively pursue through mechanization of the process, standardization, and training.

D. Perishability

1. **Perishability** means that services cannot be stored, warehoused, or inventoried.

2. One of the most important challenges in many service industries is finding ways to synchronize supply and demand. Such techniques as differential pricing during nonpeak periods may help.

3 Describe the components of service quality and the gap model of service quality

III. Service Quality

Because of the four unique characteristics of services, service quality is difficult to define and measure. Customers evaluate services along 5 dimensions.

A. These dimensions are:

1. **Reliability** is the ability to perform the service dependably, accurately, and consistently.

2. **Responsiveness** is the ability to provide prompt service.

3. **Assurance** is the knowledge and courtesy of employees and their ability to convey trust.

4. **Empathy** is caring, individualized attention to customers.

5. **Tangibles** are the physical evidence of the service.

B. The Gap Model of Service Quality

A model of service quality called the **gap model** identifies five gaps that can cause problems in service delivery.

1. Gap 1: the gap between what customers want and what management thinks customers want.

2. Gap 2: the gap between what management thinks customers want and the quality specifications that management develops to provide the service.

3. Gap 3: the gap between the service quality specifications and service that is actually provided.

4. Gap 4: the gap between what the company provides and what the customer is told it provides.

5. Gap 5: the gap between the service that customers receive and the service they want.

4 Explain why services marketing is important to manufacturers

IV. Services Marketing in Manufacturing

A. It is very difficult to make a clear distinction between manufacturing and service firms because service is often a major factor in the success of manufacturing firms.

B. The service component offers the potential for manufacturers to build a strong competitive advantage, especially in industries in which consumers perceive product offerings to be similar.

5 Develop marketing mixes for services

V. Marketing Mixes for Services

Elements of the marketing mix need to be adjusted to meet the special challenges posed by the unique characteristics of services.

A. Product (Service) Strategy

The development of a product strategy requires that the business focus on the service process itself.

1. What Is Being Processed

a. The three types of processing are people processing, possession processing, and information processing.

b. Because customers' experiences and involvement differ for each of these types of services, marketing strategies may also need to differ.

2. Core and Supplementary Services

The service product can be viewed as a bundle of activities that include the **core service**, which is the most basic benefit the customer is buying, and a group of **supplementary services** that support or enhance the core service.

3. Mass Customization

a. Customized services are more flexible and responsive to individual customer needs but also cost more.

b. Standardized services are more efficient and cost less.

c. An emerging strategy is called **mass customization** which uses technology to deliver customized services on a mass basis.

4. The Service Mix

a. Most service organizations offer more than one service. A service firm must understand it is managing a portfolio of services that include opportunities, risks, and challenges.

b. Service marketing strategy includes deciding what new services to introduce to which target markets, what existing services to maintain, and what services to eliminate from the portfolio.

B. Distribution Strategy

1. Distribution strategies for service organizations focus on convenience, number of outlets, direct vs indirect distribution, location and scheduling.

2. Management must set objectives for the distribution system and select operation type (usually direct or franchise), location, and scheduling.

C. Promotion Strategy

1. Four promotional strategies for dealing with the unique features of services are

a. Stressing tangible cues

b. Using personal information sources

c. Creating a strong organizational image

d. Engaging in postpurchase communication

D. Price Strategy

The unique characteristics of services present some special pricing challenges.

1. It is important to define the unit of service consumption.

2. For services that are composed of multiple elements, the issue is whether each element should be priced separately.

3. Several trends, such as increased competition and deregulation, have changed pricing from a passive to an active component of firms' marketing strategies in services (such as banking, telecommunications, and transportation).

4. Three categories of pricing objectives have been suggested:

a. **Revenue-oriented pricing** focuses on maximizing the surplus of income over costs.

b. **Operations-oriented pricing** seeks to match supply and demand by varying prices to ensure maximum use of productive capacity at any specific point in time.

c. **Patronage-oriented pricing** attempts to maximize the number of customers using the service.

5. In practice, a firm may use more than one objective in setting prices, often trying to blend the objectives to best fit the type of business.

6 Discuss relationship marketing in services

VI. Relationship Marketing in Services

Relationship marketing is a means for attracting, developing, and retaining customer relationships to build strong customer loyalty.

A. Relationship marketing is important in service businesses that involve a process of continuous interaction.

B. Relationship marketing can be practiced at three levels:

1. Financial: pricing incentives, such as frequent-flier programs; low customization

2. Financial and social: adds social bonds, staying in touch; medium customization

3. Financial, social, and structural: adds value-added services that are not available from other firms; high customization

7 Explain internal marketing in services

VII. **Internal marketing** means treating employees as customers and developing systems and benefits that satisfy their needs.

A. Happy employees are more likely to deliver high-quality service.

B. Internal marketing activities include competing for talent, offering a vision, training, stressing teamwork, giving employees freedom to make decisions, measuring and rewarding quality, and learning about employee needs.

8 Discuss global issues in services marketing

VIII. Global Issues in Services Marketing

The United States is the world's largest exporter of services.

A. Services that are exported include financial, credit card, construction, engineering, insurance, and leisure services.

B. The marketing mix for each service usually has to be designed to meet the specific needs of each country.

9 Describe nonprofit organization marketing

IX. Nonprofit Organization Marketing

A **nonprofit organization** is an organization that exists to achieve some goal other than the usual business goals of profit, market share, or return on investment.

Nonprofit organizations account for over 20 percent of the economic activity in the United States.

A. What Is Nonprofit Organization Marketing?

1. **Nonprofit organization marketing** is the effort of public and private nonprofit organizations to bring about mutually satisfying exchanges with their target markets.

a. Nonprofit organizations include government organizations as well as private museums, theaters, schools, churches, and other nongovernment organizations.

b. Many nonprofit organizations do not realize that their functions and activities, such as choosing a target market, setting prices, and communication, involve marketing.

c. Nonprofit organizations may believe that marketing is only appropriate in commercial, profit-seeking organizations.

10 Explain the unique aspects of nonprofit organization marketing

X. Unique Aspects of Nonprofit Marketing Strategy

Differences between business and nonprofit marketers create unique challenges for nonprofit managers in setting marketing objectives, selecting target markets, and developing appropriate marketing mixes.

A. Setting Objectives

Without a profit motive to guide them, many nonprofit organizations have a multitude of vague, diverse, intangible, and often conflicting objectives. Still, goals must be set, and measures must be established.

B. Target Markets

1. Nonprofit organizations often have apathetic or strongly opposed target markets as recipients of their services.

2. Many nonprofit organizations are pressured or required to adopt undifferentiated segmentation strategies to serve the maximum number of people or to achieve economies of scale.

3. Nonprofit organizations are often expected to complement rather than compete against private-sector organizations.

C. Product Decisions

1. Nonprofit organizations often market complex behaviors or ideas.

2. The benefit strength of many nonprofit offerings is quite weak or indirect for the individual.

3. Many nonprofit products elicit either very low involvement ("Prevent forest fires") or very high involvement ("Stop smoking").

D. Distribution Decisions

Distribution for nonprofit products is usually direct from producer to consumer.

1. A nonprofit organization's capacity for distributing its service offerings to potential customer groups when and where they are needed is typically a key variable in determining the success of those service offerings.

2. Many nonprofit services are not facility-dependent and have the freedom to go to the clients.

3. Soliciting funds and donations of goods also requires good channels of distribution

E. Promotion Decisions

Many nonprofit organizations, such as federal agencies, are explicitly or implicitly prohibited from advertising, thus limiting their promotion options.

Other nonprofit organizations may not have the resources to allow promotion.

1. Professional volunteers can assist organizations in developing and implementing promotion strategies.

2. Advertising agencies may donate their services.

3. Sales promotion activities are increasingly being used by nonprofit organizations.

4. A public service advertisement (PSA), also called a public service announcement, is an announcement that promotes programs, activities, or services of federal, state, or local governments or the programs, activities, or services of nonprofit organizations. The time and space for a PSA is donated by the media.

5. Non profit organizations may license their name and/or image to communicate to a large audience.

F. Price Decisions

Five key characteristics distinguish nonprofit organization pricing decisions from profit-sector pricing decisions.

1. Pricing objectives are generally concerned with covering costs partially or fully or equitably distributing resources.

2. Nonprofit services may have nonfinancial prices consisting of the costs of waiting time, embarrassment, or effort.

3. Indirect payment by consumers of nonprofit services through taxes is common.

4. Separation between payers (often those in a good financial situation) and users (in a poor economic situation) is common for many charitable organizations.

5. Many nonprofit organizations use below-cost pricing.

VOCABULARY PRACTICE

Fill in the blank(s) with the appropriate term or phrase from the alphabetized list of chapter key terms.

assurance
core service
credence quality
empathy
experience quality
gap model
heterogeneity
inseparability
intangibility
internal marketing
mass customization
nonprofit organization
nonprofit organization marketing
perishability
public service advertisement (PSA)
reliability
responsiveness
search quality
service
supplementary service
tangibles

1 Discuss the importance of services to the economy

1. The result of applying human or mechanical efforts to people or objects and involves a deed, performance, or effort that cannot be physically possessed is a ____________________________.

2 Discuss the differences between services and goods

2. The fundamental difference between services and goods is that services cannot be touched, seen, tasted, heard, or felt in the same manner in which goods can be sensed. This is the characteristic of ____________________________.

3. Evaluating the quality of services is more difficult than evaluating the quality of goods. Goods exhibit characteristics such as color or size that can be easily assessed before purchase. Most services, however, do not exhibit a(n) ____________________________. A characteristic that can only be assessed after use is a(n) ____________________________ and could characterize restaurant or travel services. A characteristic that cannot be easily assessed even after the purchase and experience is a(n) ____________________________. Medical and consulting services exemplify this quality.

4. Another important characteristic of services is that services are sold and then produced and consumed at the same time. This is the characteristic of ____________________________ , which also means that services cannot be produced in a centralized location and consumed in a decentralized location.

5. Because services tend to be labor-intensive and production and consumption often occur simultaneously, consistency and quality control are difficult to achieve. This ____________________________ means that service performances tend to be less standardized and uniform than goods.

6. Services cannot be stored, warehoused, or inventoried. This is the service characteristic of ____________________________, which provides an important challenge to service marketers who must balance supply with demand.

3 Describe the components of service quality, and the gap model of service quality

7. Customers evaluate service quality by five components. The ability to perform the service dependably, accurately, and consistently is known as ______________________________. The ability to provide prompt service is called ______________________________. When employees are knowledgeable and courteous, customers have great ______________________________ in the service provider. ______________________________ indicates the caring, individualized attention given to customers. Finally, the physical evidence of the service is known as ______________________________.

8. The ______________________________ identifies five gaps that can cause problems in service delivery.

5 Develop marketing mixes for services

9. The service offering can be viewed as a bundle of activities. These activities include the most basic benefit the customer is buying, which is the ______________________________. The activities also include support or enhancement services knows as the ______________________________.

10. Instead of choosing to either standardize or customize a service, a firm may incorporate elements of both by adopting an emerging strategy called ______________________________.

7 Explain internal marketing in services

11. The quality of a firm's employees is an important part of service quality. Firms that treat employees as customers and develop systems and benefits that satisfy their needs are practicing ______________________________.

9 Describe nonprofit organization marketing

12. An entity that exists to achieve some goal other than profit, market share, or return on investment is a(n) ______________________________. These entities practice ______________________________.

10 Explain the unique aspects of nonprofit organization marketing

13. An announcement, free of charge, that promotes programs, activities, or services of governments or nonprofit organizations is a(n) ______________________________.

Check your answers to these questions before proceeding to the next section.

TRUE/FALSE QUESTIONS

Mark the statement **T** if it is true and **F** if it is false.

2 Discuss the differences between services and goods

_____ 1. Services differ from products, because services are generally more intangible, inseparable, heterogeneous, and perishable.

3 Describe the components of service quality, and the gap model of service quality

_____ 2. Martha is a customer service agent at a department store. When a customer calls with a complaint about merchandise purchased at the store, she calls the customer back immediately and resolves the issue as quickly as possible. This is an example of the reliability component of services.

5 Develop marketing mixes for services

_____ 3. Lowanda is designing a promotional strategy for her consulting service. Her promotions should use personal sources and stress tangible cues of the service.

6 Discuss relationship marketing in services

_____ 4. IBM offers its employees flexible work hours, work-at-home programs, on-site childcare, and travel benefits that reward quality performance. IBM is practicing relationship marketing.

9 Describe nonprofit organization marketing

_____ 5. The only marketing mix component that nonprofit organization do not use is price.

10 Explain the unique aspects of nonprofit organization marketing

_____ 6. Target markets are generally easier to persuade in nonprofit organization marketing because they are not asked to "buy" anything.

_____ 7. The characteristics of for-profit services and nonprofit organizations suggest that their distribution channels should be direct.

_____ 8. Paid advertisements that promote programs, activities, or services of nonprofit agencies or governments are called PSAs (public service advertisements or announcements).

_____ 9. Consumers in nonprofit situations pay prices not in dollars but in other ways such as time or opportunity costs.

Check your answers to these questions before proceeding to the next section.

AGREE/DISAGREE QUESTIONS

For the following statements, indicate reasons why you may agree and disagree with the statement.

1. Employees who work for firms selling services must be trained better in customer service than employees who work for firms selling goods.

   Reason(s) to agree:

   Reason(s) to disagree:

2. The "gap model" of service quality could also be used for the marketing of goods.

   Reason(s) to agree:

   Reason(s) to disagree:

3. Service providers should use promotion to build a strong organizational image rather than advertise their products.

   Reason(s) to agree:

   Reason(s) to disagree:

4. Target markets are easier to identify for nonprofit organizations than for for-profit organizations.

   Reason(s) to agree:

   Reason(s) to disagree:

MULTIPLE-CHOICE QUESTIONS

Select the response that best answers the question, and write the corresponding letter in the space provided.

2 Discuss the differences between services and goods

_____ 1. Randy left his job marketing detergent brands at a large company in order to consult to small businesses that need marketing expertise. Consulting proves to be a much more difficult than marketing detergent largely because consulting is so:

a. intangible
b. unknowable
c. tangible
d. amortizable
e. elusive

_____ 2. Lauro has just accepted a position with Therapy-Plus, a major provider of physical therapy to the elderly. Lauro realizes that selling physical therapy services to individuals and health maintenance organizations will be challenging. What is one of the key aspects of the job Lauro can expect to find?

a. The customer will select and evaluate health care services on the basis of search qualities.
b. Customers are engaged in this as a low-involvement product.
c. The tangibility of Therapy-Plus will be high.
d. The use of the service will involve standardized service levels and not exhibit heterogeneity problems.
e. A consumer may not have the ability to judge the quality of service because of its credence quality.

_____ 3. At the Sundae Sampler, customers place their orders, watch their ice-cream sundaes being prepared, and then quickly eat the sundaes in the store before the ice cream melts. This is an example of a characteristic of services known as:

a. inseparability
b. intangibility
c. heterogeneity
d. perishability
e. dependency

_____ 4. Jorgio owns Pet Plus, a complete pet service that offers grooming, obedience training, show training for animals and handlers, boarding, special food preparation, and minor veterinary care. To maintain a quality image, Jorgio provides complete training for each employee in the company operations, objectives, and expected performance standards. What unique aspect of services is Jorgio attempting to address?

a. inventoriability
b. perishability
c. intangibility
d. inseparability
e. heterogeneity

_____ 5. Carl is frustrated with the uneven business cycles in his catering business. Because he cannot produce a product in slow times and inventory it for the busy holiday seasons, he has to suffer through some extreme peak periods and slow periods. This is because his service business exemplifies the characteristic of:

a. perishability
b. inseparability
c. intangibility
d. heterogeneity
e. insecurity

3 Describe the components of service quality, and the gap model of service quality

_____ 6. Kentucky Fried Chicken once measured its managers' success by how little chicken was thrown away at the end of the night. Customers who came into stores late at night either had to wait for chicken to be cooked or to settle for chicken that was cooked several hours before. This is an example of a gap between:
   a. what customers want and what management thinks customers want.
   b. the service quality specifications and the service that is actually provided.
   c. what the company provides and what the customer is told it provides.
   d. what management thinks what customers want and the quality specifications that management develops to provide the service.
   e. the service that customers receive and the service they want.

_____ 7. Sylvia runs a day care center for the children of professional parents. In order to enrollment in her center, Sylvia lowered tuition but had to increase the number of children that each teacher would care for. Sylvia was surprised when some of the parents pulled their children out of the center, especially when the tuition had been lowered. This is an example of a gap between:
   a. what customers want and what management thinks customers want.
   b. the service quality specifications and the service that is actually provided.
   c. what the company provides and what the customer is told it provides.
   d. what management thinks what customers want and the quality specifications that management develops to provide the service.
   e. the service that customers receive and the service they want.

_____ 8. A major airline once advertised that passengers should fly the airline because of the "respect" that airline personnel gave to passengers. However, passengers were always disappointed by the lack of respect that the personnel gave them. This is an example of a gap between:
   a. what customers want and what management thinks customers want.
   b. the service quality specifications and the service that is actually provided.
   c. what the company provides and what the customer is told it provides.
   d. what management thinks what customers want and the quality specifications that management develops to provide the service.
   e. the service that customers receive and the service they want.

4 Explain why services marketing is important to manufacturers

_____ 9. In addition to the sale of the basic computer hardware, Gateway Computer now offers services such as maintenance and troubleshooting and its own Internet service. Gateway like does this in order to:
   a. differentiate itself from other computer companies.
   b. gain loyalty from its customer base.
   c. gain a competitive advantage over other computer companies.
   d. None of the above.
   e. All of the above.

5 Develop marketing mixes for services

_____ 10. British Airways' first-class passengers sit in a semi-private pear-wood berth that converts into a bed, thus providing more comfort on long flights. This service is an example of the airline's:
   a. standardized service.
   b. core service.
   c. customized service.
   d. supplementary service.
   e. component service.

_____ 11. You are opening your first hair-styling salon and are concerned with the development of your distribution mix decision. Which of the following reflects the key issue in your distribution decision?
a. considerations of the storage of the service
b. the development of a long channel of intermediaries
c. standardizing the service
d. intensity of distribution
e. the location of your service for customer convenience

_____ 12. You have recently opened a specialty service that cleans, stores, and insures valuable jewelry. Your shop is exquisitely decorated, and the keys to the storage vaults are solid gold. Each time a customer visits the store to make a deposit or withdrawal, free champagne is offered. After each customer contact, the customers is mailed a thank-you note, handwritten on expensive stationery. The thank-you note is a(n):
a. skimming strategy
b. prestige prompt
c. tangible cue
d. intangible cue
e. pricing strategy

_____ 13. Long-distance telephone companies have more expensive rates during peak usage times (weekdays) and discount rates for slower times (evenings and weekends). This illustrates the ___________ pricing objective.
a. peak profitability
b. operations-oriented
c. patronage-oriented
d. supply-demand
e. revenue-oriented

6 Discuss relationship marketing in services

_____ 14. The Xerox corporation stays in touch with its customers with phone calls and greeting cards, periodically sends out needs-assessment questionnaires, and designs new services to meet new needs. This is an example of relationship marketing based on:
a. financial bonds
b. service delivery
c. social bonds
d. internal bonds
e. structural bonds

8 Discuss global issues in services marketing

_____ 15. Many U.S. services industries have the potential for globalization because of their existing competitive advantages. Industries that possess distinct advantages include all of the following EXCEPT:
a. insurance
b. construction and engineering
c. computer microchip manufacturing
d. credit card operations
e. leisure

9 Describe nonprofit organization marketing

_____ 16. Green America is a nonprofit organization and therefore has:
a. no need to develop an understanding of pricing
b. a marketing orientation
c. the same organizational structure as profit-making firms
d. no impact on nonbuyers
e. a need for marketing skills

10 Explain the unique aspects of nonprofit organization marketing

____ 17. What is one of the consequences of the nonprofit orientation of the Humane Society?
a. The Humane Society does not have to worry about revenues and costs.
b. The Humane Society will not make as much money as a profit-oriented firm.
c. The success of the Humane Society cannot be measured in financial terms.
d. The Humane Society is not expected to be as efficient as a for-profit firm.
e. The Humane Society's success is based on how much money is donated to it.

____ 18. Cancel Cocaine is a not-for-profit organization with the goal of providing low-cost assistance to cocaine addicts who do not have health insurance. To raise funds, it distributes pamphlets and videos to local businesses and wealthy homeowners, that encourage the donation of money. Cancel Cocaine is most likely to face the target market issue of:
a. pressure to adopt undifferentiated segmentation strategies
b. lack of bottom-line objectives
c. complementary positioning
d. apathetic or strongly opposed targets
e. benefit complexity and strength

____ 19. A local church formed a committee to provide aid to the homeless. The church applied to the city and county governments for funding but was denied when it could not adequately state what the services would include. In this case the committee failed because it had not defined the:
a. user market
b. product to be offered
c. sponsoring organizations
d. donor market
e. target market

____ 20. The Penny Press has been offering free printing to the American Cancer Society. This could be expected to provide the Penny Press with all of the following benefits EXCEPT:
a. community approval
b. goodwill
c. personal satisfaction
d. personal contacts
e. immediate financial gain

____ 21. The Project Literacy program wants to notify the community about its upcoming fund-raiser rummage sale. As an experienced marketer and director of the program, you plan to visit several local radio and televisions stations to request:
a. public service advertising
b. nonsponsored advertising
c. free telemarketing
d. freestanding inserts
e. social advertising

Check your answers to these questions before proceeding to the next section.

ESSAY QUESTIONS

1. Services have four unique characteristics that distinguish them from goods. Name and briefly define each of these four characteristics. These characteristics can make the marketing of services more difficult. What special strategies should marketers adopt for services to address each of the four characteristics? Use the example of a restaurant to help describe each of the four characteristics.

2. Discuss service strategies for each of the four P's in the marketing mix.

3. You are the marketing manager at an organization that has the goal of decreasing the incidence of smoking among teen-agers. Describe unique issues that you will encounter in each of the 4 P's in achieving your organizational goals.

CHAPTER NOTES

Use this space to record notes on the topics you are having the most trouble understanding.

ANSWERS TO THE END-OF-CHAPTER DISCUSSION AND WRITING QUESTIONS

Use this space to work on the questions at the end of the chapter.

ANSWERS TO THE END-OF-CHAPTER CASE

Use this space to work on the questions at the end of the chapter.

CHAPTER GUIDES and QUESTIONS

PART FOUR

DISTRIBUTION DECISIONS

CHAPTER 13 Marketing Channels and Logistics Decisions

LEARNING OBJECTIVES

1 Explain what a marketing channel is and why intermediaries are needed

Marketing channels are composed of members who perform negotiating functions. Some intermediaries buy and resell products; and other intermediaries aid the exchange of ownership between buyers and sellers without taking title. Nonmember channel participants do not negotiate, but function as an auxiliary part of the marketing channel structure.

Intermediaries such as retailers and wholesalers can achieve economies of scale through specialization and division of labor. Essentially, producers hire channel members to perform tasks and activities (such as transportation or selling to the consumer) that the producer is not prepared to perform or that the intermediary is better prepared to perform.

Distribution channels also aid in overcoming barriers to exchange that are created in the production process. Channel members such as wholesalers help overcome the discrepancies of quantity, assortment, time, and space by bringing a variety of items together at a convenient location and appropriate times.

2 Define the types of channel intermediaries and describe their functions and activities

Channel members can perform transactional functions, which may include the negotiation of the transportation, quantity, payment terms, and other details of the exchange process. Channel members often perform the task of physical distribution, moving products from where they are produced to where they are consumed. Channel members also assume risk-taking responsibilities as they assume possession of products. Examples of risk are obsolescence, deterioration, pilferage, bad debts, and price fluctuations.

Channel members may also perform logistical functions. Intermediaries create efficiencies by reducing the amount of contacts between producers and users. In order to overcome the discrepancies of quantity and assortment, channel members perform sorting functions, including sorting out, accumulation, allocation, and assorting.

Channel members may also perform facilitating functions, including research and gathering information about other channel members and consumers. Channel members perform promotion activities to make prospective buyers aware that products exist and to explain their features, advantages, and benefits. Sometimes channel members provide financing services for other channel members. Inventories are financed as products flow through or are held in a channel, and many wholesalers and retailers provide credit for final purchasers.

3 Describe the channel structures for consumer and business-to-business products and discuss alternative channel arrangements

One consumer channel is the direct channel, which entails producers selling directly to consumers. The direct channel is less common than other channel configurations, such as the retailer channel (producer to retailer to consumer) or the wholesaler channel (producer to wholesaler to retailer to consumer). At the end of the spectrum is the agent/broker channel in which agents or brokers bring manufacturers and wholesalers together for negotiation. Ownership then passes to one or more wholesalers and then to retailers, who sell to the ultimate consumer of the product.

Many business product producers sell directly, in large quantities, to the large manufacturers that are their customers. The channel from producer to government is also a direct channel. This arrangement is usually used because of specialized products and bidding. If a middleman is used, it is usually an industrial distributor or manufacturers' representative. Another common business-to-business channel configuration is direct to government.

4 Discuss the issues that influence channel strategy

When determining marketing channel strategy, the marketing manager must determine what market, product, and producer factors will influence the channel selected. The manager must also determine the appropriate level distribution intensity.

5 Explain channel leadership, conflict, and partnering

Power, control, conflict and partnering are the main social dimensions of marketing channel relationships. Channel power refers to the capacity of one channel member to control or influence other channel members. Channel control occurs when one channel member intentionally affects another member's behavior. Channel leadership is the exercise of authority and power. Channel conflict occurs when there is a clash of goals and methods among the members of a distribution channel. Channel partnering is the joint effort of all channel members to create a supply chain that serves customers and creates a competitive advantage. Collaborating channel partners meet the needs of consumers more effectively by ensuring the right products reach shelves at the right time and at a lower cost, boosting sales and profits.

6 Discuss logistics and supply chain management and their evolution into distribution practice

Logistics is the process of strategically managing the efficient flow and storage of raw materials, in-process inventory, and finished goods from point of origin to point of consumption. The supply chain connects all of the business entities, both internal and external to the company, that perform or support the logistics function. Supply chain management, or integrated logistics, coordinates and integrates all of these activities performed by supply chain members into a seamless process that delivers enhanced customer and economic value through synchronized management of the flow of goods and information from sourcing to consumption.

7 Discuss the concept of balancing logistics service and cost

As distribution departments have evolved, their emphasis has changed from getting the lowest transportation rates to the broader issue of minimizing total distribution costs. The main goal of physical distribution is getting the right goods to the right place at the right time for the least cost.

Unfortunately, it is difficult to maximize customer service while minimizing distribution costs. Large inventories, for instance, allow better customer service but drive up total costs. Customers are not concerned with the process of physical distribution but are concerned with the result: the availability of merchandise, timeliness of deliveries, and quality (condition and accuracy) of shipments.

8 Describe the integrated functions of the supply chain

The logistics supply chain consists of several interrelated and integrated functions: 1) procuring supplies and raw materials; 2) scheduling production; 3) processing orders; 4) managing inventories of raw materials and finished goods; 5) warehousing and materials-handling; and 6) selecting modes of transportation. Integrating and linking all of the logistics functions of the supply chain is the logistics information system.

9 Discuss new technology and emerging trends in logistics

Several technological advances and trends affect the physical distribution industry today. Technology and automation has linked intermediaries in information networks. Concern for the environment is also impacting physical distribution. Another trend is the use of third-party carriers and a quest for quality in transportation. Finally, the need to understand global physical distribution has assumed greater importance for many companies.

10 Identify the special problems and opportunities associated with distribution in service organizations

The fastest-growing part of the U.S. economy is the services sector. The same skills, techniques, and strategies used to manage inventory can be used to manage service inventory.

The major challenge is determining how to deliver the service cost effectively, and when and where the customer wants it. This involves managing three critical variables: minimizing wait times, managing service capacity, and providing delivery through distribution channels.

11 Discuss channel structure and logistics issues in global markets

Global marketing channels are important to U.S. companies that export their products or manufacture abroad. Manufacturers introducing products in foreign countries must decide what type of channel structure to use. Should the product be marketed through direct channels or through foreign intermediaries? Foreign intermediaries must be chosen very carefully because they will impact the brand image. Also, channel structures in other countries may be very different from those in the United States.

PRETEST

Answer the following questions to see how well you understand the material. Re-take it after you review to check yourself.

1. What is a marketing channel?

2. List and briefly describe three types of channel intermediaries.

3. List and briefly describe three intensities of distribution.

4. What is channel conflict? Channel leadership? Channel partnering?

5. What are the six major functions of the supply chain?

CHAPTER OUTLINE

1 Explain what a marketing channel is and why intermediaries are needed

I. What Is a Marketing Channel?

A. The term *channel* is derived from the Latin word *canalis*, which means canal. A marketing channel can be viewed as a canal or pipeline for products.

B. A **marketing channel** or **channel of distribution** is the set of interdependent organizations that facilitate the transfer of ownership as products move from producer to business user or consumer.

C. Many different types of organizations participate in marketing channels.

1. **Channel members**, also called intermediaries:

a. Negotiate with one another

b. Buy and sell products

c. Facilitate the change of ownership between buyer and seller

D. Providing Specialization and Division of Labor
Essentially producers hire channels members to perform tasks and activities (such as transportation or selling to the consumer) that the producer is not prepared to perform or that the intermediary is better prepared to perform.

E. Overcoming Discrepancies

1. A **discrepancy of quantity** is the difference between the amount of product produced and the amount an end user wants to buy..

2. A **discrepancy of assortment** is the lack of all the items necessary to receive full satisfaction from a product or products. A manufacturer may produce only one product, yet it may require the addition of several more products to actually use the first product.

3. A **temporal discrepancy** is created when a product is produced and a consumer is not ready to purchase it.

4. A **spatial discrepancy** is the difference between the location of the producer and the location of widely scattered markets.

F. Contact Efficiency

Marketing channels simplify distribution by reducing the amount of transactions required to get products from manufacturers to consumers. Retailers assemble a selection of merchandise so that one contact (buying trip) can result in the purchase of many different items.

2 Define the types of channel intermediaries and describe their functions and activities

II. Channel Intermediaries and Their Functions

A. Types of Channel Intermediaries

The primary difference separating intermediaries is whether or not they take title to the product

1. **Retailers** are firms that sell mainly to consumers. They take title to the goods which they sell.

2. **Merchant wholesalers** facilitate the movement of products and services from the manufacturer to producers, resellers, government institutions and retailers. Merchant wholesalers take title to the goods which they sell.

3. **Agents** and **brokers** facilitate the sale of a product from producer to end user by representing retailers, wholesalers, or manufacturers. Agents and brokers do not take title.

B. Channel Functions Performed by Intermediaries

1. Transactional functions include contacting buyers, promoting the product to be sold, negotiating the sale, and taking on risks associated with owning and keeping the product in inventory.

2. Logistical functions include transporting, storing, sorting out, accumulating, allocating, and assorting products.

3. Facilitating functions include research and financing.

C. Individual members can be added or eliminated from a channel, but these functions must be performed by someone. It could be the producer, wholesaler, retailer, or end user.

3 Describe the channel structures for consumer and business-to-business products and discuss alternative channel arrangements

III. Channel Structures

The appropriate configuration of channel members to move a product from the producer to the end user may differ greatly by product.

A. Channels for Consumer Products

1. One channel is the **direct channel**, which entails producers selling directly to consumers. This includes telemarketing, mail order, and catalog shopping and forms of electronic retailing.

2. The longest typical channel found for consumer products is the agent/broker channel, in which agents or brokers bring manufacturers and wholesalers together for negotiation. Ownership then passes from one or more wholesalers to retailers, and finally retailers sell to the consumer.

3. Most consumer products are sold through the retailer and the wholesaler channel

   a. A *retailer channel* is used when the retailer is large and can buy in large quantities direct from the manufacturer.

   b. A *wholesaler channel* is often used for low-cost, frequently purchased items such as candy and gum. The wholesalers purchase large quantities and break it into smaller lots which are sold to retailers.

B. Channels for Business and Industrial Products

1. Channels for business and industrial markets are usually characterized by fewer intermediaries.

2. Many producers sell directly, in large quantities, to the large manufacturers that are their customers.

3. The channel from producer to government buyers is direct. This is usually because of specialized products and bidding.

4. For standardized items of moderate or low value, an intermediary is often used. This is usually an industrial distributor or manufacturers' representative.

C. Alternative Channel Arrangements

1. Some producers select two or more channels to distribute the same products to target markets, a practice called **multiple distribution** or **dual distribution.** A variation of this practice is the marketing of similar products using different brand names.

2. Often nontraditional channel arrangements help differentiate a firm's product from the competition's.

3. Adaptive channels are flexible and responsive channel alternatives that are used when a firm identifies critical but rare customer requirements which they cannot fill.

4. **Strategic channel alliances** are sometimes formed to use another manufacturer's already established channel.

4 Discuss the issues that influence channel strategy

IV. Channel Strategy Decisions

Marketing channel strategy is substantially influenced by channel objectives, which in turn should reflect marketing objectives and organization objectives.

The marketing manager faces two important issues: what factors will influence the channel(s) and what level of distribution intensity will be appropriate.

A. Factors Affecting Channel Choice

1. Market Factors

These include target customer profiles and preferences, geographic location, size of market, and competition.

2. Product Factors

These include whether products are complex, customized, standard, the price, stage of product life cycle, and delicacy of the product.

3. Producer Factors

Producer factors that impact channel choice include size of managerial, financial, and marketing resources, number of product lines, and desire for control of marketing channels.

B. Levels of Distribution Intensity

1. **Intensive distribution** is distribution aimed at having a product available in every outlet where target consumers might want to buy it.

    a. Many convenience goods and supplies have intensive distribution.

    b. Low-value, frequently purchased products may require a long channel of distribution..

2. **Selective distribution** is distribution achieved by screening dealers to eliminate all but a few in any single area.
Shopping goods usually have selective distribution, as do some specialty products.

3. The most restrictive form of distribution, **exclusive distribution**, entails establishing one or a few dealers within a given area.

    a. Consumer specialty goods, a few shopping goods, and major industrial equipment use exclusive distribution.

    b. This limited distribution aids in establishing an image of exclusiveness for the product..

5 Explain channel leadership, conflict, and partnering

V. Channel Relationships

Social relationships play an important role in building unity among channel members.

A. Channel Power, Control, and Leadership

1. **Channel power** is a channel member's capacity to control or influence the behavior of another member's behavior.

2. **Channel control** occurs when one channel member affects another member's behavior.

3. The channel member who assumes channel leadership and exercises authority and power is the **channel leader** or **channel captain**.

B. Channel Conflict

1. Inequitable channel relationships often lead to **channel conflict**.

2. Sources of conflict may be conflicting goals, failure to fulfill the expectations of other channel members, idealogical differences, and different perceptions of reality.

3. **Horizontal conflict** occurs among channel members on the same level.

4. **Vertical conflict** occurs between different levels in a marketing channel.

C. Channel Partnering

1. **Channel partnering** or **channel cooperation** is the joint effort of all channel members to create a supply chain that serves customers and creates a competitive advantage.

2. This is in contrast to the adversarial relationships of the past.

3. Channel alliances and partnerships are created as firms try to leverage the intellectual, material and marketing resources of the channel, and make entry into far-flung markets easier and more cost effective.

6 Discuss logistics and supply chain management and their evolution into distribution practice

VI. Logistics Decisions and Supply Chain Management

A. **Logistics** describes the process of strategically managing the efficient flow and storage of raw materials, in-process inventory, and finished goods from point of origin to point of consumption.

B. The **supply chain** is the connected chain of all of the business entities, both internal and external to the company, that perform or support the logistics function.

C. **Supply chain management** or **integrated logistics** coordinates and integrates all of the activities performed by supply chain members into a seamless process.

1. Supply chain management is customer driven.

2. Supply chain management allows companies to respond with the unique product configuration and mix of services demanded by the customer.

3. Supply chain management is a *communicator* of customer demand and is a *physical flow process*.

D. The evolution of integrated logistics and supply chain management can be traced through:

1. *Physical distribution*, the movement of products to market.

2. *Logistics management* which shifted the focus to the entire system, stressing a total cost perspective.

3. The focus of logistics management changed to include the customer in the early 1970s.

4. By the end of the 1980s logistics was gaining recognition as a means to achieve differentiation for the firm.

E. Benefits of supply chain management include reduced costs in inventory management, transportation, warehousing, and packaging, improved service, and enhanced revenues.

7 Discuss the concept of balancing logistics service and cost

VII. Balancing Logistics Service and Cost

A. **Logistics service** is the package of activities performed by a supply chain member to ensure that the right product is in the right place at the right time.

B. In an effort to set service levels that maximize service while minimizing cost, logistics managers utilize the *total cost approach.* This requires trade-offs among the functional elements of the supply chain (e.g. procurement, inventory, warehousing, transportation, etc.).

C. Total Distribution Costs

1. Most distribution managers try to set their service level goals at a point that maximizes services yet minimizes cost.

2. This means that the complex interrelationship of all the physical distribution factors must be examined, along with their impact on customer service.

3. Implementing the total cost approach requires cost trade-offs because of the conflicting goals of distribution and service.

8 Describe the integrated functions of the supply chain

VIII. Integrated Functions of the Supply Chain

The **logistics information system** integrates and links all of the logistics functions of the supply chain.

The **supply chain team** orchestrates the movement of goods, services and information from the source to the consumer. They typically cut across organizational boundaries and may include participants from external members of the supply chain.

A. Sourcing and procurement links the manufacturer and the supplier. They are responsible for the vendor relations and for reducing the cost of raw materials and supplies.

B. Production scheduling may be based on pushing products into the supply chain and waiting for customers to place their orders, or by allowing customer orders to pull the production of goods and services.

The **mass customization** or **build-to-order** process uniquely tailors mass-market goods and services to the needs of the individuals who buy them.

C. **Just-in-time production (JIT)**, sometimes called *lean production*, requires manufacturers to work closely with suppliers and transportation providers to get required items to the assembly line or factory floor at the precise time they are needed for production.

D. The **order processing system** processes the requirements of the customer and sends the information into the supply chain via the logistics information system.

Electronic data interchange (EDI) uses computer technology to replace the paper documents that usually accompany business transactions.

E. An **inventory control system** develops and maintains an adequate assortment of materials or products to meet a manufacturer's or a customer's needs.

1. Managing inventory from the supplier to the manufacturer is called **materials requirement planning (MRP)** or **materials management**.

2. Systems that manage the finished goods inventory from the manufacturer to end user is commonly referred to as **distribution resource planning.**

   Enhanced versions of DRP common in the retailing and supermarket industries are *continuous replenishment* (CR), *efficient consumer response* (ECR), and *vendor managed inventory* (VMI).

F. Storage helps manufacturers manage supply and demand, and provides time utility to buyers and sellers. A **materials-handling system** moves inventory into, within, and out of the warehouse.

G. The five major modes of transportation are railroads, motor carriers, pipelines, water transportation, and airways.

   Logistics managers choose a mode on the basis of:

   1. **Cost:** total amount a specific carrier charges to move the product from the point of origin to the destination.
   2. **Transit time:** total time a carrier has possession of goods.
   3. **Reliability:** consistency with which the carrier delivers goods on time and in acceptable condition.
   4. **Capability:** ability of the carrier to provide the appropriate equipment and conditions for moving goods.
   5. **Accessibility:** carrier's ability to move goods over a specific route or network.
   6. **Traceability:** relative ease with which a shipment can be located and transferred.

9 Discuss new technology and emerging trends in logistics

IX. Trends in Physical Distribution

A. Automation

   Computer technology has boosted the efficiency of the physical distribution process dramatically in warehousing, materials management, inventory control, and transportation. Automated equipment handles and moves the goods in and out of storage, and computers keep track of inventory and location information.

B. Outsourcing Logistics Functions

   1. In **outsourcing or contract logistics**, a manufacturer or supplier turns over the entire function of buying and managing transportation to a third party.
   2. Third-party contact logistics allows companies to cut inventories, locate stock at fewer plants and distribution centers, and still provide the same level of service or better.
   3. Outsourcing may even lead to firms allowing business partners to take over the final assembly of their product or its packaging in an effort to reduce inventory costs, speed of delivery, or meet customer requirements better.

C. Electronic Distribution

Electronic distribution is the most recent development in the physical distribution arena.

1. This will include any product or service that can be distributed electronically, whether through traditional cable or through satellite transmission.

2. In the future movies, music, books, newspapers and more may use these methods.

10 Identify the special problems and opportunities associated with physical distribution in service organizations

X. Channels and Distribution Decisions for Services

The fastest-growing part of the U.S. economy is the service sector. The same skills, techniques, and strategies used to manage inventory can be used to manage service inventory.

A. Service Industries' Distribution Opportunities

1. Service distribution means getting the right service and the right people and the right information to the right place at the right time.

2. Production and consumption are simultaneous.

3. Service distribution attempts to minimize wait times

a. manage service capacity
b. improve delivery through new distribution channels

11 Discuss channel structure and decisions in global markets

XI. Channels and Distribution Decisions for Global Markets

A. Developing Global Marketing Channels

Manufacturers introducing products in global markets must determine what type of channel structure to use.

1. Channel structure abroad may not be very similar to channels in the United States.

2. Channel types available in foreign countries usually differ as well.

3. Marketers must be aware of "gray" marketing channels in which products are distributed through unauthorized channel intermediaries.

B. Global trade growth requires a well-thought-out global logistics strategy.

1. One critical issue is the legalities of trade in other countries.

2. A second factor is the transportation infrastructure.

VOCABULARY PRACTICE

Fill in the blank(s) with the appropriate term or phrase from the alphabetized list of chapter key terms.

adaptive channel
agents and brokers
channel conflict
channel control
channel leader
channel partnering
channel power
direct channel
discrepancy of assortment
discrepancy of quantity
distribution resource planning (DRP)
dual distribution (multiple distribution)
electronic data interchange (EDI)
electronic distribution
exclusive distribution
horizontal conflict
intensive distribution
inventory control system
just-in-time (JIT) inventory management
logistics
logistics information system
logistics service
marketing channel (channel of distribution)
mass customization
materials-handling system
materials requirement planning (MRP)
merchant wholesaler
order processing system
outsourcing (contract logistics)
retailer
selective distribution
spatial discrepancy
strategic channel alliance
supply chain
supply chain management (integrated logistics)
supply chain team
temporal discrepancy
vertical conflict

1 Explain what a marketing channel is and why intermediaries are needed

1. A set of interdependent organizations that ease the transfer of ownership as products move from producer to business user or consumer is a(n) ______________________.

2. Channel members are often included between producers and consumers because they help overcome discrepancies. The difference between the amount of product produced and the amount an end user wants to buy is a(n) ______________________; this is overcome by making products available in the amounts buyers desire. The lack of all the items necessary to provide full satisfaction to a buyer is a(n) ______________________; to overcome this discrepancy, channels assemble assortments of products that buyers want available at one place. When a product is produced but a consumer is not ready to purchase it, a(n) ______________________ is created; channels overcome this discrepancy by maintaining inventories. Markets are usually scattered over large geographic regions, creating a(n) ______________________; channels overcome this discrepancy by making products available in convenient locations.

2 Define the types of channel intermediaries and describe their functions and activities

3. There are three types of intermediaries that help move products from producer to the final customer. The first is ______________________ who act as salespeople representing either producers, retailers, or wholesalers and who do not take title to the goods. The second is ______________________ who facilitate the movement of product from producers to end user and who take title to the goods. These organizations specialize in warehousing, inventory management and transportation. The third type of intermediary is a

____________________, who sell directly to consumers and provide many services to these end users.

3 Describe the channel structures for consumer and business-to-business products and discuss alternative channel arrangements

4. If a producer sells directly to consumers with no intermediaries, it is using a(n) ____________________.

5. Producers often devise alternative marketing channels to move their products. Some producers select two or more different channels to distribute the same products to target markets. This practice is known as ____________________. Channel members may also decide to share resources in order to fulfill a customer's requirements and form a(n) ____________________. A producer may also to elect to use another manufacturer's already-established channel for distribution by forming a(n) ____________________.

4 Discuss the issues that influence channel strategy

6. Distribution aimed at maximum market coverage is ____________________. If products are distributed to select dealers in a particular region, ____________________ is taking place. The most restrictive form of market coverage is ____________________, which entails only one or a few dealers within a given area.

5 Explain channel leadership, conflict, and partnering

7. A channel member's capacity to control or influence the behavior of other channel members is known as ____________________. When one channel member affect another member's behavior it has exercised ____________________ and assumes a position of ____________________.

8. When a clash of goals and methods among members of a distribution channel occurs, this is known as ____________________. Conflict that occurs among channel members on the same level, such as two retailers, is known as ____________________, while conflict that occurs between different levels in a marketing channel, such as between the manufacturer and wholesaler, is known as ____________________.

9. ____________________ is the joint effort of all channel members to create a supply chain that serves customers and creates a competitive advantage.

6 Discuss logistics and supply chain management and their evolution into distribution practice

10. The process of strategically managing the efficient flow and storage of raw materials, in-process inventory and finished goods from point of origin to point of consumption is known as ____________________. The connected chain of all the business entities, both internal and external to the company, that perform the logistics function is a ____________________. All activities performed by members of the chain must be coordinated and integrated, and this process is known as ____________________.

7 Discuss the concept of balancing logistics service and cost

11. Interrelated activities performed by a member of the supply chain to ensure that the right product is in the right place at the right time is known as ______________________________.

8 Describe the integrated functions of the supply chain

12. A ______________________________ is information technology that integrates and links all of the logistics functions of the supply chain. The entire group of individuals who orchestrate the movement of goods, services and information from the source to the consumer is known as a(n) ______________________________.

13. A production method whereby products are not made until an order is placed by the customer and where products are made according to customer specifications is known as ______________________________. Another important service that suppliers offer to their customers is ______________________________, whereby the supplier delivers raw materials just when they are needed on the production line. This helps the manufacturer to reduce inventory costs.

14. The ______________________________ processes the requirements of the customer and sends the information t the supply chain via the logistics information system. These systems are becoming more automated, and many firms use ______________________________ to reduce the amount of paper documents needed for transactions, such as purchase orders and invoices.

15. The physical distribution subsystem that develops and maintains an adequate assortment of products to meet customers' demands is the ______________________________. Managing inventory from the supplier to the manufacturer is called ______________________________. Systems that manage the finished goods inventory from the manufacturer to end user are called ______________________________. A(n) ______________________________ moves inventory into, within, and out of the warehouse.

9 Discuss new technology and emerging trends in logistics

16. The most recent development in physical distribution includes any kind of product or service that can be distributed by fiber optic cable or through satellite transmission of electronic signals. This is known as ______________________________. Another trend in logistics is ______________________________ which occurs when a manufacturer uses an independent third party to manage an entire function of the logistics system, such as transportation, warehousing, or order processing.

Check your answers to these questions before proceeding to the next section.

TRUE/FALSE QUESTIONS

Mark the statement **T** if it is true and **F** if it is false.

1 Explain what a marketing channel is and why intermediaries are needed

_____ 1. It is generally better to "cut out the middleman" in order to save costs.

_____ 2. Ted stopped by the convenience store to buy a can of soda pop. He is glad that he can buy one can at a time, instead of having to buy the thousands of gallons of soda the manufacturer produces every day. For Ted, the convenience store overcomes a discrepancy of assortment.

_____ 3. Jill would like to put some kiwifruit in her fruit salad. Fortunately, she does not have to hop on an airplane and fly to New Zealand to buy one; she can buy a kiwifruit at the neighborhood grocery store. In this case, marketing channels have helped to overcome a spatial discrepancy.

2 Define the types of channel intermediaries and describe their functions and activities

_____ 4. The three basic functions that a channel intermediary provides are transactional (including physical distribution and sorting), logistical (including research and financing), and facilitating (including contacting, promoting, negotiating, and risk taking).

3 Describe the channel structures for consumer and business-to-business products and discuss alternative channel arrangements

_____ 5. Company Alpha sells cotton balls for consumers to use. Company Zeta sells cotton for the fabric industry to make clothing. It is more likely that Company Zeta will use a direct channel because the direct channel is used more often in business markets than in consumer markets.

_____ 6. Many clothing manufacturers sell both through catalogs and through retail shops. This is known as dual distribution.

_____ 7. Volvo united with Federal Express Logistics to overcome its problem getting replacement parts where needed for emergency roadside repairs. This is an example of a strategic channel alliance.

4 Discuss the issues that influence channel strategy

_____ 8. The objective of intensive distribution is to achieve mass market selling.

_____ 9. The D'Or Galleries sells its solid gold picture frames at only five stores in the United States. D'Or promotes the frames intensively to those retailers and performs much cooperative advertising. This is an example of intensive distribution.

_____ 10. The Mini-Micro company manufactures miniature microwave ovens suitable for campers, dorm rooms, and small kitchens. The company's marketing research indicates that consumers are willing to look around for miniature microwaves but may not be willing to search or travel extensively to acquire the product. Mini-Micro should use selective distribution.

5 Explain channel leadership, conflict, and partnering

_____ 11. Savings Mart is a large discount retailer that dictates to its suppliers to provide the lowest possible cost and to adhere to a just-in-time inventory system. Wholesalers and manufacturers generally yield to the authority of this giant retailer. Savings Mart exercises channel power.

_____ 12. Fred's Mart and Josephine's Foods are two independent grocery stores that are claiming that a large manufacturer is treating them unfairly because they are small. This is an example of vertical conflict within the marketing channel.

7 Discuss the concept of balancing logistics service and cost

_____ 13. The most important factor in physical distribution service is low cost.

8 Describe the integrated functions of the supply chain

_____ 14. The supply chain team is not responsible for transportation; this is the job of wholesalers.

_____ 15. U-Build-It Computer Company specializes in providing computers built-to-order to its customers. This is an example of mass customization.

_____ 16. Fiske's Disks, Inc., produces computer diskettes and has just changed its assembly method. The diskettes are produced on an assembly line, and suppliers deliver the needed parts in five small daily shipments directly to the line. Fiske's has noticed a dramatic decrease in carrying costs but a sharp increase in delivery costs because the company must pay more for the added delivery convenience. Fiske's Disks uses Just-in-Time (JIT) inventory management.

_____ 17. Nancy Gordan orders office supplies from a local retailer through her computer. The computer program is set up to transmit purchase orders and invoices between customers and the retailer. The retailer then delivers the products to her office. This is an example of electronic distribution.

9 Discuss new technology and emerging trends in logistics

_____ 18. The use of outsourcing has resulted in firms using fewer carriers, but demanding more from them.

10 Identify the special problems and opportunities associated with distribution in service organizations

_____ 19. Marketers of services do not have to address the question of logistics because there is no distribution of a service in which there are no physical products.

11 Discuss channel structure and logistics issues in global markets

_____ 20. Small firms are most likely to enter global markets because logistics are easier for smaller firms.

Check your answers to these questions before proceeding to the next section.

AGREE/DISAGREE QUESTIONS

For the following statements, indicate reasons why you may agree and disagree with the statement.

1. Marketing channels are far too complex. The best channel is direct: to go from manufacturer to retailer.

   Reason(s) to agree:

   Reason(s) to disagree:

2. When a channel leader emerges, there will inevitably be channel conflict.

Reason(s) to agree:

Reason(s) to disagree:

3. It is better to use intensive distribution, rather than selective or exclusive distribution, because a larger number of outlets means increased sales.

Reason(s) to agree:

Reason(s) to disagree:

4. There is no such thing as a marketing channel for services.

Reason(s) to agree:

Reason(s) to disagree:

MULTIPLE-CHOICE QUESTIONS

Select the response that best answers the question, and write the corresponding letter in the space provided.

1 Explain what a marketing channel is and why intermediaries are needed

_____ 1. Which of the following is NOT a function of a marketing channel?
a. To provide specialization
b. To achieve economies of scale in production
c. To divide labor
d. To provide contact efficiency
e. To overcome discrepancy of assortment

____ 2. The Hit-the-Beach Store purchases swimsuits, sunscreen products, towels, beach toys, beach chairs, and surf boards from a variety of manufacturers and brings these items to its large store in a large California coastal city. The store's goal is to provide every amenity needed for a trip to the beach, so Hit-the-Beach is aiding consumers by overcoming the:
a. discrepancy of assortment
b. discrepancy of quantity
c. spatial discrepancy
d. temporal discrepancy
e. contact discrepancy

____ 3. The Howlin' Halloween company operates its manufacturing facilities year-round, but the sales season for Halloween costumes and party decorations is usually September 1 through November 1. However, sales remain steady all year as the company sells to wholesale distributors that stock the product. The wholesale distributors are helping to overcome a:
a. discrepancy of assortment
b. temporal discrepancy
c. contact discrepancy
d. discrepancy of quantity
e. spatial discrepancy

2 Define the type of channel intermediaries and describe their functions and activities

____ 4. McKesson Drug purchases large quantities of over-the-counter and prescription pharmaceutical products from manufacturers and sells them to drug stores. McKesson takes title to the products when they are purchased. This company is a(n):
a. retailer
b. agent
c. broker
d. merchant wholesaler
e. manufacturer's representative

____ 5. Murdock and Company represents several health and beauty care products made by small manufacturers that cannot afford their own sales forces. Murdock sells the goods to distributors and retailers and does not carry brands that compete directly with each other. Murdock does not take title to the goods. This company is a(n):
a. retailer
b. distributor
c. merchant wholesaler
d. placement company
e. agent

____ 6. Lands' End sells casual and outdoor clothing on-line and through catalogs directly to consumers. Lands' End is a(n):
a. retailer
b. distributor
c. merchant wholesaler
d. broker
e. agent

3 Describe the channel structures for consumer and business-to-business products and discuss alternative channel arrangements

____ 7. The Custo-Lift Company manufactures customized, high-tech elevator shelving systems for specialized automated warehouses. For distribution, you would expect Custo-Lift to use a:
a. network of facilitating agents
b. horizontally integrated channel
c. long channel
d. vertical marketing system
e. direct channel

_____ 8. General Electric sells large home appliances both through independent retailers (department stores and discount houses) and directly to large housing-tract builders. This is an example of:
   a. intensive distribution
   b. intermediary distribution
   c. selective distribution
   d. alternative distribution
   e. dual distribution

4 Discuss the issues that influence channel strategy

_____ 9. Life Savers candy is sold in grocery stores, service stations, convenience stores, drugstores, discount stores, and vending machines. This is a(n) ___________ distribution strategy.
   a. exclusive
   b. franchising
   c. selective
   d. intensive
   e. horizontal

_____ 10. Lennox Air Conditioners carefully screens its dealers to ensure a quality dealer image and service ability. Only a few dealers are chosen in any single geographic area. This is an example of:
   a. selective distribution
   b. intensive distribution
   c. exclusive distribution
   d. dual distribution
   e. intermediary distribution

_____ 11. Rolls Royce has a restrictive policy of only establishing one or two dealers within a given large geographic area. Buyers of this type of expensive car will travel to acquire the product, so this form of distribution, ___________, is appropriate for the product.
   a. intermediary
   b. intensive
   c. selective
   d. dual
   e. exclusive

5 Explain channel leadership, conflict, and partnering

_____ 12. As the largest retailer of any kind in the world, Wal-Mart exerts great power over its suppliers, who provide the best possible prices. In the marketing channel, Wal-Mart would be considered a:
   a. channel authority
   b. channel member
   c. channel leader
   d. channel team
   e. merchant wholesaler

_____ 13. American Booksellers Association, a group of small independent bookstores, filed a suit against the large chains Barnes & Noble and Borders, alleging that they violated antitrust laws by using their buying power for deeper discounts on books. This is an example of:
   a. horizontal channel conflict
   b. lateral channel conflict
   c. retailer conflict
   d. vertical channel conflict
   e. channel conspiracy

7 Discuss the concept of balancing logistics service and cost

_____ 14. The sales representatives at Mechanix, Inc., are worried that they will lose more customer accounts if service is not faster. But the warehouse is trying to cut costs by holding less inventory and using slower transportation modes. This is an example of:
a. an inefficient warehousing system
b. the difficulties in an inventory control system
c. the need for third-party or contact warehousing
d. the complex balancing required of the physical distribution system
e. a situation in which a mechanized material handling system would work

8 Describe the integrated functions of the supply chain

_____ 15. Spiffy Spices, Inc., is trying to decide on a location for its new warehouse and should consider all of the following EXCEPT:
a. the cost and quality of industrial land
b. transportation costs and distances to customers
c. the stage of the product life cycle
d. the location of production facilities
e. the availability of transportation modes

_____ 16. The Megabit Company experiences much damage and loss when moving its computer products from the manufacturing plant to the storage facilities. The company should consider reducing the number of times an item is moved in the warehouse by installing a(n):
a. inventory control system
b. order-processing system
c. safety procedure
d. materials-handling system
e. transportation system

_____ 17. A large bank commissions a computer manufacturer to produce three hundred computers that would be tailored to the bank's specific needs. This is an example of:
a. JIT production
b. auto ID
c. EOQ
d. mass customization
e. procurement

_____ 18. The PorterCo company can no longer afford expensive storage facilities for its inventory. Furthermore, its computer products need to be built as soon as customers place orders. PorterCo should establish a(n) ____________ system.
a. UPC
b. auto ID
c. EOQ
d. JIT
e. TQM

_____ 19. The Limited clothing store can immediately transmit orders for new clothing to its suppliers in Asia. The Asian suppliers can electronically transmit back to the Limited information about in-stock items and shipping dates. This is an example of:
a. UPC
b. EDI
c. EOQ
d. JIT
e. ROP

_____ 20. Massachusetts Mining harvests timber and mines metal ores and coals. This company is most likely to use ___________ for transportation to customers.
a. railroads
b. airplanes
c. trucklines
d. water transportation
e. pipelines

_____ 21. The Prime Cut Cattle Company has determined that its chief priority for choosing a transportation mode in the United States is transit time because meats must be fresh. However, this must be tempered by practical cost considerations. Prime Cut should use ___________ to ship their meats.
a. railroads
b. airplanes
c. motor carriers
d. water transportation
e. pipelines

_____ 22. Florida Greenhouses, Inc. promises its customers fresh-cut tropical flower arrangements with premium service (at a premium price). Florida Greenhouses most likely uses:
a. railroads
b. airways
c. motor carriers
d. water transportation
e. pipelines

_____ 23. Some U.S. car buyers save money by purchasing Mercedes-Benz cars in Germany and adding U.S. required emission systems while the car is in transit. The most likely transportation mode is:
a. water carrier
b. motor carrier
c. railroad
d. air
e. ocean pipeline

9 Discuss new technology and emerging trends in logistics

_____ 24. Remex Diesel recognizes its main business strengths as efficient manufacturing procedures, an excellent dealer network, extensive market research, and an experienced marketing staff. The company has never developed a strong physical distribution system because it is not a focus of the business. It makes sense for Remex Diesel to use:
a. private warehousing
b. intermodal sourcing
c. distribution center systems
d. freight forwarding
e. contract warehousing

10 Identify the special problems and opportunities associated with distribution in service organizations

_____ 25. You are responsible for physical distribution of your company's service and should focus on:
a. ensuring the intangibility of the service so physical distribution becomes a less important factor
b. customer-oriented order processing and inventory control
c. making sure that production and consumption are simultaneous
d. setting quality standards, choosing faster transportation modes, and using safety stock
e. minimizing wait times, managing service capacity, and providing delivery through distribution channels

11 Discuss channel structure and logistics issues in global markets

____ 26. Celestial Seasonings, a United States tea manufacturer, would like to sell its herbal teas in China and Russia. Given the product and market characteristics, Celestial Seasonings should probably:
- a. use a direct-mail wholesaler
- b. establish an overseas company sales force
- c. use independent foreign intermediaries for distribution
- d. establish a vertical marketing system and act as channel leader
- e. purchase the services of an industrial distributor

Check your answers to these questions before proceeding to the next section.

ESSAY QUESTIONS

1. Channel members, such as wholesalers and retailers, are often included between producers and business users or consumers for three important reasons. Name and describe each of these reasons.

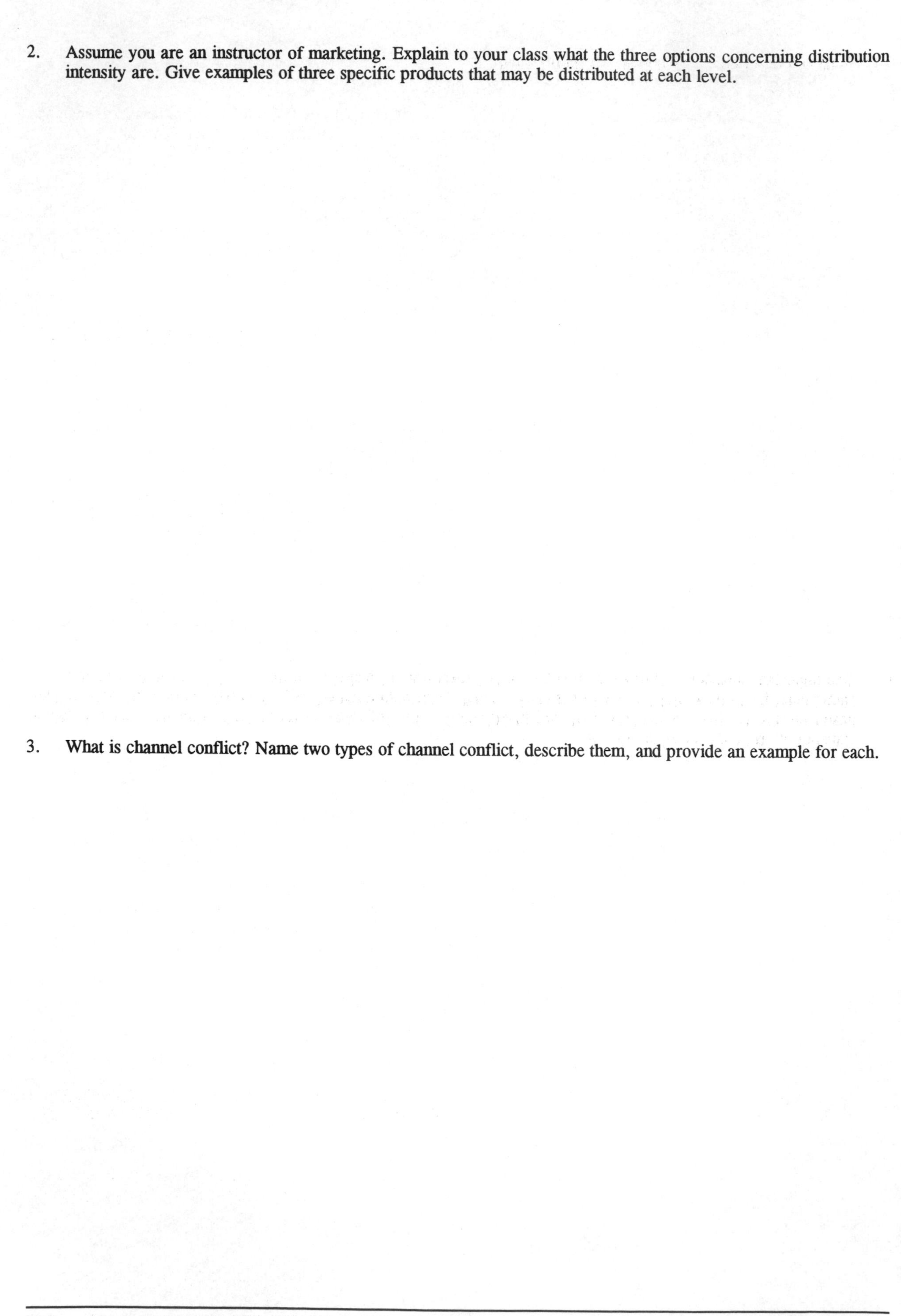

2. Assume you are an instructor of marketing. Explain to your class what the three options concerning distribution intensity are. Give examples of three specific products that may be distributed at each level.

3. What is channel conflict? Name two types of channel conflict, describe them, and provide an example for each.

4. Name and describe six functions of a supply chain.

5. One important subsystem of physical distribution is transportation. Name the different modes of transportation. For each mode, list two examples of cargo typically carried. Distribution managers select different transportation modes based on several distinct criteria. List and briefly define five of these criteria, and name the best and worst transportation mode for each criteria.

CHAPTER NOTES

Use this space to record notes on the topics you are having the most trouble understanding.

ANSWERS TO THE END-OF-CHAPTER DISCUSSION AND WRITING QUESTIONS

Use this space to work on the questions at the end of the chapter.

ANSWERS TO THE END-OF-CHAPTER CASE

Use this space to work on the questions at the end of the chapter.

CHAPTER 14 Retailing

LEARNING OBJECTIVES

1 Discuss the importance of retailing in the U.S. economy

Retailing is a huge and visible portion of marketing. Retailing is one of the largest employers, with over 18 million people employed by over 2 million U.S. retailers. The industry is dominated by a few giant organizations. Over half of all retail sales are produced by fewer than 10 percent of all retail establishments. The list of giant retailers is headed up by Wal-Mart, Kmart, and Sears.

2 Explain the dimensions by which retailers can be classified

Many different kinds of retailers exist. They can be differentiated on the basis of ownership, level of service, product assortment, and general price levels.

3 Describe the major types of retail operations.

There are several types of retailing, such as department stores, specialty stores, mass merchandisers, supermarkets, discount stores, franchising, and nonstore forms (vending, in-home selling, and direct marketing).

Housing several departments under one roof, a department store carries a wide variety of shopping and specialty goods, including apparel, cosmetics, housewares, electronics, and sometimes furniture. Each department is usually headed by a buyer who selects the merchandise mix for the department and may also choose the promotion devices and personnel.

Very similar to department stores in many respects, the mass merchandising shopping chains have wide product assortments, yet their sheer size in sales volume, promotion budgets, and number of stores sets them apart. Sears, JC Penney, and Montgomery Ward fall into this category. Mass merchandising is the strategy of using moderate to low prices on large quantities of merchandize, coupled with big promotional budgets to stimulate a high turnover of products.

A specialty store is not only a type of store but also a method of retail operations, specializing in a given type of merchandise. A typical specialty store carries a deep but narrow assortment of merchandise and offers attentive customer service and knowledgeable sales clerks.

Americans spend approximately 9 percent of their disposable income in supermarkets--large, departmentalized, self-service retailers that specialize in foodstuffs and a few nonfood items.

Discount stores are retailers that compete on the basis of low prices, high turnover, and high volume. A major force in retailing, discounters can be classified into four major categories: full-line discounters (Kmart and Wal-Mart), discount specialty stores (Office Depot), membership clubs, and off-price discount retailers (factory outlets).

4 Discuss nonstore retailing techniques

Nonstore retailing refers to shopping without visiting a store and has three major categories: vending; direct retailing; and direct marketing, which includes direct mail, catalogs, telemarketing, and electronic retailing.

The shop-at-home cable television industry has quickly grown into a billion-dollar business with a loyal customer following.

5 Define franchising and describe its two basic forms

Franchising goes back almost one hundred years and is becoming one of the most profitable business methods in the United States today. A franchise is a continuing relationship in which a franchisor grants retail operating rights to a franchisee. The franchisor originates the trade name, product, methods of operation, and so forth and, in return, receives a fee and continuing revenues from the business operation of the franchisee.

6 List the major tasks involved in developing a retail marketing strategy

Retailers must develop marketing strategies based on overall goals and strategic plans. The key tasks in retail strategy are defining and selecting a target market and developing the six P's of the retailing mix to successfully meet the needs of the chosen target market.

The first task in developing the retail strategy is to define the target market. This process begins with market segmentation. The markets are often defined on demographic, geographic, and lifestyle or psychographic dimensions.

The six P's of the retailing mix consist of those elements controlled by the retailer that are combined together in varying degrees and forms and directed as a single retailing method to the target market. The six P's include product, place, price, promotion, personnel, and presentation, and together these items project the store's image.

The presentation of a retail store to its customers helps determine the store's image. The most predominant aspect is the store's atmosphere--the store's overall impression established by the physical layout or decor.

Personnel and customer service are a prominent aspect of retailing. Most retail sales involve a customer-salesperson interaction.

7 Discuss the challenges of expanding retailing operations into global markets

Retailers are seeking to growth their business by entering global markets. The homogenization of tastes and product preferences around the world, the lowering of trade barriers, and the emergence of underserved markets has made the prospects of expanding across national borders more feasible for many retailers. Retailers wanting to expand globally should first determine what their core competency is and determine if this differentiation is what the local market wants.

8 Discuss future trends in retailing

With increased competition and slow domestic growth, mature retailers are looking for growth opportunities in the developing consumer economies of others countries. The homogenization of tastes and the lowering of trade barriers has made the prospect feasible. Additionally, retailers of the future will shift from distribution centers to marketers. Staples won't be sold in stores, but instead delivered directly to the consumer, freeing shoppers to visit stores for products they enjoy buying. Advances in technology will make it easier for consumers to obtain the products they want. Growing consumer demand for convenience will move retailers to be more solution-oriented.

PRETEST

Answer the following questions to see how well you understand the material. Re-take it after you review to check yourself.

1. What is retailing?

2. List ten types of stores.

3. List and briefly describe four types of nonstore retailing.

4. List and briefly describe the six Ps of retailing.

CHAPTER OUTLINE

1 Discuss the importance of retailing in the U.S. economy

I. The Role of Retailing

Retailing is all activities directly related to the sale of goods and services to the ultimate consumer for personal, nonbusiness use.

A. Retailing is one of the largest employers, with over 19 million people employed by over 2 million U.S. retailers.

B. The industry is dominated by a few giant organizations. Over half of all retail sales are produced by less than 10 percent of all retail establishments.

C. The list of giant retailers (in order) is headed by Wal-Mart, Sears and Kmart.

2 Explain the dimensions by which retailers can be classified

II. Classification of Retail Operations

Retailers can be differentiated on the basis of ownership, level of service, product assortment, and price.

A. Ownership

Retailers can be broadly classified by form of ownership: independent, part of a chain, or a franchise outlet.

1. **Independent retailers** are owned by a single person or partnership and not operated as part of a larger retail institution.

2. **Chain stores** are owned and operated as a group by a single organization.

3. **Franchises** are owned and operated by individuals but are licensed by a larger supporting organization.

B. Level of Service

Various services provided by retailers can be classified along a continuum from full-service (such as exclusive clothing stores) to self-service (such as factory outlets and warehouse clubs).

C. Product Assortment

Another basis for positioning or classifying stores is by the breadth and depth of their product line.

1. Specialty stores may be very narrow in product lines but offer great depth within that area.

2. Discounters such as Wal-Mart and Kmart are just the opposite, offering a wide array of products with limited depth.

D. Price

Some stores such as department stores typically charge full price, while other stores emphasize their discounting or outlet pricing.
Gross margin shows how much a retailer makes. It is expressed as a percentage of sales after the cost of goods sold is subtracted.

3 Describe the major types of retail operations

III. Major Types of Retail Operations

A. Department Stores

1. Having several departments under one roof, a **department store** carries a wide variety of shopping and specialty goods, including apparel, cosmetics, housewares, electronics, and sometimes furniture.

2. Each department is usually headed by a **buyer** who selects the merchandise mix for the department and who may also choose the promotion devices and personnel.

3. Most department stores buy directly from manufacturers and some may even have merchandise produced under the store's brand name.

4. Some department stores may have so much buying strength that they dominate small manufacturers.

5. Most large department stores are owned by national chains, with few independent stores remaining.

6. Specialty retailers have responded to the consumer cost consciousness and value-orientation.

7. To protect themselves from powerful new specialty retailers, discounters, outlets, and other price cutters, department store managers are using several strategies.

   a. Many department stores are repositioning themselves as specialty outlets with a "store-within-a-store" format, dividing departments into mini boutiques.

   b. Department stores view their high level of service as the one unique benefit they offer to consumers. Emphasizing service rather than price enables them to compete with discounters.

   c. Expansion and renovation of existing stores allows growth in new market areas and changes in merchandising directions and images.

C. Specialty Stores

1. A **specialty store** is not only a type of store but also a method of retail operations that specializes in a given type of merchandise.

2. A typical specialty store carries a deep but narrow assortment of merchandise and offers attentive customer service and knowledgeable sales clerks.

3. Price is usually of secondary importance to customers of specialty stores.

D. Supermarkets

1. Americans spend approximately 10 percent of their disposable income in **supermarkets**, which are large, departmentalized, self-service retailers that specialize in foodstuffs and limited nonfood items.

2. With razor-thin profit margins (1 to 2 percent) and the slow annual population growth (less than 1 percent), supermarkets have had to discover and exploit several demographic and lifestyle trends in order to prosper.

3. *Superstores* offer both foods, non-foods and services in a store which is usually twice the size of supermarkets.

4. Offering a wide variety of nontraditional goods and services under one roof is **scrambled merchandising**. These superstores offer one-stop shopping for many food and nonfood needs, and may include pharmacies, flower shops, bookstores, salad bars, takeout food sections, dry cleaners, photo processing, health food sections, and banking.

5. *Loyalty marketing programs* reward loyal customers with discounts or gifts.

E. Drugstores

Drugstores stock pharmacy - related products and services as their main draw.

1. Customers are most attracted by:

    a. the pharmacist

    b. convenience

    c. acceptance of third party prescription drug plans

2. Drugstores have developed value-added services in order to compete with growing competition.

3. Demographic trends support growth in the drugstore industry.

F. **Convenience stores** carry a limited line of high-turnover convenience goods and resemble miniature supermarkets.

Prices are higher than supermarkets because the stores offer so many conveniences--location, long hours, and fast service.

G. Discount Stores

Discount stores are retail stores that compete on the basis of low prices, high turnover, and high volume.

1. **Full-line discounters, also called mass merchandisers**, carry a broad assortment of nationally branded hard goods (broader than a department store) and offer customers few services. Very similar to department stores in many respects, the mass merchandising shopping chains have wide product assortments, yet their sheer size in sales volume, promotion budgets, and number of stores sets them apart.

2. **Mass merchandising** is the strategy of setting moderate to low prices on large quantities of products, coupled with big promotional budgets, to stimulate high inventory turnover.

3. Several different hybrids of the full-line discounters have appeared:

    a. **Hypermarkets** are even larger than the largest supermarket and discount store, with over 200,000 square feet. They must generate a huge volume to compensate for very low gross margins. This concept has had limited success in the United States.

4. **Supercenters** combine groceries and general merchandise goods with a wide range of services including pharmacy, dry cleaning, photo finishing, optical shops, and hair salons. Walmart operates over 300 supercenters.

    a. The *extreme-value retailer* is characterized by a narrow selection of basic merchandise and rock bottom prices.

5. **Discount specialty stores** are single-line stores offering merchandise such as sporting goods, electronics, auto parts, office supplies, or toys.

    These stores offer a nearly complete selection of one line of merchandise and use self-service, discount prices, high volume, and high-turnover merchandise to their advantage.

    a. They are often termed **category killers** because they so heavily dominate their narrow segment.

6. **Warehouse membership clubs** offer a limited selection of brand-name appliances, household items, and groceries, usually in bulk on a cash-and-carry basis to members only.

7. **Off-price discount retailers** sell at prices that are 25 percent or more below traditional department store prices because these retailers pay cash for their stock, and usually don't ask for return privileges.

    a. The off-price retailers purchase goods at cost or less from manufacturers' overruns, bankruptcies, irregulars, and unsold end-of-season output.

    b. Merchandise styles and brands usually change from month to month as availability fluctuates.

    c. **Factory outlets** are a type of off-price retailer with one big difference: They are owned and operated by manufacturers and carry one line of merchandise (their own).

    d. Factory outlets sell discontinued merchandise, factory seconds, and canceled orders.

    e. Factory outlets attempt to avoid direct competition with their normal retail outlets by locating in out-of-the-way places.

H. Restaurants

Restaurants straddle the line between a retailing establishment and a service establishment.

1. Tangible products are food and drink.

2. Service elements are food preparation and food service.

3. Food away from home accounts for anywhere from 25 percent to 50 percent of the household food budget.

4. More restaurants are now competing directly with supermarkets by offering take-out and delivery in an effort to capture more of the home replacement market.

4 Discuss nonstore retailing techniques

IV. Nonstore Retailing

Nonstore retailing refers to shopping without visiting a store and is currently growing faster than in-store shopping because of the consumer demand for convenience.

A. **Automatic Vending** is a form of nonstore retailing that uses automated machines ffer products for sale.

1. Vending accounts for $21 billion worth of sales each year.

2. Trends include branching out into different types of merchandise and a debit card system.

B. **Direct retailing** occurs in a home setting and includes door-to-door sales and party plan selling, and the trend is toward party-plans rather than cold door-to-door canvassing. The sales of direct retailers have suffered as more women have entered the work force, and the direct retailers have had to become more creative in reaching women.

C. **Direct marketing**, sometimes called **direct-response marketing**, describes a variety of techniques such as telephone selling, direct mail, catalogs, and newspaper, television, or radio ads that invite the shopper to buy from their homes.

Direct marketing seeks an immediate response from the consumer who is in his or her home.

1. Direct Mail

Direct mail can be the most or least efficient retailing method, depending on the quality of the mailing list and the effectiveness of the mailing piece. Direct mail can be very precise in targeting customers according to demographics, geographics, and even psychographics.

With a technique called predictive modeling, direct-mailers can pick out individuals most likely to buy their products using census statistics, lifestyle and financial information, and past-purchase and credit history.

Direct mailers have suffered an image problem because of the volume of junk mail that is of no interest to the recipients and the scams that announce "you have won a prize!"

2. Catalogs and Mail Order

Over 13 billion catalogs are mailed annually, with the average U.S. household receiving one every five days.

Improved customer service and quick delivery has boosted customer confidence in mail order.

3. Telemarketing

Telemarketing is the use of the telephone to sell directly to consumers. It consists of outbound sales calls, usually unsolicited, and inbound calls, that is, taking orders through toll-free 800 numbers or fee-based 900 numbers.

4. Electronic Retailing

a. Shop-At-Home Networks
The shop-at-home cable television industry has quickly grown into a billion dollar business with a loyal customer following. Merchandise is displayed and demonstrated on the screen and credit-card orders are taken over toll-free phone lines.

b. Online retailing
Online retailing is a two-way interactive service offered to users with personal computers.

Users of online retailing "subscribe" to information and shopping services. Prodigy, CompuServe, and America Online are the most popular electronic shopping and information services.

5 Define franchising and describe its two basic forms

V. Franchising

A. A *franchise* is the right to operate a business or to sell a product

B. The **franchiser** originates the trade name, product, methods of operation, and so forth, and in return receives a fee and continuing revenues from the business operation of the **franchisee**.

C. Franchising offers many benefits to a person wanting to own and manage a business:

1. an opportunity to become an independent businessperson with relatively little capital

2. a product or service that has already been established in the marketplace

3. technical training and managerial assistance by the franchisor

4. quality-control standards enforced by the franchisor that will aid in ensuring product uniformity throughout the franchise system

D. The franchiser obtains benefits also:

1. company expansion with limited capital investment

2. motivated store owners

3. bulk purchasing of inventory

E. There are two basic forms of franchises: product and trade name franchising, and business format franchising.

6 List the major tasks involved in developing a retail marketing strategy

VI. Retail Marketing Strategy

The key tasks in retail strategy are defining and selecting a target market and developing the six P's of the retailing mix to successfully meet the needs of the chosen target market.

Retailers control the six P's of the **retailing mix**: the four P's of the marketing mix plus personnel and presentation.

A. Defining a Target Market

1. The first task in developing a retail strategy is to define the target market.

2. This process begins with market segmentation. The markets are often defined by demographics, geographics, and lifestyles or psychographics.

B. Choosing the Retailing Mix

The six P's of the retailing mix are all variables that can be used in a marketing plan. The six P's include product, place, price, promotion, personnel, and presentation, and together these items project the store's image.

1. The first element in the retailing mix is the **product offering**, also called the product assortment or merchandise mix.

   a. The retailer must determine what to sell, and this is based on research of the target market, past sales, fashion trends, customer requests, and other sources.

   b. The width and depth of the product assortment must be determined. *Width* refers to the assortment of products offered; *depth* refers to the number of different brands offered within each assortment.

   c. Computer technology, through *electronic data interchange* (EDI), is allowing retailers to respond quickly to fashion trends and new merchandise opportunities. EDI aids in producing and buying merchandise. This leads to *efficient consumer response* (ECR) a philosophy of streamlining the way products move to the consumer.

2. Retail promotion strategy includes advertising, public relations, publicity, and sales promotion.

   a. The goal of the retailer's promotional strategy is to help position the store in consumers' minds.

   b. Media advertising for retailers is generally concentrated at the local level.

   c. Publicity and public relations are a very important part of the promotional mix.

3. The third element in the retail mix is site location. Selecting a proper site is extremely important.

   a. A location is a large and long-term commitment of resources.

   b. Growth goals often imply that more new locations should be established.

c. The environment may change over time, and the value of a location may deteriorate.

d. A decision must be made on whether or not to locate as a freestanding unit or to become a shopping mall tenant.

1) Freestanding stores are often used by large retailers such as Target or Kmart and sellers of shopping goods such as automobiles and furniture. They are "destination" stores.

2) Shopping centers began after World War II in response to the migration of the U.S. population to the suburbs. The first *strip centers* were typically located along a busy street. Small *community shopping centers* contained offered a broader variety of goods and contained 75,000 to 300,000 square feet of retail space. Huge *regional malls* are entirely enclosed and have *anchor* or *generator stores* located on opposite ends of the mall.

4. Another important element in the retail mix is price.

a. Because many retail prices are based on the cost of the merchandise, an essential part of the pricing element of the retail mix is efficient and timely buying.

b. Price is also a large part of a retail store's positioning strategy and classification.

c. A trend has been *everyday low pricing (EDLP)*, a move away from sales and discounts.

5. The presentation of a store determines the store's image.

a. The predominant aspect is the store's **atmosphere**, the store's mood or feeling as established by the physical layout, decor, and surroundings.

b. The main determinants of a store's atmosphere are:

1) employee type and density

2) merchandise type and density

3) fixture type and density

4) sound

5) odors

6) visual factors

6. Personnel and customer service are a prominent aspect of retailing. Most retail sales involve a customer-salesperson relationship, if only briefly.

a. An important task for retail salespeople is personal selling, persuading shoppers to buy.

b. Most salespeople are trained in two common selling techniques: trading-up and suggestive selling.

1) Trading-up means convincing customers to buy a higher-priced item than they originally intended.

2) Suggestion selling is a common practice used by most retailers, which seeks to broaden the customer's original purchase with related items.

7 Discuss the challenges of expanding retailing operations into global markets

VII. Global Retailing

Mature retailers are looking for growth opportunities in the growing consumer economics of other countries.

A, Several events have made expansion across national boundaries more feasible.

1. Spread of communication networks and mass media has homogenized tastes and product preferences.

2. The lowering of trade barriers and tariffs.

3. High growth potential in underserved markets.

B. Prerequisite for going global include:

1. A secure and profitable position domestically.

2. A long-term perspective.

3. A global strategy that meshes with the retailers overall corporate strategy.

C. Retailers need to skillfully make adjustments where necessary.

8 Describe future trends in retailing

VIII. Trends in Retailing

Future trends in retailing include the use of entertainment, a shift toward providing greater convenience and the emergence of customer management programs.

A. Entertainment

Adding entertainment to the retail environment is one of the most popular strategies in retailing. Entertainment may include anything that makes shoppers have a good time.

B. Convenience and Efficiency

Today's consumer is looking for ways to shop quicker and more efficiently. Consumers are visiting stores less often. New technology may offer many innovative solutions for consumers.

C. Customer Management

Through customer management strategies, leading retailers are intensifying their efforts to identify, satisfy, retain and maximize the value of their best customers. *One-to-one marketing* is the use of database technology to manage customer relationships.

1. *Customer relationship marketing* (CRM) originated out of the need to more accurately target a fragmented customer base.

2. Retailers are taking active measures to develop loyalty programs that identify and reward their best customers.

3. *Clienteling* strongly emphasizes personal contact on the part of managers and sales associates with customers.

VOCABULARY PRACTICE

Fill in the blank(s) with the appropriate term or phrase from the alphabetized list of chapter key terms.

atmosphere
automatic vending
buyer
category killers
chain stores
convenience store
department store
direct marketing (direct response marketing)
direct retailing
discount store
drugstore
factory outlet
franchise
franchisee
franchiser
full-line discount store
gross margin
hypermarket
independent retailers
mass merchandising
nonstore retailing
off-price retailer
private-label brands
product offering (product assortment)
retailing
retailing mix
scrambled merchandising
specialty discount store
specialty store
supercenter
supermarket
telemarketing
warehouse membership club

1 Discuss the importance of retailing in the U.S. economy

1. All activities directly related to the sale of goods and services to the ultimate consumer for personal, nonbusiness use or consumption can be defined as ______________________________.

2 Explain the dimensions by which retailers can be classified

2. Retailers can be broadly classified by form of ownership. Retailers than are owned by a single person or partnership are ______________________________. Retailers that are owned and operated as a group by a single firm are ______________________________.

3. One way to classify retailers is according to price. One measure of this is expressed as a percentage of sales less the cost of goods sold. This measure is ______________________________.

3 Describe the major types of retail operations

4. One method of retail operations carries a wide variety of shopping and specialty goods in several departments under one roof. This defines a(n) ______________________________. The head of each department, or the ______________________________, selects the merchandise mix and may be responsible for promotion and personnel.

5. As retailers focus more carefully on segmentation and tailor their merchandise to specific target markets, stores may carry a deeper but narrower assortment of merchandise. This type of store would be a(n) ______________________________.

6. A large, departmentalized, self-service retailer that specializes in wide assortments of foodstuffs and limited nonfood items is a(n) ______________________________. Sometimes these retailers offer a wide variety of nontraditional goods and services, which is called ______________________________. A miniature version of this retailer type is a(n) ______________________________, which carries only a limited line of high-turnover goods. ______________________________ carry a variety of pharmaceutical products—including over-the-counter (OTC) and prescription drugs—and cosmetics, health and beauty care items, and specialty products.

7. A retail chain that competes on the basis of low prices, high turnover, and high volume is a(n) ______________________________. A type of retailer that offers consumers very limited service and carries a much broader assortment of well-known, nationally-branded "hard goods" are known as ______________________________. These retailers use moderate to low prices on large quantities of merchandise, coupled with big promotional budgets, to stimulate high turnover of products, a strategy known as ______________________________.

8. A hybrid form of a full-line discounter combines a supermarket and general merchandise discount store in a large space. This is a ____________________. Similar to this type is a retailer about half the size, which is a(n) ____________________.

9. Stores that offer a nearly complete selection of single-line merchandise and heavily dominate their narrow merchandise segment are called ____________________ or are also known as ____________________.

10. Limited-service retailers that carry bulk items and sell to members are called ____________________.

11. If the store sells manufacturers' overruns and irregular merchandise at prices far below that of department stores, the store would be a(n) ____________________. If the retailer carries one line of discount merchandise, its own, it is called a(n) ____________________.

4 Discuss nonstore retailing techniques

12. Shopping without visiting a store is termed ____________________. A low-profile form of this retailing is ____________________, which involves consumers making purchases from machines.

13. Another form of retailing involves representatives visiting the customer to sell, which is known as ____________________. Other techniques used to get consumers to purchase from their home, office, or other nonretail setting. These techniques are known as ____________________ or ____________________ and include direct mail, catalogs, and electronic retailing, as well as ____________________, which involves a systematic use of telephones in the selling process.

5 Define franchising and describe its two basic forms

14. A continuing relation in which a company grants operating rights (trade name, product, operation methods) to another party is a(n) ____________________. The individual or business granting the business rights is called the ____________________, and the individual or business granted the right to operate and sell the product or service is called the ____________________.

6 List the major tasks involved in developing a retail marketing strategy

15. Product, place, promotion, price, personnel, and presentation make up the ____________________. This first element in this grouping is known as the merchandise mix, product assortment, or ____________________. Retailers can also choose to sell their own propriety brand name product called ____________________.

16. The presentation of a retail store to its customers helps determine the store's image. The predominant aspect of a store's presentation is its ____________________, which refers to how a store's physical layout, decor, and surroundings convey an overall impression.

Check your answers to these questions before proceeding to the next section.

TRUE/FALSE QUESTIONS

Mark the statement **T** if it is true and **F** if it is false.

1 Discuss the importance of retailing in the U.S. economy

_____ 1. Lands' End sells casual clothing through catalogs mailed to homes and online service. Because Lands' End does not have a store front, it is not considered a retailer.

2 Explain the dimensions by which retailers can be classified

_____ 2. Sales minus the cost of goods sold will give a retailer's net income.

3 Describe the major types of retail operations

_____ 3. Plingers is a retailing establishment. At Plingers, Mario is the buyer for ladies' shoes and develops promotions for the shoe lines as well as purchasing them. Tina is the buyer for all luggage products; she has similar promotional responsibilities and also is responsible for her personnel. Plingers is most likely a department store.

_____ 4. Each Aames Store has several departments, including housewares and menswear. Because there are many Aames stores and each store has huge sales volumes, the store would be a department store rather than a mass merchandiser.

_____ 5. The Sock Stop is a retail store with a narrow assortment of merchandise (socks and hosiery) and with great depth in its product line. The Sock Stop is a specialty store.

_____ 6. Jon Schommer owns a neighborhood pharmacy. He is considering offering other profitable items that customer sometimes ask for, including snack foods, diapers, cosmetics, greeting cards, and magazines. This would be an example of multibrand merchandising.

_____ 7. Factory outlets are called "category killers" because they dominate the competition in their narrow merchandise segment.

4 Discuss nonstore retailing techniques

_____ 8. Cindee sells cosmetics through her web site and via a 1-800 number. Cindee practices direct-response marketing.

5 Define franchising and describe its two basic forms

_____ 9. Michelle's makes a unique frozen custard and sells this famous custard at a local parlor. Michelle's has agreed to provide individuals in neighboring counties with the special frozen custard recipes, trade name, parlor store plan, and management principles, in return for a percentage of the new revenues. Michelle's is a franchisor.

6 List the major tasks involved in developing a retail marketing strategy

_____ 10. The six Ps of retailing are product, price, promotion, place, packaging, and presentation.

8 Discuss future trends in retailing

_____ 11. In the future, the Swamp Supermarket can expect to focus on marketing produce items as part of a "solutions" package, because staple goods will be home-delivered rather than shopped for.

Check your answers to these questions before proceeding to the next section.

AGREE/DISAGREE QUESTIONS

For the following statements, indicate reasons why you may agree and disagree with the statement.

1. With communication technology, traditional store retailing will yield to nonstore retailing and eventually disappear.

   Reason(s) to agree:

   Reason(s) to disagree:

2. Given the dominance of large retail chains, the small independent store generally has no purpose.

   Reason(s) to agree:

   Reason(s) to disagree:

3. U.S. retailers should always adapt their products to local market when they enter foreign countries.

   Reason(s) to agree:

   Reason(s) to disagree:

MULTIPLE-CHOICE QUESTIONS

Select the response that best answers the question, and write the corresponding letter in the space provided.

1 Discuss the importance of retailing in the U.S. economy

_____ 1. Which of the following statements about retailing is NOT true?
a. Retail sales represent about 25 percent of the gross domestic product of the U.S.
b. Large retail operations represent over 75 percent of all retail sales in the U.S.
c. The retailing industry employs about 20 percent of the nation's workers.
d. Large retail operations employ about 40 percent of the nation's retail workers.
e. The nation's top retailer is Wal-Mart.

2 Explain the dimensions by which retailers can be classified

_____ 2. A store's gross margin is:
a. the final profit after all expenses are subtracted from revenue.
b. the cost of merchandise from suppliers.
c. the marketing expenses as a percentage of sales.
d. the net sales as a percentage of gross sales.
e. the percentage of sales after cost of goods is subtracted.

3 Describe the major types of retail operations

_____ 3. The Perfect Pet Parlor is a chain of stores selling a large selection of pet food and pet accessories at a high margin with excellent sales support. The store would most likely be classified as a:
a. hypermarket
b. convenience store
c. warehouse membership club
d. specialty store
e. mass merchandiser

_____ 4. Waygro is a large, departmentalized, self-service retailer that specializes in wide assortments of foodstuffs and limited nonfood items. Waygro is a:
a. mass merchandiser
b. convenience store
c. warehouse membership club
d. discount store
e. supermarket

_____ 5. Albertsons, Kroger, Winn-Dixie, and other supermarkets offer a variety of nontraditional goods and services such as video rental, flower shops, dry cleaning, and banking. This practice is called:
a. convenience merchandising
b. retail wheeling
c. scrambled merchandising
d. trade-up positioning
e. specialty service

_____ 6. The Muffy Mart utilizes a strategy of setting low prices on large quantities of products and then uses daily television advertisements to stimulate a high turnover of inventory. Muffy Mart also offers a wide variety of product lines. Muffy Mart is a:
a. mass merchandiser
b. factory outlet
c. convenience store
d. wheel of retailing
e. specialty store

_____ 7. Smart-Sam's has opened a new store, even larger than its largest supermarket. This new store is over 200,000 square feet and combines a supermarket and discount department store. The store must generate a volume of over $1,000,000 per week in sales just to break even. This new store is a:
a. mass merchandiser
b. warehouse membership club
c. hypermarket
d. discount store
e. factory outlet

_____ 8. SoundSensation offers stereo equipment and accessories. It has a deep assortment and low prices. The store is operated on a self-service, no-frills concept. SoundSensation is a:
a. specialty store
b. factory outlet
c. warehouse membership club
d. discount specialty store
e. discount store

_____ 9. Jill buys many of the office supplies for her income tax preparation business at a retailer that stocks a limited selection of items, which are sold in bulk on a cash-and-carry basis to members only. She browses through a huge store that has a warehouse atmosphere. She buys computer paper, pencils, a small copy machine, and a television for her waiting area. Jill is shopping at a(n):
a. off-price discount retailer
b. factory outlet
c. warehouse membership club
d. industrial supply warehouse
e. hypermarket

_____ 10. The Waverly Wear clothing company has to find a way to dispose of its overrun and unsold end-of-season clothes. As a marketing consultant, you suggest that they either open a factory outlet store or sell the merchandise to a(n):
a. off-price discount retailer
b. hypermarket
c. department store
d. supermarket
e. mass merchandiser

_____ 11. The American Tourister Luggage Company has decided that the most profitable way to dispose of out-of-season and irregular stock would be to open a store and sell its own merchandise in a remote location. This is the retail strategy of:
a. hypermarkets
b. mass merchandisers
c. discount stores
d. factory outlets
e. bargain basements

4 Discuss nonstore retailing techniques

_____ 12. Which of the following is NOT an example of nonstore retailing?
a. An Avon sales person sells cosmetics in an office setting.
b. L.L. Bean sells clothing through catalog sales.
c. A famous chef sells her pasta sauce in her restaurant.
d. The QVC network sells jewelry through television.
e. PC Flowers sells flower arrangements on-line.

_____ 13. BigFoote sells its hunting and hiking boots through catalogs in the mail. This retailing technique is known as:
a. online retailing
b. direct marketing
c. franchising
d. vending
e. in-home retailing

_____ 14. The Super Shoppe has decided to display products on a cable television channel and encourage shoppers to call a toll-free number to purchase the merchandise with a credit card. This form of retailing is called a(n):

a. in-store electronic shopping technique
b. online method
c. electronic point of sale
d. shop-at-home network
e. catalog viewing

5 Define franchising and describe its two basic forms

_____ 15. Brigitte has decided to buy a franchise pet grooming business rather than develop her own independent business. She probably chose the franchise for all of the following reasons EXCEPT:

a. a management training program
b. obtaining a well-known product or service name
c. the individual can try personal, innovative product and service ideas in the business
d. the established image of the franchise
e. product uniformity, which is ensured by quality-control standards

6 List the major tasks involved in developing a retail marketing strategy

_____ 16. The retailing mix consists of all of the following EXCEPT:

a. packaging
b. product
c. presentation
d. place
e. price

_____ 17. The Greenhouse Exotique has decided that consumers will be drawn to its offering of rare and specialty plants. The store's manager believes that consumers will be willing to drive out of their way to buy these plants. The company needs to keep its overhead costs (such as rent) low and wants to avoid locating near competing nurseries and plant stores. For a location the Greenhouse Exotique will probably choose a:

a. factory outlet
b. freestanding store
c. strip center
d. shopping center
e. regional mall

_____ 18. Alistair wants to open a small, specialty toy store and is considering locating it in a regional shopping mall. He should know that:

a. parking is usually inadequate
b. the leases required are usually inexpensive
c. his store could be the mall anchor
d. the mall atmosphere and other stores will help attract shoppers
e. there is usually a problem with image because malls have no unified image, as a shopping center does

_____ 19. Zepps and Ablards are two anchor stores and therefore:

a. are most likely large department stores that are located at opposite ends of a mall to create a heavy pedestrian traffic flow
b. are the stores within the mall that sell services rather than products
c. probably specialize in high-priced items like furniture
d. are supermarkets that are located within shopping malls
e. are retail stores that "drop off" to freestanding locations

_____ 20. Beach Bums is a new, trendy store that specializes in swimwear and beach accessories. The store is decorated in neon colors, is full of potted palm trees, has sand on the floor, and plays beach music in the background. These factors are used to create the store's:
  a. cultural impact
  b. merchandise mix
  c. target strategy
  d. atmosphere
  e. promotional strategy

_____ 21. Assume you are the manager of a retail store and you are currently deciding your store's service level. Which of the following is NOT an issue for you to consider for retail service level?
  a. browsing patterns of the consumers
  b. the service offered by the competition
  c. type of merchandise handled
  d. socioeconomic characteristics of the target market
  e. cost of providing the service

7 Discuss the challenges of expanding retailing operations into global markets

_____ 22. Which of the following statements would NOT explain why retailing has become global?
  a. Communication technology has homogenized tastes and preferences throughout the world.
  b. Trade barriers and tariffs have been lowered.
  c. Many markets in the underdeveloped world have vast sales potential.
  d. Retailers around the world have been merging and creating global retail chains.
  e. Marketers have become very savvy about how to conduct market research in other countries.

8 Discuss future trends in retailing

_____ 23. Zepp is interested in global retailing. Zepp is most likely to be successful in this endeavor with a:
  a. product and trade name franchise
  b. megamall with multiple department store anchors
  c. mail-order catalog
  d. specialty store chain
  e. hypermart

Check your answers to these questions before proceeding to the next section.

ESSAY QUESTIONS

1. There are several types of retail stores, each offering a different product assortment, service level, and price level, according to the shopping preferences of its customers. Name eight types of retailers. For each type, indicate the level of service, price level, and width of product assortment.

2. Retail strategy involves the six P's of the retailing mix. Name each of the six P's, and briefly define elements of each P that are unique to retailing.

3. You are the owner of a new gourmet coffee shop that will compete against large chains such as Starbucks. Name three trends in retailing and describe how you could use each one to compete as a small independent retailer in such a competitive business.

CHAPTER NOTES

Use this space to record notes on the topics you are having the most trouble understanding.

ANSWERS TO THE END-OF-CHAPTER DISCUSSION AND WRITING QUESTIONS

Use this space to work on the questions at the end of the chapter.

ANSWERS TO THE END-OF-CHAPTER CASE

Use this space to work on the questions at the end of the chapter.

CHAPTER GUIDES and QUESTIONS

PART FIVE

INTEGRATED MARKETING COMMUNICATIONS

CHAPTER 15 Marketing Communication and Personal Selling

LEARNING OBJECTIVES

1 Discuss the role of promotion in the marketing mix

Promotion is communication by marketers that informs, persuades, and reminds potential buyers of a product in order to influence an opinion or elicit a response. Promotional strategy is a plan for the optimal use of the elements of promotion: advertising, personal selling, sales promotion, and public relations. The main function of promotion is to convince target customers that the goods and services offered provide a differential advantage over the competition.

2 Discuss the elements of the promotional mix

A combination of the various promotion tools is called the promotional mix. The four major tools that make up the promotional mix include advertising, personal selling, sales promotion, and public relations.

3 Describe the communication process

The communication process begins when the sender has a thought or idea and wants to share it with one or more receivers. The source then encodes this message and sends it, via a channel, to the receiver(s) for decoding. In turn, the sender receives feedback from the receiver(s) as to whether the message was understood.

4 Explain the goals and tasks of promotion

Promotion seeks to modify behavior and thoughts in some way and to reinforce existing behavior. There are three tasks for promotion: informing, persuading, and reminding the target market.

Informative promotion explains a good's or service's purpose and benefits. Promotions that inform the consumer are typically used to increase demand for a general product category or introduce a new good or service.

Persuasive promotion is designed to stimulate a purchase or an action. Promotions that persuade the consumer to buy are essential during the growth stage of the product life cycle, when competition becomes fierce.

Reminder promotion is used to keep the product and brand name in the public's mind. Promotions that remind are generally used during the maturity stage of the product life cycle.

5 Discuss the AIDA concept and its relationship to the promotional mix

The hierarchy of effects model outlines the four basic stages in the purchase decision-making process: awareness, interest, desire, and action. These stages are initiated and propelled by promotional activities. The promotional mix needs to recognize and fit the customer's stage in the AIDA model.

6 Describe the factors that affect the promotional mix

Many factors can affect the promotional mix. These factors include the nature of the product, product life cycle stage, target market characteristics, the type of buying decision involved, the availability of funds, and the feasibility of either a push or pull strategy.

7 Describe personal selling

Personal selling allows salespeople to thoroughly explain and demonstrate a good or service. Salespeople have the flexibility to tailor a sales pitch to the particular needs and preferences of individual customers. Personal selling is more efficient than some other promotion methods because salespeople target qualified prospects and avoid wasting effort on unlikely buyers. Personal selling affords greater managerial control over promotion costs. Finally, personal selling is the most effective method of closing a sale.

8 Discuss the key differences between relationship selling and traditional selling

Relationship selling is the practice of building, maintaining and enhancing interactions with customers in order to develop long-term satisfaction through mutually beneficial partnerships. Traditional selling, on the other hand, is transaction-focused. That is, the salesperson is most concerned with making one-time sales and moving on to the next prospect. Salespeople practicing relationship selling spend more time understanding a prospect's needs and developing solutions to meet those needs.

9 List the steps in the selling process

The selling process consists of seven basic steps:

1. Generating sales leads
2. Qualifying sales leads
3. Making the sales approach
4. Making the sales presentation
5. handling objections
6. Closing the sale
7. Following up

10 Describe the functions of sales management

Sales managers set overall company sales objectives and individual salespeople's quotas. They establish a sales force structure using geographic-, product-, function-, or market-oriented variables. Size of the sales force, compensation plans, recruiting, training, motivation, and evaluation are also functions of the sales manager.

PRETEST

Answer the following questions to see how well you understand the material. Re-take it after you review to check yourself.

1. List and briefly describe four elements of the promotion mix.

2. List and briefly describe the four steps of the communication process.

3. What is AIDA?

4. List the seven steps of the selling process.

CHAPTER OUTLINE

1 Discuss the role of promotion in the marketing mix

I. The Role of Promotion in the Marketing Mix

Promotion is communication by marketers that informs, persuades, and reminds potential buyers of a product in order to influence an opinion or elicit a response.

A. **Promotional strategy** is a coordinated plan for the optimal use of the elements of promotion: advertising, personal selling, sales promotion, and public relations.

B. The main function of promotion is to convince target customers that the goods and services offered provide a differential advantage over the competition.

C. A **differential advantage** is a set of unique features of a company and its products that are perceived by the target market as significant and superior to the competition

2 Discuss the elements of the promotional mix

II. The Promotional Mix

A combination of the various promotional tools is called the **promotional mix**. It includes advertising, personal selling, sales promotion, and public relations.

A. Personal Selling

Personal selling is a situation in which two people communicate in an attempt to influence each other in a purchase situation.

1. Traditional methods of personal selling include a planned, face-to-face presentation to one or more prospective buyers for the purpose of making a sale.

2. More current notions on the subject of personal selling emphasize the relationship that develops between a salesperson and a buyer.

B. Advertising

Advertising is any form of paid communication in which the sponsor or company is identified. It may be transmitted via many different media and does not provide direct feedback.

1. Cost per contact is very low because advertising can reach such a large number of people.

2. The total cost to advertise, however, is typically very high

C. Sales Promotion

Sales promotion consists of all marketing activities—other than personal selling, advertising, and public relations—that stimulate consumer purchasing and dealer effectiveness.

1. Promotion is used to improve the effectiveness of other ingredients in the promotional mix.

2. Sales promotion can be aimed at either the end consumers, trade customers, or a company's employees. Examples include free samples, contests, bonuses, trade shows, and coupons.

D. Public Relations

Public relations is the marketing function that evaluates public attitudes, identifies areas within the organization that the public may be interested in, and executes a program of action to earn public understanding and acceptance.

1. A good public relations program can generate favorable publicity.

2. **Publicity** is public information about a company, good, or service appearing in the mass media as a news item. Although publicity does not require paid media space or time, a cost is generated by public relations employees who organize and distribute news.

3 Describe the communication process

III. Marketing Communication

Communication is the process by which we exchange or share meanings through a common set of symbols.

A. Categories of Communication

1. **Interpersonal communication** is direct, face-to-face communication between two or more people.

2. **Mass communication** refers to communicating to large audiences, usually through a mass medium such as television or newspaper.

3. For effective communication between two communicators (sender and receiver), common understanding or overlapping frames of reference are required.

B. The Communication Process

The communication process involves both a sender and a receiver, and begins when the sender has a thought or idea and wants to share it with one or more receivers.

1. The **sender** is the originator, or source, of the message in the communication process.

   Encoding is the conversion of the sender's ideas and thoughts into a message, usually in the form of words or signs.

2. Transmission of a message requires a **channel**, such as a voice, gesture, newspaper, or other communication medium.

   Although transmitted, the message may not be received by the desired target audience. **Noise** is anything that interferes with, distorts, or slows the transmission of information.

3. The **receiver** is the person who decodes the message.

   a. **Decoding** is the interpretation of the language and symbols sent by the source through a channel.

   b. Even though the message is received, it may not be properly understood and decoded.

4. In interpersonal communication, the receiver's response to a message is direct **feedback** to the source.

   Mass communicators are cut off from direct feedback and must rely on market research or sales trends for indirect feedback.

C. The Communication Process and the Promotional Mix

The four elements in the promotional mix differ in their ability to affect the target audience. The elements also differ in how they interact with the communication process.

D. Integrated Marketing Communications

1. Companies are adopting the concept of **integrated marketing communications (IMC)**.

2. IMC is the method of carefully coordinating all the promotional activities to product a consistent, unified message that is customer-focused.

3. Marketing managers work out the roles of the various promotional elements in the marketing mix and monitor results.

4 Explain the goals and tasks of promotion

IV. The Goals and Tasks of Promotion

Promotion seeks to modify behavior and thoughts in some way and to reinforce existing behavior. Thus, the goal of promotion is to *inform*, *persuade*, and *remind*.

A. Informative promotion is generally more prevalent during the early stages of the product life cycle, when it can increase demand for a product category.

1. Informative promotion explains the purpose and benefits of a good or service.

2. More complex products often require informative promotion that explains technical benefits.

B. Persuasion, the second promotional task, is simply attempting to motivate a consumer to purchase or use more of a product.

1. Persuasion normally becomes the primary promotion goal when the product enters the growth stage of the product life cycle.

2. The aim of persuasion is to convince the customer to buy the company's brand rather than the competitor's.

C. Reminder promotion is used to keep the product and brand name in the public's mind.

1. This type of advertising is common during the maturity stage of the product life cycle.

2. The purpose of these ads is to trigger memory.

5 Discuss the AIDA concept and its relationship to the promotional mix

V. Promotional Goals and the AIDA Concept

A. A classic model for reaching promotional goals is called the **AIDA concept**, standing for Attention-Interest-Desire-Action. It outlines the stages of consumer involvement with a promotional message.

B. The AIDA concept assumes that promotion propels consumers along four steps in the purchase-decision process.

1. The advertiser must first make the target market *aware* that the product exists.

2. The next step is to create an *interest* in the product.

3. The *desire* to purchase the product is the third step in the process.

4. *Action* is the final step in the purchase decision process.

The promoter's task is to determine where on the purchase ladder most of the target consumers are located and design a promotion plan to meet their needs.

C. AIDA and the Promotional Mix

Each promotional tool is more effective at certain stages of the hierarchy of effects model.

1. Advertising is most effective in creating awareness.

2. Personal selling is most effective at creating customer interest for a product and for creating desire.

3. Sales promotion is most effective in creating strong desire and purchase intent.

4. Public relations has the greatest effect in building awareness about the company, good, or service.

6 Describe the factors that affect the promotional mix

VI. Factors Affecting the Promotional Mix

A. Nature of the Product

1. The characteristics of the product influences the promotional mix.

a. Industrial or business products are often expensive, complex, and customized, requiring personal selling.

b. Consumer products are promoted mainly through advertising to create brand familiarity.

2. As the costs and risks of the product increase, so does the need for personal selling. Risks can be financial or social.

B. Stage in the Product Life Cycle

The product's stage in its life cycle can also affect the promotional mix.

1. During the *introduction* stage in the product life cycle, both advertising and publicity are very important in informing the target audience that the product is available.

2. During the *growth stage* of the product life cycle, the promotional strategy is designed to build and maintain brand loyalty. Advertising and public relations continue to be major elements, but sales promotion can be reduced.

3. As a product reaches the *maturity stage* of its life cycle, increased competition mandates the emphasis of persuasive and reminder advertising and the increased focus on sales promotion.

4. During the *decline stage*, personal selling and sales promotion may be maintained but other forms of promotion, especially advertising, are reduced.

C. Target Market Characteristics

The characteristics of the target market influence the blend of promotion tools.

Widely scattered markets, highly informed buyers, and brand-loyal repeat purchasers generally respond to a blend of advertising and sales promotion with less personal selling.

D. Type of Buying Decision

The type of buying decision—whether routine or complex—also affects the promotional mix.

1. Advertising and sales promotion are most effective for routine decisions.

2. For decisions which are neither routine nor complex, advertising and public relations help establish awareness.

3. Personal selling is used in complex buying situations.

E. Available Funds

1. Available (or unavailable) funds may be the most important factor in determining the promotional mix.

2. A lack of money may force a firm to rely on publicity or commission-only manufacturers' sales agents.

3. When funds are available a firm will generally try to optimize its return on promotion dollars while minimizing the *cost per contact*, the cost of reaching one member of the target market.

F. Push and Pull Strategies

1. Some manufacturers employ a **push strategy**, which uses aggressive personal selling and trade advertising to convince wholesalers and retailers to carry and sell the merchandise.

2. At the other extreme, a **pull strategy** stimulates consumer demand with consumer advertising and special promotions and thus obtains product distribution.

7 Describe personal selling

VII. Personal Selling

A. **Personal selling** is direct communication between a sales representative and one or more prospective purchasers, for the purpose of making a sale. This can be accomplished through a face-to-face, personal sales call or over the telephone, called telemarketing.

B. Advantages of Personal Selling

1. Personal selling can be used to provide a detailed explanation or demonstration of the product.

2. The message can be varied by the salesperson to fit the motivations and interests of each prospective customer.

3. Personal selling can be directed to specific qualified prospects.

4. Personal selling costs can be controlled by adjusting the size of the sales force.

5. Personal selling is most effective in obtaining a sale and gaining a satisfied customer.

C. Certain customer and product characteristics indicate that personal selling might work better than other forms of promotion. In general, personal selling is more important if the product has a high value, the product is custom-made, the product is technically complex, there are few customers, and customers are concentrated.

8 Discuss the key differences between relationship selling and traditional selling

VIII. Relationship Selling

A. Traditional selling, on the other hand, is transaction-focused. That is, the salesperson is most concerned with making one-time sales and moving on to the next prospect.

B. **Relationship selling**, or **consultive selling**, is a multi-stage process that emphasizes personalization and empathy as key ingredients in identifying prospects and developing them as long-term, satisfied customers.

Salespeople practicing relationship selling spend more time understanding a prospect's needs and developing solutions to meet those needs.

C. Advances in electronic commerce on the Internet are threatening the buyer-seller relationship. Purchasing supplies and materials through a computer screen negates the importance of the face-to-face encounter with the salesperson.

9 List the steps in the selling process

IX. Steps in the Selling Process

The **sales process** or **sales cycle** is the set of steps a salesperson goes through to sell a particular product or service.
The steps of selling follow the AIDA concept

Traditional selling and relationship selling follow the same basic steps. The difference is the relative importance placed on key steps.

A. Generating Leads

1. **Lead generation**, or **prospecting**, is the identification of those firms and people most likely to buy the seller's offerings.

2. Sales leads come from advertising and other media, favorable publicity, direct mail and telemarketing, cold calling, Internet Web sites, client referrals, salesperson networking, trade shows and conventions, and internal company records.

   a. A **referral** is a recommendation from a customer or business associate.

   b. **Networking** is finding out about potential clients from friends, business contacts, coworkers, acquaintances, and fellow members in professional and civic organizations.

   c. **Cold calling** occurs when the salesperson approaches potential buyers without any prior knowledge of the prospects' needs or financial status.

B. Qualifying Leads

The next step is **lead qualification**, which determines the prospects who have:

1. A recognized need

2. Buying power (ability and authority to purchase)

3. Receptivity and accessibility

Often the task of lead qualification is handled by a telemarketing group or a sales support person who *prequalifies* the lead for the salesperson.

With more and more companies setting up web sites on the Internet, qualifying online leads has also received some attention.

C. Approaching the Customer and Probing Needs

1. Prior to approaching the customer, the salesperson should learn as much as possible about the prospects' organization and its buyers. This process is called the **preapproach**.

2. During the approach, the salesperson's ultimate goal is to conduct a **needs assessment** in which he finds out as much as possible about the prospects' situation.

   a. The consultative salesperson must be an expert on his or her product or service.

   b. The salesperson should know more about the customer than they know themselves.

   c. The salesperson must know as much about the competitor's company and products as he or she knows about their own. *Competitive intelligence* includes identifying the competitors, their products and services, advantages and disadvantages, strengths and weaknesses.

   d. The salesperson should be involved in active research concerning the industry.

3. Creating a *customer profile* during the approach helps salespeople optimize their time and resources.

D. Developing and Proposing Solutions

A **sales proposal** is a written document or professional presentation that outlines how the company's product or service will meet or exceed the client's needs. The **sales presentation** is the face-to-face explanation of a product's benefits to a prospective buyer; it is the heart of the selling process.

1. The quality of both the sales proposal and presentation can make or break the sale.

2. The selling presentation can be enhanced by allowing customers to touch or hold the product, using visual aids, and emphasizing important selling points of the product.

3. Technology has become an important part to presenting solutions.

E. Handling Objections

Objections should be viewed positively as a request or need for more information. Anticipating objections is the best way to prepare for them.

F. Closing the Sale

1. At the end of the presentation, the salesperson attempts to close the sale. This requires skill and courage on the part of the salesperson.

2. **Negotiation** often plays a key role in the closing of the sale. The salesperson offers special concessions at the end of the selling process and uses it in closing the sale. Examples include price cuts, free installation, free service, or trials.

3. Accepted closing techniques may differ greatly from country to country.

4. Rarely is a sale closed on the first call.

G. Following Up

Most businesses rely on repeat sales, and repeat sales depend on thorough **follow-up**.

Salespeople must ensure that

1. Delivery schedules are met.

2. Goods or services perform as promised.

3. The buyer's employees are properly trained to use the products.

10 Describe the functions of sales management

X. Sales Management

Sales management is one of marketing's most critical areas. It has several important functions.

A. Defining Sales Objectives and the Sales Process

1. Overall sales force objectives are usually stated in terms of desired dollar sales volume, market share, or profit level.

2. Individual salespeople are also assigned objectives in terms of quotas. A **quota** is simply a statement of sales objectives, usually based on sales volume but sometimes including key accounts, new accounts, and specific products.

   A sales manager needs to formally define the specific procedures salespeople go through to do their jobs, examine the sales process in their business.

B. Determining the Sales Force Structure

1. Sales departments are most commonly organized by geographic regions, by product lines, by marketing function performed, by market or industry, or by individual client or account.

2. Market or industry based structure and key account structures are gaining in popularity with today's emphasis on relationship selling.

C. Compensating and Motivating the Sales Force

Compensation planning is one of the sales manager's toughest jobs.

a. The **straight commission** system provides salespeople with a specified percentage of their sales revenue as income. No compensation is received until a sale is made. This system encourages salespeople to spend as much time as possible selling and may make them reluctant to perform nonselling activities.

b. The **straight salary** system compensates salespeople with a stated salary regardless of sales productivity. It may provide little incentive to produce but is useful in sales situations that require spending a great deal of time on prospecting, doing paperwork, training customers, and performing other nonselling tasks.

c. *Combination systems* offer a base salary plus an incentive, usually a bonus based on sales. This system provides selling incentives while allowing managers to control the activities of their sales forces..

d. Sales incentives: recognition at ceremonies, premiums, awards, merchandise, vacations, and cash bonuses are often used to motivate salespersons.

e. An effective sales manager inspires his or her salespeople to achieve their goals through clear and enthusiastic communications.

D. Evaluating the sales force requires regular feedback. Typical performance measures include sales volume, contribution to profit, calls per order, sales or profits per call or percentage of calls achieving specific goals.

VOCABULARY PRACTICE

Fill in the blank(s) with the appropriate term or phrase from the alphabetized list of chapter key terms.

advertising
AIDA concept
channel
cold calling
communication
decoding
differential advantage
encoding
feedback
follow-up
integrated marketing communication (IMC)
interpersonal communication
lead generation
lead qualification
mass communication
needs assessment
negotiation
networking
noise
personal selling
preapproach
promotion
promotional mix
promotional strategy
publicity
public relations
pull strategy
push strategy
quota
receiver
relationship selling (consultative selling)
sales presentation
sales process (sales cycle)
sales promotion
sales proposal
sender
straight commission
straight salary

1 Discuss the role of promotion in the marketing mix

1. Communication by marketers that informs, persuades, and reminds potential buyers of a product is ______________________________. A plan for the optimal use of advertising, personal selling, sales promotion, and public relations is the ______________________________. This vital part of the marketing mix helps communicate a set of unique features of a company and its products that are superior to the competition; this is the company's ______________________________.

2 Discuss the elements of the promotional mix

2. A combination of promotional tools makes up the ______________________________, which refers to the amount of funds allocated to each promotional tool and to the managerial emphasis placed on each technique. There are four major tools that make up this combination. The first is a form of impersonal, one-way mass communication paid for by the sponsor and transmitted by many different media; this tool is ______________________________. A second tool involves a planned face-to-face presentation to one or more prospective purchasers for the purpose of making a sale; this tool is ______________________________. A third promotional tool includes marketing activities that stimulate consumer purchasing and dealer effectiveness; this tool is called ______________________________ and may include such activities as coupons, contests, bonuses, and samples. The fourth promotional tool performs functions such as evaluating public attitudes and executing programs that earn public understanding and acceptance; this tool is known as ______________________________, and a solid program utilizing this tool can generate favorable ______________________________, which is public information appearing in the mass media as a news item.

3 Describe the communication process

3. The process by which individuals exchange or share meanings through a common set of symbols is called ______________________________. This can be divided into two major categories. The direct, face-to-face category is called ______________________________, while the category directed to large audiences is called ______________________________.

4. The communication process begins with an originator, or source of a thought or idea, who is called the ______________________________. When these ideas and thoughts are converted into a message, usually in the form of words or signs, ______________________________ has taken place. Then the message is transmitted through a(n) ______________________________, such as a voice on a radio ad or a printed coupon. The message is communicated to a(n) ______________________________ who will interpret the language and symbols of the message. This process in known as ______________________________. When a receiver responds to a message, this gives direct ______________________________ to the source. Anything that interferes with, distorts, or slows this communication process is called ______________________________.

5. The method of carefully coordinating all promotional activities to produce a consistent, unified message that is customer focused is ____________________________.

5 Discuss the AIDA concept and its relationship to the promotional mix

6. One model for reaching promotional objectives is called the ____________________________, which is an acronym for a sequential process.

6 Describe the factors that affect the promotional mix

7. Manufacturers may use aggressive personal selling and trade advertising to convince a wholesaler or retailer to carry and sell their merchandise; this approach is known as a(n) ____________________________. Alternatively, the manufacturer may wish to promote heavily to the end consumer to stimulate demand; this approach is the ____________________________.

8 Discuss the key differences between relationship selling and traditional selling

8. The practice of building, maintaining, and enhancing interactions with customers in order to develop long-term satisfaction through mutually beneficial partnerships is called ____________________________.

9 List the steps in the selling process

9. The set of steps a salesperson goes through to sell a product or service is called the ____________________________.

10. The first step in the selling process is the identification of those firms and people most likely to buy the seller's offerings; this is called ____________________________ and can be achieved with several different methods. With one method, the salesperson approaches potential buyers without any knowledge of the prospects' needs or financial status. This unsolicited sales method is ____________________________. A salesperson can also use a recommendation, or ____________________________, from a customer or business associate. The method of using friends, business contacts, co-workers, and acquaintances as a means of meeting new people who could become potential clients is called ____________________________.

11. The second step in the selling process is ____________________________, which determines the prospects who have the authority to buy and can afford to pay for the product or service.

12. During the ____________________________, the salesperson learns as much as possible about the prospect's organization and its buyers. When determining the customer's specific needs and wants and the range of options a customer has for satisfying them, the salesperson in performing a(n) ____________________________.

13. Once the salesperson has identified customer needs, he/she will put together a written document that outlines how the companies product or service will meet the customer's needs, otherwise known as a(n)

_______________________________. Once this has been completed, the salesperson will make the _______________________________, which is the formal meeting with the client.

14. At the end of the sales presentation, the salesperson can close the sale if the prospect's objections have been handled properly. Sometimes a salesperson will withhold a special concession until the end of the selling process and use it in closing the sale; this strategy is called _______________________________.

15. One of the most important aspects of the selling process is the last step: the _______________________________. This activity can help repeat sales.

10 Describe the functions of sales management

16. The first task in sales force management is to set sales objectives. Individual salespeople may be assigned objectives in the form of a statement of sales volume, financial achievements, or account goals. This type of stated objective is a(n) _______________________________.

17. There are three basic compensation methods for salespeople. The first method provides salespeople with a specified percentage of their sales revenue, but no compensation is received until after the sale is made; this is a(n) _______________________________ system. The second method compensates people with a stated dollar figure regardless of sales productivity; this is the _______________________________ system. Finally, to achieve the advantages of both of the first two systems, a combination system can be used.

Check your answers to these questions before proceeding to the next section.

TRUE/FALSE QUESTIONS

Mark the statement **T** if it is true and **F** if it is false.

1 Discuss the role of promotion in the marketing mix

_____ 1. Communication by marketers with the intent of informing, persuading, and reminding potential buyers of a product is known as advertising.

2 Discuss the elements of the promotional mix

_____ 2. Zippy, Inc., uses trade magazines, top forty radio, and local newspapers for its promotions. Zippy uses mass communication.

3 Describe the communication process

_____ 3. The Addle Ad Agency conducted research on a popular television ad and found that people were so caught up in the flashy pictures and catchy music that the message in the ad was not noticed. This is an example of noise in the communications process.

_____ 4. Once an advertisement is seen by a member of the target market, the message is understood as intended.

4 Explain the goals and tasks of promotion

_____ 5. Promotion can have the basic tasks to inform, persuade, or remind the consumer about a product, even though different types of promotion may be used (personal selling, advertising, sales promotion, or public relations).

5 Discuss the AIDA concept and its relationship to the promotional mix

_____ 6. AIDA stands for Attention-Interest-Desire-Action.

_____ 7. Of the four types of promotion, sales promotion is the most effective at getting potential customers to purchase the product (the "action" of AIDA).

6 Describe the factors that affect the promotional mix

_____ 8. Tia is trying to develop a promotional mix for her firm's new product, a high-technology solar heating/cooling system. Because her product is complex and carries high economic risk, she should concentrate on personal selling.

_____ 9. Totter Toys uses personal selling and sales promotions to encourage intermediaries to carry and sell its toy products. Totter Toys is using a pull strategy.

9 List the steps in the selling process

_____ 10. Generating, qualifying, and approaching sales leads are all important steps in the personal-selling process, but the sales presentation is the heart of the process.

_____ 11. Seth is approaching the end of his sales presentation. He is prepared to offer free training and a 10% price discount if the client does not seem to want to make a purchase. Seth is using negotiation to close the sale.

_____ 12. The last selling duty of a salesperson is to close the sale.

10 Describe the functions of sales management

_____ 13. The Squinch Squeegee Company is using the workload approach to determine the number of salespeople it should hire. The company's ideal number of customers is 500, and each customer should be visited every month. Because Squinch's squeegees are not complex, the average time for each sales call is half an hour. If each salesperson can give 1,500 hours of selling time per year, Squinch should hire a total of two salespeople.

_____ 14. Edward earns a 20 percent commission on every sale he makes to supplement his base income of $20,000. This is an example of a straight commission system.

Check your answers to these questions before proceeding to the next section.

AGREE/DISAGREE QUESTIONS

For the following statements, indicate reasons why you may agree and disagree with the statement.

1. Advertising is the most important element of the promotional mix.

   Reason(s) to agree:

   Reason(s) to disagree:

2. Advertising alone cannot make customers complete the AIDA process.

   Reason(s) to agree:

   Reason(s) to disagree:

3. Personal selling is the opposite of true marketing.

   Reason(s) to agree:

   Reason(s) to disagree:

4. The selling process as described in the textbook cannot be applied to all products and selling situations.

   Reason(s) to agree:

   Reason(s) to disagree:

MULTIPLE-CHOICE QUESTIONS

Select the response that best answers the question, and write the corresponding letter in the space provided.

1 Discuss the role of promotion in the marketing mix

_____ 1. The Racer's Bicycle Company manufactures racing bicycles that are clearly superior to the competition in terms of weight, durability, and aerodynamics. Racer's promotion emphasizes this:
a. unique selling proposition
b. comparative advantage
c. special benefit
d. differential advantage
e. promotional theme

2 Discuss the elements of the promotional mix

_____ 2. A local exterminator uses its own fleet of cars to communicate its service. The cars are painted yellow and are made to resemble a mouse. This is a form of:
a. personal selling
b. advertising
c. sales promotion
d. public relations
e. publicity

_____ 3. Julie Gourmand is a chef in a new downtown restaurant. She has sent out press releases to the major local media and has invited food critics to dine in her restaurant. Julie is engaging in:
a. public relations
b. personal selling
c. sales promotion
d. advertising
e. publicity

_____ 4. Sally's Security Service has sent a representative to most of the large local corporations to introduce the firm and its services and to explain the rates for corporate customers for after-hours security. This is a form of:
a. mass communication
b. implicit communication
c. public relations
d. personal selling
e. telcmarketing

_____ 5. Most large warehouse membership clubs give away samples of the packaged foods sold in their stores. This is a form of:
a. public relations
b. personal selling
c. sales promotion
d. advertising
e. publicity

_____ 6. Mammoth Oil Company uses radio, television, magazine, and newspaper advertising for its promotion. This form of communication is known as:
a. reference
b. factual
c. mass
d. interpersonal
e. public

3 Describe the communication process

_____ 7. All of the following are noise in the communication process EXCEPT:
a. three competing ads on the same page of a magazine
b. music, flashing lights, and hot temperatures in the store dressing room
c. important news stories in the newspaper with bold headlines
d. two people with a shared frame of reference
e. a crying child and ringing telephone while you are watching television

_____ 8. PiperCo places its messages on local radio stations and in the yellow pages of the phone book. These communications are directed at the ___________, the person who will decode the message.
a. sender
b. communicator
c. encoder
d. channeler
e. receiver

_____ 9. Rad-Rays manufactures sunglasses targeted at college students. Its last advertising campaign was written by a retirement-age manager who did not attend college. The ad campaign generated poor response. This demonstrates how important it is to have:
a. common frames of reference
b. common social classes
c. several copywriters for advertisements
d. similar ethnic backgrounds
e. messages that are always pretested

4 Explain the goals and tasks of promotion

_____ 10. The Micro-Blaze is a surgeon's tool for microscopic eye surgery and has been on the market for eight years. The advertising agency is writing the ad copy for the Micro-Blaze and has decided on a(n) ___________ format because the product is technical.
a. informative
b. persuasive
c. reminder
d. talkative
e. influencer

_____ 11. Quinn's Quiches has started a new ad campaign aimed at changing negative perceptions of cholesterol-laden eggs (a primary ingredient in quiche). The ads state that eggs are a good source of protein and that Quinn's Quiches are preferred by famous athletes. The message ends by asking viewers to buy Quinn's Quiches for their next special dinner. This promotion has the task of:
a. informing
b. persuading
c. reminding
d. suppressing
e. rewarding

_____ 12. Crest toothpaste should use ___________ promotion to keep the brand name in the public's mind.
a. influence
b. reminder
c. informative
d. persuasive
e. humorous

_____ 13. Icellee's Ice Cream is in the maturity stage of the product life cycle. Icellee should probably use ____________ promotion.
- a. reminder
- b. amusement
- c. informative
- d. persuasive
- e. influence

5 Discuss the AIDA concept and its relationship to the promotional mix

_____ 14. AIDA stands for:
- a. attention, interest, desire, action
- b. attention, intention, desire, action
- c. awareness, interest, desire, action
- d. awareness, interest, demand, attention
- e. awareness, intention, desire, attention

_____ 15. You are the public relations manager for an environmentally friendly paint. You know that public relations will have its greatest impact in the ____________ stage of the AIDA model.
- a. awareness
- b. attention
- c. desire
- d. demand
- e. action

_____ 16. Displays in grocery stores, coupons, premiums, and trial-size packages are most useful at the ____________ stage the AIDA model.
- a. awareness
- b. attention
- c. desire
- d. demand
- e. action

6 Describe the factors that affect the promotional mix

_____ 17. As Freddie's Flour plans its promotion for its new oat-blend flours, all of the following factors can be expected to affect the promotional mix EXCEPT that it is:
- a. just being introduced
- b. targeted at restaurants that will use the flour for their baking
- c. a healthy product
- d. packaged in five-pound bags
- e. targeted at home users

_____ 18. Lubri-Care used to market its motor oils heavily to car owners. But after years of television and newspaper advertising, Lubri-Care has decided to focus on personal selling and sales promotion to auto supply stores and large discount stores such as Wal-Mart. This new strategy is a:
- a. sales promotion strategy
- b. foot-in-the-door strategy
- c. pull strategy
- d. push strategy
- e. public relations strategy

_____ 19. Chewees Gum uses television advertising, couponing, and free store samples to increase sales. This is a ___________ strategy.
a. push
b. pull
c. personal selling
d. reinforcing
e. publicity

7 Describe personal selling

_____ 20. The newly formed Industrial Intruder burglar alarm company needs more clients. Ike, the founder of the firm, has decided to spend three days per week visiting other businesses and introducing himself and his firm. Ike is engaged in:
a. personal selling
b. advertising
c. public relations
d. sales promotion
e. research and development

_____ 21. Brown's sells customized oil field drilling equipment in Texas and Oklahoma. You would expect Brown's to rely on ___________ to promote their products.
a. publicity
b. advertising
c. personal selling
d. sales promotion
e. word-of-mouth

9 List the steps in the selling process

_____ 22. The Chem-Gro Company has developed a new type of liquid chemical fertilizer for large lawn areas. The company has purchased mailing lists of landscape contractors, corporate lawn care services, and turf companies to send out brochures with a detachable card that can be sent in for more information. Chem-Gro is involved in:
a. qualification
b. the sales approach
c. a sales presentation
d. the contact procedure
e. lead generation

_____ 23. When Sammy began selling Singer sewing machines, he called friends, relatives, former business acquaintances, and members of his neighborhood club. He asked them if they knew anyone who was looking for a deal on a new sewing machine. This technique is known as:
a. networking
b. cold calling
c. quota driving
d. qualifying
e. following up

_____ 24. The SnoGo snowblower salesperson is demonstrating how the snowblower works, letting the potential customer try it personally. This is the ___________ stage of the sale.
a. leading
b. follow-up
c. sales presentation
d. sales approach
e. closing

_____ 25. The employees at Compu-Consulting focus on solving problems for their clients rather than trying to sell more services. They concentrate on finding a variety of quality suppliers of computers, equipment, office supplies, and training programs, depending upon the client's needs. This is a(n) ____________ approach.

a. stimulus-response
b. need-satisfaction
c. canned
d. prepared
e. adaptive

_____ 26. As Sherri is trying to conclude the sale of the aluminum house siding system to the Renaud family, she finally offers them free gutter cleaning services and six pairs of window shutters for half-price. This is a closing technique called:

a. summative
b. assumptive
c. follow-up
d. adaptive
e. negotiation

10 Describe the functions of sales management

_____ 27. At the Elica Electronics Store, all salespeople are paid a percentage of their individual sales, and there is no guaranteed minimum pay. This is a(n):

a. straight commission plan
b. salary
c. hourly wage
d. combination pay plan
e. bonus plan

_____ 28. The salespeople at Victory Veterinarian Supply spend most of their time filling out information reports, calling on small customers, and dispensing product information. Because of this, Victory decided on a:

a. bonus system
b. percentage of sales plan
c. straight salary system
d. combination pay plan
e. straight commission

_____ 29. Perkle's Restaurant Supply would like to reduce its employee turnover and offer some incentive to sell more volume. The company is considering various pay plans, and you would recommend a(n):

a. combination pay plan
b. straight commission system
c. hourly wage
d. percentage of sales without salary
e. guaranteed salary

Check your answers to these questions before proceeding to the next section.

ESSAY QUESTIONS

1. The promotional mix is made up of a blend of four promotional tools. Name and describe each of these four tools and give specific examples of each tool.

2. Draw a diagram that illustrates the communication process. Then briefly describe each of the steps in the communication process.

3. The ultimate objective of any promotion is a purchase or some other activity. One model for reaching promotional objectives is called the AIDA concept. Using the stages of the AIDA model, describe the activities of a salesperson selling an executive cellular telephone.

4. The nature of the promotional mix depends upon several types of factors. Name five of these factors, and describe how changes in those factors affect the mix of promotional elements.

5. You are a sales representative for a large computer company that sells computers to business firms. Name the seven steps of the selling process and describe each step as it related to selling computers.

CHAPTER NOTES

Use this space to record notes on the topics you are having the most trouble understanding.

ANSWERS TO THE END-OF-CHAPTER DISCUSSION AND WRITING QUESTIONS

Use this space to work on the questions at the end of the chapter.

ANSWERS TO THE END-OF-CHAPTER CASE

Use this space to work on the questions at the end of the chapter.

CHAPTER 16 Advertising, Sales Promotion and Public Relations

LEARNING OBJECTIVES

1 Discuss the effect advertising has on market share, consumers, brand loyalty, and perception of product attributes

Advertising can help increase or maintain brand awareness and, subsequently, market share. New products use a proportionately large amount of advertising to build brand awareness. Brands with large market share tend to use advertising as a reminder mechanism to maintain that share. Advertising can seldom change strongly held consumer values, but it may transform a negative attitude toward a product into a more positive one. Advertising ultimately influences consumer purchases. Advertising of a brand may increase brand loyalty and increase purchases by brand-loyal customers.

Advertising can also change the importance of a brand's attributes to consumers. By emphasizing different brand attributes, advertisers can customize their appeal to respond to changing consumer needs or emphasize an advantage over competing brands.

2 Identify the major types of advertising

The major types of advertising are institutional advertising and product advertising. Institutional advertising is not product-oriented; rather its purpose is to foster a positive company image among the general public, investment community, customers, and employees. Product advertising is designed primarily to promote goods or services and is classified into three categories: pioneering, competitive, and comparative.

3 Describe the advertising campaign process

An advertising campaign is a series of related advertisements focusing on a common theme and common objectives. The advertising campaign process consists of several important steps. First, the campaign objectives are established. Next, creative decisions are made. Then various media are evaluated and selected. Finally, the overall campaign is assessed through various forms of testing.

4 Describe media evaluation and selection techniques

Managers select advertising media based on three main variables: characteristics of the target market, audience selectivity, and geographic selectivity. Media mix decisions usually depend on cost per thousand (CPM), reach, and frequency factors.

5 Define and state the objectives of sales promotion

The primary objectives of sales promotion are to increase trial purchasing of products, consumer inventories, and repurchasing. Promotion is also designed to support advertising activities.

6 Discuss the most common forms of consumer sales promotion

Consumer sales promotion includes coupons, premiums, contests and sweepstakes, sampling, and point-of-purchase displays.

7 List the most common forms of trade sales promotion

Trade sales promotion includes some consumer sales promotions, plus trade allowances, push money, training programs, free merchandise, store demonstrations, and meetings, conventions, and trade shows.

8 Discuss the role of public relations in the promotional mix

Public relations is a component of the promotional mix. A company fosters good publicity to enhance its image and promote its products. An equally important aspect of public relations is crisis management, or managing bad publicity in a way that is least detrimental to a firm's image.

PRETEST

Answer the following questions to see how well you understand the material. Re-take it after you review to check yourself.

1. Name and briefly describe three types of product advertising.

2. List the four steps in creating an advertising campaign.

3. List at least seven different types of media.

4. List six types of consumer sales promotion tools and six types of trade sales promotion tools.

5. List six public relations tools.

CHAPTER OUTLINE

1 Discuss the effect advertising has on market share, consumers, brand loyalty, and perception of product attributes

I. Effects of Advertising

Advertising is any form of impersonal, paid communication in which the sponsor or company is identified. The amount of advertising spending in the United States increases annually, with the estimated expenditures now exceeding $187 billion per year.

- Although the amount of advertising spending is quite large, the industry itself is very small.

- Only about 272,000 people are employed in the advertising departments of manufacturers, wholesalers, retailers, advertising agencies, and the various media services, such as radio, television, newspaper, magazines, and direct-mail firms.

- The five largest U.S. advertisers—Procter & Gamble, Philip Morris, and General Motors, each spend well over $2 billion annually on advertising.

A. Advertising and Market Share

1. New brands with a small market share tend to spend proportionately more for advertising and sales promotion than those with a large market share.

2. After a certain level of advertising is reached, diminishing returns set in. This phenomenon, known as the **advertising response function**, explains why well-known brands can spend proportionately less on advertising than new brands can.

3. Advertising requires a certain minimum level of exposure to measurably affect purchase habits.

B. Advertising and the Consumer

1. According to estimates, Americans are exposed to hundreds of advertisements a day from the various types of advertising media.

2. Attitudes and values are deeply rooted within an individual's psychological makeup. Advertising seldom succeeds in changing an attitude that stems from a person's moral code or culture. But advertising does attempt to change attitudes toward brands and to create an attitude toward the advertisement itself.

C. Advertising and Brand Loyalty

Advertising is used to reinforce and increase brand usage by regular purchasers of the product.

D. Advertising and Product Attributes

By emphasizing certain factors, advertising can affect the relative importance consumers place on various product attributes.

2 Identify the major types of advertising

II. Major Types of Advertising

The two major types of advertising are institutional advertising and product advertising. **Product advertising** touts the benefits of a specific good or service. **Institutional advertising** is used if the goal of the campaign is to build the image of the company or the industry.

A. Institutional Advertising

1. Institutional advertising has four important audiences: the public, the investment community, customers, and employees.

2. A unique form of institutional advertising called **advocacy advertising** is a way for corporations to express their views on controversial issues.

3. Many advocacy campaigns react to criticism or blame, media attacks, or impending legislation.

B. Product Advertising

1. **Pioneering advertising** is intended to stimulate primary demand for a new product or product category.

2. The goal of **competitive advertising** is to influence demand for a specific brand; it is often used when a product enters the growth phase of the product life cycle.

3. **Comparative advertising** compares two or more specifically named or shown competing brands on one or more specific product attributes.

   a. Advertisers often make taste, price, and preference claims in reference to the competition.

   b. Before the 1970s, advertisements were not allowed to show or identify the competition by name. But the Federal Trade Commission changed its stance to allow direct comparisons, prohibiting only false descriptions.

   c. These ads attract attention, enhance purchase intentions, increase consumers' message perception, generate positive attitudes, and increase personal relevance.

3 Describe the advertising campaign process

III. Steps in Creating an Advertising Campaign

The **advertising campaign** is a series of related advertisements for a particular product—focusing on a common theme, slogan, and set of advertising appeals—that extends for a defined time period.

A. Determine Campaign Objectives

1. The first step is to determine the **advertising objective**, the specific communication task a campaign should accomplish for a specified target audience during a specified period of time.

2. The DAGMAR approach (Defining Advertising Goals for Measured Advertising Results) is one method that stresses defining the objective as a percent of change.

B. Make Creative Decisions

1. Creative decisions cannot be completed without knowing which **medium**, or message channel, will be used to convey the message to the target market.

2. Marketers strive to identify product benefits, not product attributes, that will be the message to the consumers.

3. Marketing research and creative intuition are usually used to list the perceived benefits of a product and to rank these benefits.

4. The next step is to develop possible advertising appeals. An **advertising appeal** identifies a reason a person should purchase a product.

a. Advertising campaigns can focus on one or more appeals, which are developed by the creative people in the advertising agency.

b. Typical appeals are profit, health, love, fear, convenience, and fun.

5. The next step is to evaluate the proposed appeals. An appeal needs to be desirable, exclusive, and believable.

6. The dominant appeal for the campaign will be the **unique selling proposition**, and it usually becomes the campaign slogan.

7. The fourth and final step is executing the advertising message. Message execution is the way the advertisement will be portrayed. Examples of message execution style include fantasy, humor, demonstration, and slice of life.

a. Executional styles for foreign advertising are often quite different from those we are accustomed to in the United States.

b. Global advertising managers are increasingly concerned with the issue of standardization vs. customization.

4 Describe media evaluation and selection techniques

IV. Make Media Decisions

Advertisers spend over $187 billion annually on the various types of media advertising (national, regional, and local). Advertising media are channels that advertisers use in mass communication.

A. Six major types of advertising media are available: newspapers, magazines, radio, television, outdoor advertising, and the Internet. Marketers can also use alternative media to reach their target market.

1. Newspaper advertising has the advantage of geographic flexibility and timeliness. Newspapers reach a very broad mass market.

Cooperative advertising is an arrangement in which the manufacturer and retailer split the costs of advertising the manufacturer's brand.

2. Magazines are often targeted to a very narrow market. Although they may offer a very high cost per contact, the cost per potential customer may be much lower.

3. Radio can be directed to very specific audiences, has a large out-of-home audience, has low unit and production costs, is timely, and can have geographic flexibility.

4. Television can be divided into networks (ABC, NBC, CBS, and Fox), independent stations, cable channels, and direct broadcast satellite television.

   a. Cable television is the largest growth market; over two thirds of all U.S. households subscribe to cable. It can be useful for reaching specific markets.

   b. Television reaches a huge market, but both the advertising time and production costs are very expensive.

   c. The **infomercial** is a thirty minute or longer advertisement. They are popular because of the cheap air time and the relatively small production costs.

5. Outdoor advertising is a flexible, low-cost medium that may take a variety of forms, such as billboards, skywriting, ads in and on modes of transportation, and so on. It reaches a broad and diverse market.

6. With ad revenue approaching $1 billion, the Internet has established itself as a solid advertising medium.

   a. Popular Internet sites and search engines generally sell advertising space, called "banners." Banner ads are judged to be as or more effective for boosting brand and advertising awareness.

   b. The challenges posed by online media include more consumer control and measuring effectiveness.

7. Advertisers are looking for new media vehicles to help promote their products. Some of these include facsimile (fax) machines, video shopping carts, electronic "place-based" media, interactive computer advertising, and cinema and video advertising.

B. Promotional objectives and the type of advertising a company plans to use strongly affect the selection of media.
Several methods exist for evaluating the media mix.

1. The **media mix** is the combination of media to be used for a promotional campaign. Media decisions are typically based on cost per thousand, reach, and frequency.

   a. **Cost per contact** is the cost of reaching one member of the target market.

   b. **Reach** is the number of target consumers exposed to a commercial at least once over a period of time, such as four weeks.

   c. **Frequency** measures the intensity of coverage in a specific medium. Frequency is the number of times an individual is exposed to a brand message during a specific time period.

   d. Another consideration is **audience selectivity**, the medium's ability to reach a precisely defined market.

C. After selecting the media, a **media schedule**—which designates the vehicles, the specific publications or programs, and the dates and times—must be set. There are three basic types of media schedules:

1. With a **continuous media schedule**, advertising runs steadily throughout the advertising period.

2. With a **flighted media schedule**, the advertiser schedules ads heavily every other time period (such as every other month or every two weeks).

3. A **pulsing media schedule** combines continuous scheduling with flighting, resulting in a base advertising level with heavier periods of advertising.

4. A **seasonal media schedule** is used for products that are sold more during certain times of the year.

D. Evaluating an advertising campaign can be the most demanding task facing advertisers.

5 Define and state the objectives of sales promotion

V. Sales Promotion

Sales promotion is an activity in which a short-term incentive is offered to consumers or channel members to induce the purchase of a particular good or service.

Consumer sales promotion is aimed at the ultimate consumer of a good. **Trade sales promotion** is directed to members of the marketing channel, such as wholesalers and retailers.

A. The objectives of sales promotion center around immediate purchase. Specific objectives may be to

1. Increase trial

2. Boost consumer inventory

3. Encourage repurchase

4. Support and increase the effectiveness of advertising

B. Sales promotion may also encourage brand switching in some instances (coupons) and brand loyalty in others (*frequent-buyer clubs*).

6 Discuss the most common forms of consumer sales promotion

VI. Tools for Consumer Sales Promotion

Consumer sales promotion tools are used to create new users of the product, as well as to entice current customers to buy more.

A. A **coupon** is a certificate that entitles consumers to an immediate price reduction when they purchase the item.

1. Coupons are effective for product trial and repeat purchase. They are also useful for increasing the amount of product a customer will buy.

2. Approximately 268 billion coupons are distributed through free-standing newspaper inserts annually, but only about two percent are used.

3. In-store couponing is most likely to affect customer-buying decisions.

4. **Rebates** offer the purchaser a price reduction but the reward is not as immediate as the rebate form and proof-of-purchase must be mailed in.

B. A **premium** is an extra item offered, usually with proof of purchase, to the consumer. Sometimes it may be a small item, such as a T-shirt or coffee mug, or it may be free air travel or hotel stays. Frequent-buyer clubs and programs offer premiums.

C. **Loyalty marketing programs** or **frequent-buyer programs**, reward loyal consumers for making multiple purchases.

D. Contests and sweepstakes are promotions that give away prizes and awards. A *contest* is based on some skill or ability, but *sweepstakes* rely on chance and luck.

E. **Sampling** is a way to reduce the amount of risk a consumer perceives in trying a new product. For sampling to be beneficial, the new product must have benefits that are clearly superior to existing products and must have a unique new attribute that the consumer must experience to believe in.

Distributing samples to specific location types where consumers regularly meet for a common objective or interest, is one of the most cost-efficient methods of sampling.

F. A **point-of-purchase display** is a promotional display set up at the retailer's location to build traffic, advertise the product, or induce impulse buying.

Point-of-purchase displays include shelf talkers, shelf extenders, ads on grocery carts and bags, end-aisle and floor-stand displays, in-store audio messages, and audiovisual displays.

7 List the most common forms of trade sales promotion

VII. Tools for Trade Sales Promotion

Trade promotions push a product through the distribution channel. When selling to members of the distribution channel, manufacturers use many of the same sales promotion tools used in consumer promotions. Several tools, however, are unique to intermediaries.

A. A **trade allowance** is a price reduction offered by manufacturers to intermediaries, such as wholesalers or retailers, in exchange for the performance of specified promotion activities.

B. Intermediaries receive **push money** as a bonus for pushing the manufacturer's brand. Often the push money is directed at the retailer's salespeople.

C. Sometimes a manufacturer will provide free training for the personnel of an intermediary if the product is complex.

D. Another trade promotion is free merchandise offered in lieu of quantity discounts.

E. In-store demonstrations are sometimes provided by manufacturers as a sales promotion for retailers.

F. Trade association meetings, conferences, and conventions are an important aspect of trade promotion and an opportunity to meet and interact with current and potential customers.

8 Discuss the role of public relations in the promotional mix

VIII. Public Relations

Public relations evaluates public attitudes, identifies issues that may elicit public concern, and executes programs to gain public understanding and acceptance.

Publicity is the effort to capture media attention.

Public relations campaigns strive to achieve and maintain a corporation's positive image in the eyes of the public.

- The first task of public relations management is to set objectives that fit with the corporation's overall marketing program.
- Public relations tools include press relations, product publicity, corporate communication (internal and external), public affairs, lobbying, employee and investor relations, and crisis management.

A. Major Public Relations Tools

1 New Product Publicity

Public relations can help differentiate new products by creating free news stories about the product and its uses, garnering valuable exposure.

2. Product Placement

Marketers can garner publicity by making sure their products appear at special events or in movies and on television shows.

3. Consumer Education

Free seminars and demonstrations help develop more knowledgeable and loyal consumers.

4. Event Sponsorship

Sponsoring events and charities is a popular method of getting positive exposure for the company or a product.

5. Issue Sponsorship

Companies also build awareness and loyalty by supporting their customers' favorite issues.

6. Internet Web Sites

These sites are an excellent vehicle to post news releases on products, product enhancements, strategic relationships, and financial earnings. It can also be site for feedback and a self-help desk.

C. Managing Unfavorable Publicity

1. **Crisis management** is used by public relations managers to handle the effects of bad publicity.

2. It is imperative that a firm have plans for fast and accurate communication in times of emergency.

VOCABULARY PRACTICE

Fill in the blank(s) with the appropriate term or phrase from the alphabetized list of chapter key terms.

advertising appeal
advertising campaign
advertising objective
advertising response function
advocacy advertising
audience selectivity
comparative advertising
competitive advertising
consumer sales promotion
continuous media schedule
cooperative advertising
cost per contact
coupon
crisis management
flighted media schedule
frequent buyer program
infomercial
institutional advertising
loyalty marketing program
media mix
media schedule
medium
pioneering advertising
point-of-purchase display
premium
product advertising
pulsing media schedule
push money
reach
rebate
sampling
seasonal media schedule
trade allowance
trade sales promotion
unique selling proposition

1 Discuss the effect advertising has on market share, consumers, brand loyalty, and perception of product attributes

1. After a certain level of advertising is reached, diminishing returns set in. This is known as the ______________________________.

2 Identify the major types of advertising

2. If the goal of an advertising campaign is to build the image of the company or the industry, the advertiser will use ______________________________. A unique form of this advertising is a means for corporations to express their viewpoints on various controversial issues; this is ______________________________. In contrast, if the advertiser wishes to enhance the sales of a specific product, brand, or service, ______________________________ is used, which can be implemented in one of three forms. Advertising intended to stimulate primary demand for a new product or product category is

____________________________. Advertising used to influence demand for a specific brand of a good or service is ____________________________. Finally, if the advertising compares two or more specifically named or shown competing brands on one or more specific product attributes, ____________________________ is being used.

3 Describe the advertising campaign process

3. A series of related advertisements focusing on a common theme, slogan, and a set of advertising appeals is the ____________________________. The first step in developing this series is to set goals for specific communications tasks that should be accomplished for a specified target audience during a specified period of time. These goals are ____________________________.

4. To make creative decisions, the marketer must know which message channel, or ____________________________, will be used.

5. Part of making creative decisions involves identifying a reason a person should purchase a product or service; this is the ____________________________ and could take the form of a profit motive, fear, convenience, or concern for health. The dominant form chosen becomes the ____________________________, which then becomes the campaign slogan.

4 Describe media evaluation and selection techniques

6. An arrangement under which the manufacturer and the retailer split the cost of advertising the manufacturer's brand is called ____________________________.

7. A relatively new version of direct-response advertising is in the form of a thirty-minute extended television promotion that resembles a talk show. This is a(n) ____________________________.

8. The cost of reaching one member of a market with advertising is called ____________________________. Another factor that marketing managers evaluate is the volume of advertising that will be conducted with various media, or the ____________________________. The degree of coverage of a total audience through a particular medium is ____________________________, or the number of different target consumers exposed to a commercial at least once during a specific period. The intensity of a specific medium coverage is measured by ____________________________; this is the number of times an individual is exposed to a given message during a specific time period. The ability of a medium to reach a precisely defined market is ____________________________.

9. The designation of vehicles, specific publications or programs, and the insertion dates of the advertising is the ____________________________, which has several basic types. Products that are advertised on a reminder basis use a(n) ____________________________, which allows the advertising to run steadily throughout the advertising period. If the advertiser schedules ads heavily every other month to achieve a certain impact with increased frequency and reach, a(n) ____________________________ has been used. A variation of this is ____________________________, which combines continuous scheduling with flighting. Finally,

certain times of the year, such as holidays, the flu season, or summertime call for a(n) ______________________.

5 Define and state the objectives of sales promotion

10. A promotional activity in which a short-term incentive is offered to a consumer to induce the purchase of a particular product or service is ______________________. If this is targeted at intermediaries or channel members, it is known as ______________________.

6 Discuss the most common forms of consumer sales promotion

11. There are many forms of sales promotion. A certificate given to consumers entitling them to an immediate price reduction when they purchase the item is a(n) ______________________. A gift item offered, usually with a proof-of-purchase, to the consumer is a(n) ______________________. If consumers are rewarded for brand loyalty, a ______________________ is being used. One of the fastest forms of this type of sales promotion is when consumers are rewarded for making multiple purchases, also known as a(n) ______________________. When a consumer can try a product for free, ______________________ is being used. Finally, a promotional display set up at the retailer's location to build traffic, advertise the product, or induce impulse buying is a(n) ______________________.

7 List the most common forms of trade sales promotion

12. A price reduction offered by manufacturers to intermediaries is a(n) ______________________. If intermediaries receive a bonus for moving the product through the distribution channel, the intermediary is receiving ______________________.

8 Discuss the role of public relations in the promotional mix

13. When a company receives bad publicity, public relations executives handle the effects of the bad publicity with ______________________.

Check your answers to these questions before proceeding to the next section.

TRUE/FALSE QUESTIONS

Mark the statement **T** if it is true and **F** if it is false.

1 Discuss the effect advertising has on market share, consumers, brand loyalty, and perception of product attributes

_____ 1. Georgia has noticed that the brand that she manages has started to experience declining sales, even though her firm has sustained a high level of advertising. This decline can be explained by the advertising response function.

2 Identify the major types of advertising

_____ 2. A small high-tech company has just launched a new writing pen that has a memory, much like a small computer. The advertising campaign focuses on the innovativeness of the new product. This is an example of pioneering advertising.

_____ 3. The advertisements for Dippy Chips include the names of the major competitors in the potato chip market. The ads depict kids evaluating the crunchiness of three competing potato chips. This is an example of competitive advertising.

3 Describe the advertising campaign process

_____ 4. A print ad depicts a picture of a frozen dessert accompanied by information about the dessert: "two servings, only 300 calories per serving, no cholesterol, and 100 percent of the USRDA of vitamins and minerals." This is an example of an advertisement selling the product's benefits.

_____ 5. An advertisement for an energy-efficient light bulb tells viewers that they will save $10 per month on their utility bills. This is an example of a profit motive appeal.

4 Describe media evaluation and selection techniques

_____ 6. The Tuffy Tools firm manufactures hand tools. Rather than pay for all its advertising by itself, the management of Tuffy Tools has decided to give each of the hardware stores that carries Tuffy Tools some money to be used for the retailer's advertisements. The understanding is that Tuffy Tools will be featured in the retailers' ads. This is an example of cooperative advertising.

_____ 7. A private college would like to evaluate how effectively its advertising dollars are being spent. The college's marketing department takes the total amount of advertising spending and divides it by the number of people who are exposed to the advertisements. The college is calculating the reach.

_____ 8. The two local newspapers in Dippyville reach very different audiences. Most farm workers read the *Daily Moo*, while most of the town's doctors and lawyers read the *Eye-Cue News*. Each of these newspapers displays high audience selectivity.

_____ 9. A local greeting card store only runs newspaper advertisements the week before major card-sending holidays, such as Valentine's Day, Mother's Day, Halloween, and New Year's. This is an example of a pulsing schedule.

5 Define and state the objectives of sales promotion

_____ 10. Northeast Airlines has a loyal customer base that is based on the airlines' consistently good service. Northeast should give price-off packages to its customers to encourage further loyalty.

6 Discuss the most common forms of consumer sales promotion

_____ 11. Crunchy-Worms cereal includes a free worm toy inside every box. Consumers can order the rest of the worm family by mailing the proofs-of-purchase from five boxes of cereal to the manufacturer for each additional worm. This is an example of a premium.

7 List the most common forms of trade sales promotion

_____ 12. At the annual HAI event, helicopter manufacturers and vendors of helicopter-related products gather together in a convention center and display their products in booths to customers and potential customers. This is an example of a trade show.

8 Discuss the role of public relations in the promotional mix

_____ 13. Publicity can help gain exposure for and position new products.

Check your answers to these questions before proceeding to the next section.

AGREE/DISAGREE QUESTIONS

For the following statements, indicate reasons why you may agree and disagree with the statement.

1. It is difficult to prove whether or not advertising has an impact on sales.

   Reason(s) to agree:

   Reason(s) to disagree:

2. There is never truth in advertising; it is inherently biased.

   Reason(s) to agree:

   Reason(s) to disagree:

3. Sales promotion should be used only conjunction with advertising.

   Reason(s) to agree:

   Reason(s) to disagree:

4. Public relations is the least important of all the elements of the promotion mix.

Reason(s) to agree:

Reason(s) to disagree:

MULTIPLE-CHOICE QUESTIONS

Select the response that best answers the question, and write the corresponding letter in the space provided.

1 Discuss the effect advertising has on market share, consumers, brand loyalty, and perception of product attributes

_____ 1. Pat used to work at a carpet store but now works at the Victory Toys Store. Pat is surprised at the high advertising-to-sales ratio at the store; advertising dollars are 11 percent of sales. This ratio is high because:
a. certain industries traditionally spend a large amount of sales dollars on advertising, usually because of product-related factors
b. Victory is outspending the rest of the industry in order to catch up in market share
c. Victory is an inefficient advertiser
d. all advertising work is done by agencies rather than by Victory
e. Victory is the industry leader and has to stay ahead of other toy stores

_____ 2. The Barbasol brand of shaving cream has a large market share but spends proportionally less on advertising than Zip, a shaving cream brand with a small market share. This is the case because:
a. certain industries have a practice of spending a small amount of dollars, relative to sales, on advertising
b. beyond a certain volume of promotion, diminishing returns set in
c. there is no minimum level of exposure for advertising to have an effect on sales
d. advertising will not stimulate economic growth for the industry
e. the firms with large market share do not understand the advertising-to-sales relationship

_____ 3. The new advertisements for Icy-Kreem focus on the fact that the product is fat-free and cholesterol-free. This advertising highlights:
a. changing a negative attitude
b. brand loyalty
c. product attributes
d. benefits
e. needs of the target markets

2 Identify the major types of advertising

_____ 4. A large oil company runs advertisements that show how its employees are involved in the community. This is an example of _______________ advertising.
a. pioneering
b. institutional
c. comparative
d. competitive
e. product

_____ 5. Future Power, Inc., has begun an advertising campaign to manufacturers. The campaign promotes the advantages of using alternative battery technologies (such as the new GaAs battery) for toys, computers, and electric cars. This is ___________ advertising.
a. comparative
b. innovative
c. institutional
d. competitive
e. pioneering

_____ 6. A rental car company with the second largest market share runs advertisements showing how its customer service is superior to that of the largest competitor. This is an example of ___________ advertising.
a. comparative
b. combat
c. competitive
d. institutional
e. pioneering

3 Describe the advertising campaign process

_____ 7. The first step in the advertising campaign decision process is to:
a. make media decisions
b. evaluate the campaign
c. determine campaign objectives
d. develop advertising copy
e. make creative decisions

_____ 8. A financial services advertisement targets busy professional parents by showing them in a variety of situations during a typical frantic day. This is an example of a ___________ execution.
a. spokesperson
b. demonstration
c. fantasy
d. lifestyle
e. product symbol

_____ 9. In the advertisement for a bathroom cleaner, animated, talkative scrubbing bubbles are used to show how hard the bubbles work to clean. The lively scrubbing bubbles appear in all advertisements for this product and are depicted on the packaging. This is an example of a ___________ executional style.
a. fantasy
b. lifestyle
c. spokesperson
d. product symbol
e. scientific evidence

4 Describe media evaluation and selection techniques

_____ 10. A lawn and garden store has an arrangement with a lawn mower manufacturer that 50 percent of the cost of all radio and newspaper advertisements placed by the store will be paid for by the manufacturer. This is ___________ advertising.
a. comparative
b. cooperative
c. institutional
d. corporate
e. competitive

_____ 11. The Flow-Bee hair cutting vacuum cleaner attachment is shown during a half-hour television spot. The product is demonstrated on volunteer studio audience members, testimonials are given by Flow-Bee owners, and viewers are encouraged to order the product through an 800 telephone number. Flow-Bee is using the ____________ form of direct-response advertising.

a. megamercial
b. ad expander
c. extended sales pitch
d. prolonged ad
e. infomercial

_____ 12. Peter's Perfumes is currently deciding how much space or time will be placed in each medium that the company has selected. Peter's Perfumes is determining its:

a. marketing mix
b. promotional plan
c. media mix
d. advertising campaign
e. reach objectives

_____ 13. The manufacturer of Furry's Ferret Food estimates that the product's new advertising campaign will reach 500,000 people and will cost $750,000 if the company uses newspaper ads, and $1 million if it uses television. The company's decision seems to rest on:

a. cost per contact
b. flexibility
c. noise level
d. life span
e. geographic selectivity

_____ 14. Because Calvin's Crystals wants to spend its promotional budget on advertisements that will have a long life span, it should use ____________ advertising.

a. magazine
b. radio
c. television
d. newspaper
e. outdoor

_____ 15. Kay's Catering has set up a seasonal plan for advertising, with dates selected in January, September, and December on three local radio stations, the city newspaper, and bus transit signs. This is a:

a. message execution plan
b. media profile
c. media schedule
d. reach program
e. frequency timetable

_____ 16. Ricky's Rib Restaurant runs ads for one week, every other month, in the entertainment section of the newspaper. No other forms of advertising are used. This is a ____________ scheduling plan.

a. continuous
b. flighting
c. pulsing
d. seasonal
e. duplication

5 Define and state the objectives of sales promotion

_____ 17. Sales promotion:

a. inspires long-term brand loyalty
b. is only directed to the ultimate consumer market
c. is more difficult to measure than advertising
d. is a smaller percentage of the promotion budget than advertising
e. offers a short-term incentive to buy

6 Discuss the most common forms of consumer sales promotion

_____ 18. Campbell's Soup is offering a wooden recipe box with soup recipes to consumers who send in five proofs-of-purchase. This is an example of a:
a. prize
b. coupon
c. premium
d. contest
e. free merchandise sample

_____ 19 The Dashing Dish Soap firm has decided to use free samples as a promotion technique because:
a. its dish soap is similar to other dish soaps
b. it is an inexpensive promotional tool
c. this is an increasingly popular technique with manufacturers
d. this allows the consumer to try the product risk-free
e. trial-size containers are a form of advertising

7 List the most common forms of trade sales promotion

_____ 20. The Marble Manufacturing firm has offered a price reduction to the Fun-n-Games chain of toy stores as long as the stores take responsibility for sorting and bagging marbles by color and size and for setting up special marble displays. The price reduction is also a:
a. sales discount
b. point-of-purchase discount
c. merchandise guideline
d. trade allowance
e. form of push money

8 Discuss the role of public relations in the promotional mix

_____ 21. Capital Finance has mailed its stockholders a brochure that included data from the firm on new customer services, management changes, and the firm's financial situation. This is an example of:
a. lobbying
b. corporate selling
c. a sales promotion
d. public relations
e. stockholder marketing

_____ 22. The Clearly Can Corporation sponsored a citywide litter pickup contest and sent all the proceeds from recycling the collected cans to a local charity. The event was so popular that the local newspaper took pictures for the front page, and every local news station carried a story about Clearly Can's community activities. The newspaper pictures and the news stories exemplify:
a. lobbying
b. a sales promotion
c. an attempt to build advertising credibility
d. public relations aimed at the community at large
e. publicity

Check your answers to these questions before proceeding to the next section.

ESSAY QUESTIONS

1. There are two major types of advertising. Name and define each of these two types, including special forms of each type.

2. You are responsible for developing several advertisements for Less-U, a fat substitute product with no calories to be used in place of butter, margarine, or shortening in cooking. Name five common advertising appeals. For each appeal, give an example of a Less-U advertisement message using that appeal. Then name six common advertising executional styles. Briefly describe how you would design six different Less-U advertisements (one for each executional style).

3. Advertising media are channels that advertisers use in mass communication. Name and describe the five basic media vehicles. Cite at least two advantages and two disadvantages of each basic medium in your descriptions. Then name and describe three examples of new media forms.

4. You are the sales promotion manager for Steri-Flor, a new brand of floor disinfectant/cleanser. Your company uses a pull strategy, and you are responsible for recommending sales promotion tools to accomplish this strategy. Name and define the consumer promotional tools, and for each tool describe specific sales promotion activities you would recommend for Steri-Flor.

5. You are the sales promotion manager for Steri-Flor, a new brand of floor disinfectant/cleanser. Your company uses a push strategy, and you are responsible for recommending sales promotion tools to accomplish this strategy. Name and define the promotional tools unique to the trade (not consumer sales promotion tools). For each tool describe specific sales promotion activities you would recommend for Steri-Flor.

6. You are the public relations manager for a chemical company. What are some common public relations tools you might use on a regular basis? What types of actions would you take if a crisis occurred (such as a chemical spill) and there was the possibility of unfavorable publicity?

CHAPTER NOTES

Use this space to record notes on the topics you are having the most trouble understanding.

ANSWERS TO THE END-OF-CHAPTER DISCUSSION AND WRITING QUESTIONS

Use this space to work on the questions at the end of the chapter.

ANSWERS TO THE END-OF-CHAPTER CASE

Use this space to work on the questions at the end of the chapter.

CHAPTER GUIDES and QUESTIONS

PART SIX

PRICING DECISIONS

CHAPTER 17 Pricing Concepts

LEARNING OBJECTIVES

1 Discuss the importance of pricing decisions to the economy and to the individual firm

Pricing allocates goods and services among consumers, governments, and businesses. Pricing is essential to the firm because it creates revenue, the basis of all business activity. In setting prices, marketing managers strive to find a level high enough to produce a satisfactory profit.

2 List and explain a variety of pricing objectives

Pricing objectives are commonly classified into three categories:

- Profit-oriented pricing aims for profit maximization, satisfactory level of profit, or target return on investment. With a goal of profit maximization, the firm tries to generate as much revenue as possible in relation to cost.
- Sales-oriented pricing aims to attain or maintain a certain market share or to maximize dollar or unit sales.
- Status quo pricing aims to match competitors' prices.

3 Explain the role of demand in price determination

Demand is a key determinant of price. A typical demand-price relationship is inverse—that is, when price is lowered, demand and sales increase; when price is increased, the quantity demanded falls. There is an exception in some prestige products, for which a price increase actually increases demand.

Marketing managers must consider demand elasticity when setting prices. Elasticity of demand is the degree to which quantity demanded fluctuates with changes in price. If consumers are sensitive to changes in price, demand is elastic. Conversely, if they are insensitive to prices changes, demand is inelastic. Thus an increase in price will result in lower sales for an elastic product and in little or no change in sales for an inelastic product.

4 Describe cost-oriented pricing strategies

Wholesalers and retailers commonly use markup pricing to cover their expenses and obtain a profit. In markup pricing an additional amount is added to the manufacturer's original price. Similar to markup pricing, formula pricing sets prices using a predetermined formula based on variable and fixed costs. Another pricing technique is to maximize profits by setting the price at a level in which marginal revenue (MR) equals marginal cost (MC). Some firms determine a break-even point in sales units and dollars and use this amount as a reference point for adjusting price. Finally, with target return pricing, a firm calculates the break-even point plus the additional amount that will equal its desired return on investment.

5 Demonstrate how the product life cycle, competition, distribution and promotion strategies, customer demand, the Internet and extranets, and perceptions of quality can affect price

The price of a product normally changes as it moves through the life cycle and as demand for the product and competitive conditions change. Management often sets a high price at the introductory stage of the life cycle and lowers the price over time. High prices tend to attract competition, and competition usually drives prices down as each firm tries to gain market share.

Adequate distribution for a new product can sometimes be obtained by offering a larger than usual profit margin to wholesalers and retailers. Price is also used as a promotional tool to attract customers. Special low prices often attract new shoppers and entice existing customers to buy more.

Quality perceptions can also influence pricing strategies. A firm trying to project a prestigious image often charges a premium price for a product. Consumers tend to equate high prices with high quality.

PRETEST

Answer the following questions to see how well you understand the material. Re-take it after you review to check yourself.

1. List and briefly describe three categories of pricing objectives.

2. Explain the difference between elastic demand, inelastic demand, and unitary elasticity. Describe their effect on pricing decisions.

3. List and briefly describe three pricing methods that use cost as a determinant.

4. List five alternative determinants of price.

CHAPTER OUTLINE

1 Discuss the importance of pricing decisions to the economy and to the individual firm

I. The Importance of Price

A. What Is Price?

1. **Price** is the perceived value of a good or service, most commonly expressed in dollars and cents. Price can also be expressed in terms of other goods.

2. Price is related to the perceived value at the time of the transaction and is based on the amount of expected satisfaction to be received from the good or service, not the actual satisfaction.

B. The Importance of Price to Marketing Managers

Prices are the key to company revenues.

1. Prices charged to customers multiplied by the number of units sold equals **revenue** for the firm. Revenue pays for every activity of the firm. Whatever is left over after paying for company activities is **profit**.

2. If a price is too high in the minds of consumers, sales will be lost. If a price is too low, revenues may not meet the company's goals for return on investment.

3. Trying to set the right price can be one of the most stressful and pressure-filled tasks for a marketing manager.

a. The high rate of new product introductions leads buyers to continually reevaluate the price of a new item against the value of existing products.

b. The increased availability of bargain-priced dealer and generic brands puts overall downward pressure on prices.

c. Many firms with a large market share try to maintain or regain share by cutting prices when competition heats up.

d. In the business-to-business market, buyers are becoming more efficient and better informed.

e. Competition in general is increasing.

2 List and explain a variety of pricing objectives

II. Pricing Objectives

Companies set pricing objectives that are specific, attainable, and measurable. These goals require periodic monitoring to determine the effectiveness of the strategy.

A. Profit-Oriented Pricing Objectives

1. *Profit maximization* means setting prices so total revenue is as large as possible relative to total costs for a given item. Competitors' prices and the product's perceived value mediate profit-oriented pricing.

2. Another goal of profit-oriented pricing is satisfactory profits, a reasonable level of profits that is satisfactory to stockholders and management.

3. The most common of the profit objectives is **target return on investment (ROI)**, sometimes called the firm's return on total assets. ROI measures the effectiveness of management in generating profits with its available assets:

$$\text{Return on investment} = \frac{\text{Net profit after taxes}}{\text{Total assets}}$$

B. Sales-Oriented Pricing Objectives

1. **Market share** refers to a company's product sales as a percentage of total sales for that industry. Market share can be expressed in dollars of sales or units of product. Comparisons vary depending on the terms used.

2. Rather than strive for market share, some companies try to maximize dollar or unit sales. A firm may use this strategy in an attempt to generate a maximum amount of cash in the short run or to sell off excess inventory, but this strategy may produce little or no profit.

C. Status Quo Pricing Objectives

Status quo pricing seeks to maintain existing prices or simply to meet the competition. This strategy requires little planning other than monitoring competitors' prices.

3 Explain the role of demand in price determination

III. The Demand Determinant of Price

The price that is set depends on not only pricing goals but also the demand for the good or service, the cost to the seller for that good or service, and other factors.

A. The Nature of Demand

1. **Demand** is the quantity of a product that will be sold in the market at various prices for a specified period. Ordinarily, the quantity demanded increases as the price decreases and decreases as the price increases.

   a. The *demand curve* graphs the demand for a product at various prices. The line usually curves down and to the right.

   b. The *demand schedule* is a chart that shows quantity demanded at selected prices.

2. **Supply** is the quantity of product that will be offered to the market by suppliers at various prices for a specified period. This is represented by the *supply curve*.

3. Competitive market prices are determined by a combination of supply and demand. The *supply schedule* show the amount of product suppliers will produce at different prices.

4. When supply and demand are equal, a state called **equilibrium** is achieved. At equilibrium there is no inclination for prices to rise or fall.

B. Elasticity of Demand

Elasticity of demand refers to consumers' responsiveness or sensitivity to changes in price.

1. **Elastic demand** occurs when consumers are sensitive to price changes, whereas **inelastic demand** means that an increase or decrease in price will not significantly affect demand for a product.

   a. If price goes down and revenue goes up, demand is elastic.

   b. If prices goes down and revenue goes down, demand is inelastic.

   c. If price goes up and revenue goes up, demand is inelastic.

   d. If price goes up and revenue goes down, demand is elastic.

   e. If price goes up or down and revenue remains the same, the demand elasticity is unitary.

   A marketing manager needs to know whether a product has elastic or inelastic demand to estimate the effect of a price change on sales.

2. **Unitary elasticity**, means an increase in sales exactly offsets a decrease in price so that total revenue remains the same.

3. Factors That Affect Elasticity

   Elasticity of demand is affected by

   a. The availability of substitute goods and services

   b. The price relative to a consumer's purchasing power

   c. Product durability

   d. The existence of other product uses

4 Describe cost-oriented pricing strategies

IV. The Cost Determinant of Price

Some companies price their products largely or solely on the basis of costs. This method can lead to overpricing and lost sales or to underpricing and lower returns on sales than necessary. Costs usually serve as a floor below which a good must not be priced.

A. Types of Costs

1. **Variable costs** are those that deviate with changes in the level of output—for example, the cost of materials.

2. **Fixed costs**, such as rent and executive salaries, do not change as output is increased or decreased.

3. It is helpful to calculate costs per unit or average costs. **Average variable cost (AVC)** is total variable costs divided by quantity of output. **Average total cost (ATC)** is total costs divided by output.

4. **Marginal cost** is the change in total costs associated with a one-unit change in output.

5. All these costs have a definite relationship and can be represented by curves on a cost-quantity graph.

B. Markup Pricing

Markup pricing, the most popular method used by wholesalers and retailers, is adding to cost an amount for profit and for expenses not previously accounted for. The total determines the selling price.

1. Retailers tend to discuss markup in terms of its percentage of the retail price. For example, an item that costs $1 and is marked up to $2 is marked up 50 percent from the retailer's point of view.

2. The maintained markup (or **gross margin**) is the difference between the cost and the final selling price.

3. **Keystoning** refers to markups that are double the cost.

4. The biggest appeal of markup pricing is its simplicity.

C. Profit Maximization Pricing

Profit maximization occurs when marginal revenue equals marginal cost.

1. **Marginal revenue (MR)** is the extra revenue associated with selling an additional unit of output.

2. As long as the revenue of the last unit produced and sold is greater than the cost of the last unit produced and sold, the firm should continue manufacturing.

D. Break-Even Pricing

1. **Break-even analysis** determines what sales volume must be reached for a product before the company breaks even and no profits are earned.

$$\text{Break-even quantity} = \frac{\text{Total fixed costs}}{\text{Fixed cost contribution}}$$

Fixed cost contribution (per unit) = Price per unit - Average variable cost

2. The advantage of break-even analysis is that a firm can quickly discover how much it must sell to cover costs and how much profit can be earned if higher sales volume is obtained. It is a simple formula requiring only cost information.

3. The disadvantages of break-even analysis include the fact that it ignores demand and that some costs are difficult to categorize as fixed or variable.

5 Demonstrate how the product life cycle, competition, distribution and promotion strategies, customer demands, the Internet and extranets and perceptions of quality can affect price

V. Other Determinants of Price

A. The Stage of the Product's Life Cycle

As the product moves through its life cycle, the demand for the product and the competitive conditions usually change, affecting price.

1. Prices are usually high during the introductory stage to recover development costs and take advantage of high demand originating in the core of the market.

2. Prices generally begin to lower and stabilize as the product enters the growth stage and competitors enter the market. Economies of scale lead to lower prices.

3. Maturity brings further price decreases, because competition is strong and weaker firms have been eliminated.

4. The decline stage may bring even more price decreases as firms attempt to salvage the last vestiges of demand. But when only one firm is left in the market, the prices may actually rise again as the product becomes a specialty good.

B. The Competition

1. High selling prices can attract other firms to enter a profitable market, usually at a slightly lower price.

2. When a firm enters a market, it has to decide whether to price at, below, or above market prices.

3. A firm can price its product below the market price to gain quick market share.

4. The new competitor can price above the market price if it has a distinct competitive advantage.

5. The new competitor may choose to enter a market at the "going market price" and avoid crippling price wars, assuming it will succeed through nonprice competition.

C. Distribution Strategy

1. Adequate distribution depends on convincing distributors to carry the product. This can be accomplished by:

a. Offering a larger-than-usual profit margin

b. Giving dealers a large trade allowance

2. Manufacturers have been losing control of the distribution channel to wholesalers and retailers. Some distributors engage in **selling against the brand** where well-known brands are priced higher than the distributor's own private-label brand.

3. Purchasing goods through unauthorized channels allows wholesalers and retailers to obtain higher-than-normal margins.

4. Manufacturers try to maintain price control by using:

a. Exclusive channels

b. Pre-priced packaging

D. The Impact of the Internet and Extranets

1. The Internet connects buyers and sellers, allowing buyers to quickly and easily compare product and prices while sellers are able to tailor products and prices.

2. Companies are creating private networks, or **extranets**, that link them with their suppliers and customers. This allows them to:

a. Monitor inventory costs and demand

b. Adjust prices instantly

E. Promotion Strategy

Price is often used as a promotional tool. Sales and coupons can increase consumer interest.

F. Demands on Large Customers

Large customers of manufacturers often make specific pricing demands

1. Guaranteed profit margins

2. Ticketing, packing, and shipping requirements

G. Relationship of Price to Quality

1. Consumers tend to rely on a high price as a predictor of good quality when they have great uncertainty about the purchase decision. This reliance on price as an indicator of quality seems to exist for all products, but exists more strongly for some items than for others.

2. Marketing managers can use high prices to enhance the image of their product in some cases. This is a **prestige pricing** strategy.

3. Consumers expect dealer or store brands to be cheaper than national brands. But if the savings are too great, consumers tend to believe that the dealer brand is inferior in quality.

VOCABULARY PRACTICE

Fill in the blank(s) with the appropriate term or phrase from the alphabetized list of chapter key terms.

average total cost (ATC)
average variable cost (AVC)
break-even analysis
demand
elastic demand
elasticity of demand
fixed cost
inelastic demand
keystoning
marginal cost (MC)
marginal revenue (MR)
market share
markup pricing
prestige pricing
price
price equilibrium
profit
profit maximization
return on investment (ROI)
revenue
selling against the brand
status quo pricing
supply
unitary elasticity
variable cost

1 Discuss the importance of pricing decisions to the economy and to the individual firm

1. The perceived value of a good or service that is exchanged for something else is ______________________________.

2. Prices charged to customers multiplied by the number of units sold equals the ______________________________ for the firm, which is the lifeblood of the organization. Whatever is left over of this (if anything) after paying for company operations is ______________________________.

2 List and explain a variety of pricing objectives

3. The most common of the profit objectives is sometimes called the firm's return on total assets. The ______________________________ measures the overall effectiveness of management in generating profits with its available assets. A sales-oriented pricing objective could be based on a company's product sales as a percentage of total sales for that industry, also known as ______________________________. Another pricing objective seeks to maintain the existing prices or simply meet the competition; this is a(n) ______________________________ objective.

3 Explain the role of demand in price determination

4. One determinant of price is the quantity of a product that will be sold in the market at various prices for a specified time period; this is ______________________________, and the quantity of a product that people will buy depends on its price. Another determinant of price is the quantity of a product that will be offered to the market by suppliers at various prices for a specified time period; this is the concept of ______________________________. When these two determinants of price are the same, a state called ______________________________ is achieved.

5. Consumers' responsiveness to prices or sensitivity to changes in prices refers to ______________________________. When consumers are sensitive to price changes, ______________________________ occurs. Conversely, when an increase or a decrease in price does not significantly affect demand, ______________________________ occurs. If an increase in sales exactly offsets the decrease in price so that total revenue remains the same, the situation is called ______________________________, which is a rare phenomenon.

4 Describe cost-oriented pricing strategies

6. Economic costs are important determinants of price. A cost that deviates with changes in the level of output is a(n) ______________________________. In contrast, a cost that does not change as output increases or decreases is a(n) ______________________________. There are several useful calculated cost variables. Total variable costs divided by output is the ______________________________. Total costs divided by output is the ______________________________. The change in total costs associated with a one-unit change in output is the ______________________________.

7. The most popular method used by wholesalers and retailers to establish a sales price is adding an amount to the cost for profit and expenses not previously accounted for; this is ______________________________. Markups are often based on experience. A standard markup for retailers may be 100 percent over cost, a tactic known as ______________________________.

8. The additional revenue associated with selling an additional unit of output is called ______________________________. When this is equal to marginal cost, ______________________________ occurs, which is one type of pricing objective.

9. To determine what sales volume must be reached for a product before total costs equal total revenue and no profits or losses are achieved, a marketer would use ______________________________.

5 Demonstrate how the product life cycle, competition, distribution and promotion strategies, and perceptions of quality can affect price

10. Stocking well-known brand items at high prices in order to sell store brands at discounted prices is known as ______________________________.

11. Charging a high price to help promote a high-quality image is called ______________________________.

Check your answers to these questions before proceeding to the next section.

TRUE/FALSE QUESTIONS

Mark the statement **T** if it is true and **F** if it is false.

1 Discuss the importance of pricing decisions to the economy and to the individual firm

_____ 1. Price is defined as what the customer pays for a good or service.

2 List and explain a variety of pricing objectives

_____ 2. Target return on investment (ROI) is the most common pricing objective used by firms.

_____ 3. PorterCo has a goal of cash maximization, which is an appropriate long-term pricing objective for most firms.

_____ 4. Status quo pricing objectives indicate that prices may change because prices are adjusted to meet the competition.

3 Explain the role of demand in price determination

_____ 5. When demand for a product has unitary elasticity, a firm will lose revenue if it decreases the price.

_____ 6. For a product like butter, the price is small relative to a consumer's purchasing power, and there are several alternate uses for the product. If this information is true, butter probably has elastic demand.

4 Describe cost-oriented pricing strategies

_____ 7. Markup pricing is one of the most common pricing methods used by intermediaries.

_____ 8. Break-even analysis determines what sales volume must be reached for a product before the company's total revenue equals total costs.

5 Demonstrate how the product life cycle, competition, distribution and promotion strategies, and perceptions of quality can affect price

_____ 9. When a firm enters an industry in which products are in the maturity phase of the product life cycle, the firm generally faces a decision of whether to price at the market level or above the market.

_____ 10. Consumers use price as an indicator of the quality of a product, especially when consumers have a lot of knowledge about the product.

Check your answers to these questions before proceeding to the next section.

AGREE/DISAGREE QUESTIONS

For the following statements, indicate reasons why you may agree and disagree with the statement.

1. Of all the possible pricing objectives, profit maximization is the most important.

   Reason(s) to agree:

   Reason(s) to disagree:

2. Unitary elasticity is just theoretical; in practice, demand must either be elastic or inelastic.

   Reason(s) to agree:

   Reason(s) to disagree:

3. All pricing strategies should be somewhat cost-oriented.

   Reason(s) to agree:

   Reason(s) to disagree:

4. The higher a product or service is priced, the higher its quality perception.

   Reason(s) to agree:

   Reason(s) to disagree:

MULTIPLE-CHOICE QUESTIONS

Select the response that best answers the question, and write the corresponding letter in the space provided.

1 Discuss the importance of pricing decisions to the economy and to the individual firm

_____ 1. The Cool Shades Company prices its sunglasses at $30 per pair. This year, it has sold 10,000 pairs. Each pair of sunglasses costs the company $10 to produce, and the company has paid out $100,000 in marketing, R&D, and other expenses. All of the following statements about the company are true EXCEPT:
a. The company's revenue is $300,000.
b. The company is losing money on these sunglasses.
c. The company's profit is $100,000.
d. The company's cost of goods is $100,000.
e. The company is pricing higher than the break-even price.

2 List and explain a variety of pricing objectives

_____ 2. The Foxy Furniture Firm has a pricing policy of setting prices so that the retail price is as high as the market will tolerate. However, the company constantly strives to keep costs at an industry low. The pricing policy is best described as:
a. market share pricing
b. status quo pricing
c. demand oriented
d. sales maximization
e. profit maximization

_____ 3. Bartyl's Beer determines its prices based on maintaining revenues and expenses at acceptable levels. This would lead to pricing based on:
a. satisfactory profits
b. stable sales levels
c. profit maximization
d. market share
e. consumer demand

_____ 4. The Kandy Korner managed to exceed its target ROI for the current fiscal year. The following results were found on its financial statements:

Gross revenues:	$250,000	Assets:	$500,000
Gross profits:	$100,000	Liabilities:	$200,000
Net profits:	$ 50,000	Equity:	$300,000

What was the actual ROI for The Kandy Korner?
a. 20 percent
b. 50 percent
c. 33.33 percent
d. 10 percent
e. none of the above

_____ 5. A Japanese electronics firm has launched a new CD player in the competitive U.S. market. The firm prices its products very low in order to gain a foothold in the market. This pricing objective is:
a. increasing market share
b. maximizing dollar sales
c. satisfactory profits
d. profit maximization
e. target ROI

____ 6. Kmart has lowered the price of a GE coffeemaker to $19.88. Target and Wal-Mart lower their prices on the coffeemaker to $19.97 and $19.85 the day after Kmart's price change. This is:
   a. cost-plus pricing
   b. target return pricing
   c. market share pricing
   d. predatory pricing
   e. status quo pricing

3 Explain the role of demand in price determination

____ 7. Ben's Bagel company used to price its bagels at 25¢ each. At this price, the company sold an average of 5,000 bagels per day. The company recently decided to double its price to 50¢, and the company watched its demand fall to 1,500 per day. The demand for bagels is most likely:
   a. inelastic
   b. unitary
   c. equilibrium
   d. elastic
   e. profitable

____ 8. When Pinta-Painting Co. first started, it charged $750 per house and could not keep up with all the calls. Pinta-Painting raised the price to $1,000 per house, and now all eight painters are steadily busy but no longer forced to work fourteen-hour days. The $1,000 per house price is probably a(n):
   a. supply schedule
   b. price elasticity
   c. producer surplus
   d. equilibrium price
   e. inelastic price

____ 9. When Unicorn Software lowered the price of its investment software package from $800 to $200, demand doubled from five units sold per month to ten units per month. However, total revenue dropped. This is an example of:
   a. substitute goods
   b. unitary elasticity
   c. inelastic demand
   d. consumer shortage
   e. elastic demand

____ 10. Gwin has lowered the price of his custom oven mitts from $6 to $5. He previously sold 300 oven mitts per month and now sells 360 per month. He is experiencing:
   a. unitary elasticity
   b. inelastic demand
   c. elastic demand
   d. consumer surplus
   e. producer shortage

4 Describe cost-oriented pricing strategies

____ 11. When the PinkBall company manufactures 1,000 pink balls, the cost of pink dye is $14. When output is raised to 2,500 balls, the cost of the dye is $35. The cost of the pink dye is a(n):
   a. inventory cost
   b. demand cost
   c. fixed cost
   d. marketing cost
   e. variable cost

_____ 12. The Custom-Natural Pizza store has daily sales that range from 50 to 250 pizzas. Which of the following is the BEST example of one of their fixed costs?
a. payment on leased equipment
b. flour and tomato sauce
c. electricity and gas for baking
d. paper products
e. cash register tape

_____ 13. Output at the Pine Playhouse company changed from fifteen to sixteen playhouses, and the total costs changed from $27,000 to $28,500. What was the marginal cost for the company?
a. $1,800
b. $1,781
c. $1,500
d. $27,000
e. $28,500

_____ 14. The Glass 'n' Brass Lamp Shoppe figures the price of a lamp by doubling the cost of the lamp (or taking a 100% markup). Which pricing technique is the store using?
a. turnover pricing
b. keystoning
c. marginal revenue pricing
d. target ROI pricing
e. maximum revenue pricing

_____ 15. R.J.'s Health Foods buys a brand of granola cereal for $1.50 and adds 49 cents to the cost to bring the retail price to $1.49. What pricing technique is the store using?
a. marginal revenue pricing
b. keystoning
c. turnover pricing
d. markup pricing
e. break-even pricing

_____ 16. The Lollygag Lollipop Company has the following revenues and costs:

Sales price per lollipop:	$0.50
Variable costs per lollipop:	$0.30
Total fixed costs (annual):	$50,000

What is the annual break-even quantity for the company?
a. 50,000
b. 250,000
c. 100,000
d. 166,667
e. none of the above

_____ 17. The manager of the Toothsome Toothbrush company has calculated the sales volume at which the company's costs equal revenue. The manager announced at the quarterly sales meeting that 150,000 toothbrushes, at an average of $8 per brush, must be sold during next quarter to reach this point. Which important factor has been excluded from this analysis?
a. fixed and variable cost determination
b. break-even analysis
c. target return pricing
d. market share
e. consumer demand

5 Demonstrate how the product life cycle, competition, distribution and promotion strategies, and perceptions of quality can affect price

_____ 18. Klean Detergent is the market leader of clothes detergents and is in the maturity phase of its life cycle. There are many competitors in the market that challenge Klean's leadership. Which of the following statements is most likely true about Klean's pricing?
a. As a market leader, Klean does not need to worry about price.
b. Klean uses price promotion in order to maintain its market leadership.
c. Klean must price lower than the other brands to maintain its leadership.
d. Klean can increase its price since it is in the maturity stage.
e. Klean should maintain a high price to maintain a high quality image.

_____ 19. HeartGood's Eggs decided to offer a much larger than customary profit margin to grocery wholesalers and retailers on their new low-cholesterol eggs. This pricing strategy is designed to facilitate all of the following EXCEPT:
a. encouraging retailers to advertise this high-margin item
b. giving dealers an incentive to promote the new product
c. developing wide and convenient distribution
d. maximizing profit margin for the producer
e. encouraging trial by consumers if priced low by retailers

_____ 20. When an innovative new electronics product is introduced into the market, its initial pricing is usually high. This is most likely due to:
a. The firm's distribution strategy
b. The firm's promotion strategy
c. The lack of relevant competition
d. The product's quality image
e. High product costs

_____ 21. Tommy knows little about computer diskettes and does not want to spend the time to learn about them. However, he needs to buy diskettes to use for a consulting project. Not wanting to make a poor choice, he is likely to:
a. intuitively make the right choice
b. buy an expensive brand of diskette, guessing that the price is related to quality
c. avoid making a decision by not buying anything
d. revise his goals and buy a computer instead
e. buy the cheapest diskettes because most consumers feel that price is not directly related to quality

_____ 22. The advertisements for Ramarro Car claim that it is the most expensive car in the world. This is an example of a:
a. target return pricing strategy
b. market share pricing strategy
c. prestige pricing strategy
d. maintained markup pricing strategy
e. profit maximization pricing strategy

_____ 23. The Goldstar Conveyor Corporation builds quality conveyor systems for warehouses, with innovative components and superior durability. The corporation has managed to keep its price lower than its competitors. However, Goldstar's sales have been disappointing. For a fast and simple remedy, Goldstar should:
a. reeducate the potential consumers about its products
b. lower the quality of its products
c. raise prices because of consumer expectations
d. emphasize the low price in all the advertising
e. look for a different product to manufacture

Check your answers to these questions before proceeding to the next section

ESSAY QUESTIONS

1. List the three categories of pricing objectives and the specific objectives included in each category. Then describe disadvantages of using each type of pricing objective.

2. When a marketing manager sets a price to meet pricing goals, the price established depends on several factors. Name and describe these determinants of price.

3. As a product moves through its life cycle, the demand for the product and the competitive conditions tend to change. For each stage in the product life cycle, discuss pricing strategies appropriate for that stage. Then describe how price interacts with the other three P's of the marketing mix.

PROBLEMS

1. Last quarter the Xylo company sold 1,000 strapples for $1 each, the Yeti Company sold 600 strapples at $5 each, and the Zeta Company sold 400 strapples for $2.50 apiece. Assuming the three companies are the only firms competing in the strapple market, calculate unit and dollar market share for each company for last quarter.

2. In the following scenarios, calculate answers related to the way retailers tend to calculate markups.

   a) A coffeemaker is sold for $25. The retailer added $5 to the original cost. What is the markup percentage?

   b) The cost of a calculator is $4 and the retailer applies a markup of $6. What is the markup percentage?

   c) A bookstore retailer marks up all products by 25 percent. If a book costs the retailer $15, what will the final selling price be?

   d) A gourmet foods retailer marks up all products by 75 percent. If the selling price of a specialty cheese is $12, what was the cost to the retailer?

3. Calculate the break-even quantity for TV-Terry.

TV-Terry Financial Information	
Salaries	$ 60,000
Promotion	80,000
Research and development	90,000
Equipment	20,000
Store lease	50,000
Total fixed costs	$300,000
TV-Terry selling price	$500
Variable cost	300

CHAPTER NOTES

Use this space to record notes on the topics you are having the most trouble understanding.

ANSWERS TO THE END-OF-CHAPTER DISCUSSION AND WRITING QUESTIONS

Use this space to work on the questions at the end of the chapter.

ANSWERS TO THE END-OF-CHAPTER CASE

Use this space to work on the questions at the end of the chapter.

CHAPTER 18 Setting the Right Price

LEARNING OBJECTIVES

1 Describe the procedure for setting the right price

Setting the right price on a product is a process consisting of (1) establishing pricing goals; (2) estimating demand, costs, and profits; (3) selecting a pricing policy to help determine a base price; and (4) fine-tuning the base price with pricing tactics.

2 Identify the legal and ethical constraints on pricing decisions

Laws enacted in many states to ban unfair trade practices make it illegal to sell below cost. Federal laws (the Sherman Act and the Federal Trade Commission Act) prohibit price fixing, an agreement between two or more firms on a particular price. The Robinson-Patman Act makes it illegal for firms to discriminate between two or more buyers in terms of price. Finally, federal laws also prohibit predatory pricing, which is attempting to drive the competition out of business by lowering prices drastically and then raising them after the competition is gone.

3 Explain how discounts, geographic pricing, and other special pricing tactics can be used to fine-tune the base price

These pricing techniques enable marketing managers to adjust for competition in certain markets, meet ever-changing government regulations, take advantage of unique demand situations, and meet promotional and positioning goals.

4 Discuss product line pricing

Product line pricing maximizes profits for an entire product line. When setting product line prices, marketing managers determine what type of relationship exists among the products in the line: complementary, substitute, or neutral. Managers must also consider joint (shared) costs among products of the same product line.

5 Describe the role of pricing during periods of inflation and recession

During periods of economic inflation, marketing managers may employ cost-oriented and demand-oriented tactics. Cost-oriented tactics consist of dropping products with a low profit margin, using delayed-quotation pricing, and using escalator pricing. Demand-oriented tactics include price shading and increasing demand through cultivation of selected customers, unique offerings, and systems selling.

To stimulate demand during a recession, marketers use value pricing, bundling, and unbundling. Marketers also strive to cut costs by removing unprofitable products from product lines, implementing new technologies, cutting payrolls, and pressuring suppliers for reduced prices.

PRETEST

Answer the following questions to see how well you understand the material. Re-take it after you review to check yourself.

1. List the five steps in setting the right price.

2. List and briefly describe four illegal issues regarding price.

3. List five types of discounts, rebates, and allowances.

4. List nine types of special pricing tactics.

CHAPTER OUTLINE

1 Describe the procedure for setting the right price

I. How to Set a Price on a Product

A. Establish Pricing Goals

The first step in setting a price is to establish pricing goals, which may be profit-oriented, sales-oriented, or status quo. These goals are derived from the firm's overall objectives.

B. Estimate Demand, Costs, and Profits

1. After establishing pricing goals, managers should estimate revenues at a variety of prices.

2. Next, corresponding costs should be determined and profit estimated. This information can help determine which price can best meet the firm's pricing goals.

C. Choose a Price Strategy

1. A **price strategy** defines the initial price and gives direction for price movements over the product life cycle.

   The price policy is set for a specific market segment, based on a well-defined positioning strategy.

2. The degree of freedom a company has in setting a price strategy depends on market conditions and other elements of the marketing mix. Three basic policies for setting a price on a new good or service are price skimming, penetration pricing, and status quo pricing.

   a. **Price skimming** means charging a high introductory price, often coupled with heavy promotion. As the product progresses through its life cycle, the firm may lower the price to reach successively larger markets.

      1) Price skimming is successful when demand is relatively inelastic, when a product is legally protected, when it represents a technological breakthrough, or when production is limited because of technological difficulties, shortages, or a lack of skilled craftspeople.

      2) A successful skimming strategy enables management to recover product development costs quickly. And management can lower the price if it is perceived as too high by the customers.

      3) Services use skimming policies too.

   b. **Penetration pricing** sets a relatively low price for a product as a way to reach the mass market in the early stages of the product life cycle.

      1) Penetration pricing is designed to capture a large market share, resulting in lower production costs.

2) A successful penetration pricing strategy can block entry into the market by competitors, because they cannot gain a large enough share of the market to be cost-effective.

c. **Status quo pricing** means maintaining existing prices or simply meeting the competition. Sometimes this policy can be the safest route to long-term survival if the firm is comparatively small.

2 Identify the legal and ethical constraints on pricing decisions

II. The Legality and Ethics of Price Strategy

A. Unfair Trade Practices

In over half the states, **unfair trade practice acts** put a lower limit on wholesale and retail prices. Selling below cost in these states is illegal. Wholesalers and retailers are usually required to take a certain minimum percentage markup on their combined merchandise cost and transportation cost.

B. Price Fixing

Price fixing is an agreement between two or more firms on the price they will charge for a good or service.

1. Price fixing is in violation of the Federal Trade Commission Act and the Sherman Act.

2. Price fixing is illegal *per se*; that is, it is illegal no matter how reasonable or beneficial the results may be.

C. Price Discrimination

Price discrimination, the practice of charging different prices to different customers for the same product, is prohibited by the 1936 Robinson-Patman Act.

There are three possible defenses for the seller charged with price discrimination:

1. Accounting for cost differential

2. Responding to changing market conditions

3. Meeting the competition

D. Predatory Pricing

Predatory pricing is the practice of charging a very low price for a product with the intent of driving competitors out of business or out of a market. Once competitors have been driven out, the firm raises its prices.

1. This practice is illegal under the Sherman Act and the Federal Trade Commission Act.

2. Proving the use of this practice is difficult and expensive.

3 Explain how discounts, geographic pricing, and other special pricing tactics can be used to fine-tune the base price

III. Tactics for Fine-Tuning the Base Price

A. The **base price** is the general price level at which the company expects to sell the good or service and is either above the market (price skimming), below the market (price penetration), or at the market (status quo).

B. Fine-tuning techniques are short-run approaches that do not alter the general price level but allow the firm to adjust for competition in certain markets, meet ever-changing government regulations, take advantage of unique demand situations, and meet promotional and positioning goals.

C. Discounts, Allowances, Rebates, and Value Pricing

1. When buyers receive a lower price for purchasing in multiple units or above a specified dollar amount, they are receiving a **quantity discount**, the most common form of discount.

    a. A **cumulative quantity discount**, a deduction from list price that applies to the buyer's total purchases made during a specific period, is intended to encourage customer loyalty. The amount of the discount depends on the total amount purchased during the period, and the discount increases as the quantity purchased increases.

    b. A deduction from list price that applies to a single order is a **noncumulative quantity discount**. This discount encourages larger orders.

2. A **cash discount** is a price reduction offered to a consumer, industrial user, or marketing intermediary in return for prompt payment of a bill. It saves the seller carrying costs and billing expenses, and it reduces the risk of bad debt.

3. A **functional discount** (or **trade discount**) is a discount to a wholesaler or retailer for performing channel functions.

4. A **seasonal discount** is a price reduction for buying merchandise out of season. It shifts the storage function forward to the purchaser.
5. A **promotional allowance** (trade allowance) is a payment to a dealer for promoting the manufacturer's products. It is both a pricing tool and a promotional device.

6. A **rebate** is a cash refund given for the purchase of a product during a specified period. A rebate is a temporary inducement that can be taken away without altering the basic price structure.

7. **Value-based pricing** is a pricing strategy that has grown out of the quality movement. Since the firm is customer driven, it starts with the customer, considers the competition, and then determines the appropriate price.

    a. **Trade loading** occurs when a manufacturer temporarily lowers the price to induce wholesalers and retailers to buy more goods than can be sold in a reasonable time, thereby loading inventory with idle products.

b. **Everyday low prices (EDLP)** is a pricing tactic that attacks the trade-loading problem by significantly lowering prices and eliminating functional discounts. Procter & Gamble has led the way in implementing an EDLP policy.

D. Geographic Pricing

1. **FOB origin pricing**, also called FOB factory or FOB shipping point pricing, is a pricing tactic that requires the purchaser to absorb the freight costs from the shipping point.

   a. FOB means the goods are placed "free on board" a carrier.

   b. At that point, title passes to the buyer, who pays the transportation charges.

2. In **uniform delivered pricing**, the seller pays the actual freight charges and bills every purchaser an identical, flat freight charge, regardless of the buyer's location.

3. **Zone pricing** is a modification of uniform delivered pricing that divides the market into segments or zones and charges a flat freight rate to all customers in a given zone.

4. In **freight absorption pricing**, the seller pays all or part of the actual freight charges and does not pass them on to the purchaser.

5. A **basing-point price** requires the seller to designate a location as a basing point and charge all purchasers the freight cost from that point, regardless of the city from which the goods are shipped. Often multiple basing points are used.

E. Special Pricing Tactics

Special pricing tactics are used to stimulate demand for specific products, to increase store patronage, and to offer a wider variety of merchandise at a specific price point.

1. A merchant using a **single-price tactic** offers all goods and services at the same price (or perhaps two or three prices). Single-price selling removes price comparisons from the buyer's decision-making process.

2. **Flexible pricing** (or **variable pricing**) charges different customers different prices for essentially the same merchandise bought in equal quantities. It allows the seller to meet the price of the competition or to close a sale with a price-conscious customer.

3. In **professional services pricing**—used by lawyers, doctors, counselors, and so forth—fees are based on the solution of a problem or performance of an act rather than on the actual time involved. For example, a doctor may have a set fee for a certain type of operation.

4. **Price lining** is the practice of offering a product line with several items placed in the line at specific price points. This series of prices for a type of merchandise creates a price line.

5. **Leader pricing** (or **loss-leader pricing**) is an attempt to attract customers by selling a product near cost or even below cost, hoping customers will buy other products while in the store.

6. **Bait pricing**, a deceptive and illegal practice, tries to get customers into a store through false or misleading price advertising and then uses high-pressure sales tactics to persuade customers to buy more expensive merchandise.

7. **Odd-even pricing** (or **psychological pricing**) means using odd-numbered prices to denote bargains and even-numbered prices to imply quality.

8. **Price bundling** is marketing two or more goods or services in a single package for a special price. The opposite approach, **unbundling**, reduces the add-ons that come with the basic product and charges for each separately..

9. **Two-part pricing** involves two separate charges to consume a single good or service. An annual membership fee supplemented by a court fee for each use of a tennis club is an example.

4 Discuss product line pricing

IV. Product Line Pricing

Product line pricing is setting prices for an entire line of products, which is a broader concern than setting the right price on a single item.

A. Relationships among Products

One of several types of relationships may exist among the various products in a line:

1. The products may be *complementary*, meaning that an increase in the sale of one good causes an increase in demand for the complementary good.

2. Two products in a line may be *substitutes* for one another. If buyers purchase one item in the line, they are less likely to purchase a second item in the line.

3. A *neutral* relationship may exist between two products, with the demand for one product not related to demand for the other.

B. Joint Costs

Joint costs are costs shared in the manufacturing and marketing of a product line. These costs can complicate the issue of product pricing, because the assignment of a portion of these costs to each product may be somewhat subjective.

5 Describe the role of pricing during periods of inflation and recession

V. Pricing during Difficult Times

A. Inflation

When the economy is characterized by high inflation, special cost-oriented or demand-oriented pricing tactics are often necessary.

1. Cost-Oriented Tactics

a. One popular tactic is the removal of products with a low profit margin from the product line.

1) This tactic can backfire if the product has been selling at high volume and contributing sizable profit, even at a very small margin.

2) Even a product with a low profit margin may help the firm to gain certain economies of scale in production, or its removal may alter the price/quality image of the entire line.

b. **Delayed-quotation pricing** is a popular pricing tactic for industrial installations.

1) The price is not set on the product until the item is either finished or delivered.

2) Long production lead times have forced this policy on many firms during periods of inflation.

c. **Escalator pricing** is similar to delayed-quotation pricing in that the final selling price reflects cost increases incurred between the times when the order is placed and delivery is made. An escalator clause allows for price increases based on the cost-of-living index or some other formula.

2. Demand-Oriented Tactics

a. **Price shading** involves the use of discounts by salespeople to increase demand for one or more products in a line.

b. Some firms cultivate selected demand from affluent organizations or consumers. Others concentrate on customers who favor performance over price.

c. Buyers may tolerate higher prices if the seller has designed distinctive goods or services that uniquely fit the buyers' activities.

d. Companies may pass on higher costs by shrinking products sites and keeping prices the same.

e. A buyer's dependence on the selling firm may be heightened by selling entire systems that include feasibility studies, installation, and training.

B. Recession

Periods of reduced economic activity call for special marketing tactics.

1. Value Pricing

a. Value pricing stresses to customers that they are getting a good value for their limited funds.

b. Although lower-priced products offer lower profit margins, volume increases can offset slimmer margins.

2. Bundling or Unbundling

   a. If features are added to a bundle, consumers may perceive the offering as having greater value.

   b. Companies can unbundle offerings and lower base prices to stimulate demand.

VOCABULARY PRACTICE

Fill in the blank(s) with the appropriate term or phrase from the alphabetized list of chapter key terms.

bait pricing
base price
basing-point pricing
cash discount
cumulative quantity discount
delayed-quotation pricing
escalator pricing
everyday low prices (EDLP)
flexible pricing (or variable pricing)
FOB origin pricing
freight absorption pricing
functional discount (or trade discount)
joint costs
leader pricing (or loss-leader pricing)
noncumulative quantity discount
odd-even pricing (or psychological pricing)
penetration pricing
predatory pricing
price bundling
price fixing
price lining
price shading
price skimming
price strategy
product line pricing
promotional allowance
quantity discount
rebate
seasonal discount
single-price tactic
trade loading
two-part pricing
unbundling
unfair trade practice acts
uniform delivered pricing
zone pricing

1 Describe the procedure for setting the right price

1. A definition of the initial price and the intended direction for price movements over the product life cycle is the ______________________________.

2. There are three basic policies for setting a price on a new good or service. The policy that charges a high introductory price is ______________________________. The policy that charges a relatively low price as a way to reach the mass market is ______________________________. Finally, a firm may choose simply to meet the competition, which is a status-quo policy.

2 Identify the legal and ethical constraints on pricing decisions

3. Some pricing decisions are subject to government regulation. State laws that put a lower limit on wholesale and retail prices are ______________________________. An agreement between two or more firms on the price they will charge for a product or service is ______________________________ and violates the Sherman

Act and the Federal Trade Commission Act. If a company charges a very low price for a product with the intent of driving competitors out of business or out of a market, the firm is practicing ______________________________, which is illegal under the Sherman Act and the Federal Trade Commission Act.

3 Explain how discounts, geographic pricing, and other special pricing tactics can be used to fine-tune the base price

4. The general price level at which the company expects to sell the good or service is the ______________________________. This price may be lowered through the use of discounts, allowances, and rebates.

5. There are several different ways to lower the base price. Merchants will often offer a discount to buyers that pay promptly, a pricing tactic known as a(n) ______________________________. The most common form of discount is that which offers buyers a lower price for purchasing multiple units or above a specified dollar amount; this is a(n) ______________________________ and can take two forms. The first form is a deduction from list price that applies to the buyer's total purchases made during a specific period; this is a(n) ______________________________. In contrast, a deduction from list price that applies to a single order rather than to the total volume of orders is a(n) ______________________________.

6. Another type of discount compensates intermediaries for performing a service within a distribution channel; this is called a(n) ______________________________. A form of this type of price reduction occurs when a manufacturer temporarily lowers the functional discount to induce wholesalers and retailers to buy more goods than can be sold in a reasonable time; this is ______________________________. As a response to the costly inventory problems caused by this practice, some manufacturers have initiated ______________________________.

7. A price reduction for buying merchandise at an unpopular time is a(n) ______________________________. A payment to a dealer for promoting the manufacturer's products is known as a(n) ______________________________. Finally, a cash refund given for the purchase of a product during a specified period is called a(n) ______________________________.

8. Sellers have the option of using several different geographic pricing tactics. A price tactic that requires the purchaser to absorb the freight costs from the shipping point is ______________________________, also called FOB factory or FOB shipping point. If the seller pays the actual freight charges and bills every purchaser an identical, flat freight charge, ______________________________, or "postage stamp pricing" is being used. A modification of this pricing tactic divides the total market into geographic segments and charges a flat freight rate to all customers in a given segment; this is called ______________________________. In another pricing tactic, the seller pays all or part of the actual freight charges and does not pass them on to the purchaser; this is ______________________________. If a seller designates a specific location as a point from which freight costs are charged, regardless of the actual freight cost, ______________________________ is being used.

9. There are a number of special pricing tactics. If a merchant offers all goods and services at the same price, a(n) ______________________________ is being used. If different customers pay different prices for essentially the same merchandise bought in equal quantities, ______________________________ is being practiced. If a seller offers a line of merchandise at specific price points, the tactic is called ______________________________.

10. Another special pricing tactic is an attempt by the marketing manager to induce store patronage through selling a product near or below cost; this is ______________________________. If store patronage is enhanced by misleading or false advertising, the illegal practice of ______________________________ is being used. If the manager elects to end prices in an odd number or an even number, ______________________________ is being used.

11. If two or more goods are marketed in a single package for a special price, the tactic is ______________________________. Alternatively, if services that normally come with a product are charged for separately, ______________________________ is taking place. Finally, if the pricing tactic involves two separate charges to consume a single product or service, ______________________________ is being used.

4 Discuss product line pricing

12. Setting prices for an entire line of products is ______________________________, which involves broader concerns than price tactics for individual products. One unique problem is that products in a product line may share marketing and manufacturing costs. These shared costs are ______________________________, which must be assigned or allocated to the products.

5 Describe the role of pricing during periods of inflation and recession

13. During periods of inflation, several pricing tactics may be used. One tactic does not set the price on the product until the item is either finished or delivered; this is called ______________________________. A similar form of this pricing tactic takes the form of a contractual clause stating that the final selling price will reflect cost increases incurred between the time an order is placed and delivery is made; this is ______________________________. Another tactic involves the use of discounts by salespeople to increase demand for one or more products in a line; this is called ______________________________.

Check your answers to these questions before proceeding to the next section.

TRUE/FALSE QUESTIONS

Mark the statement **T** if it is true and **F** if it is false.

1 Describe the procedure for setting the right price

_____ 1. The first step to setting price is to estimate product demand and costs.

_____ 2. A powerful and prestigious food company wants to introduce yet another brand of children's cereal, similar to the many others already on the market. This manufacturer would have a great amount of freedom in choosing a price for this new cereal.

_____ 3. It makes the most sense to use price skimming as a price policy when demand is relatively inelastic in the upper ranges of the demand curve.

_____ 4. Gary's Gas Station is located across the street from another competitive gas station. For weeks, the two have been in a price war. When Gary decreases his gas price by two cents per gallon, his competitor follows him. This is an example of status quo pricing.

2 Identify the legal and ethical constraints on pricing decisions

_____ 5. If the presidents of AT&T, MCI, and Sprint got together and decided what price they would all charge for their long-distance services, the presidents would be engaged in price discrimination.

3 Explain how discounts, geographic pricing, and other special pricing tactics can be used to fine-tune the base price

_____ 6. Promotional allowances given to retailers by manufacturers are usually passed down to consumers in the form of a temporary discounted price.

_____ 7. Functional discounts, noncumulative quantity discounts, and promotional allowances are examples of rebates given to the trade.

4 Discuss product line pricing

_____ 8. The Crockette Corporation uses uniform delivered pricing, which is a way of legally discriminating against buyers that are located close to the point of shipping because these buyers pay the same amount as buyers located far from the point of shipping.

_____ 9. The My-Tool-ich company is setting the prices for its entire line of power tools. The firm's goals are to achieve maximum profit for the entire product line rather than for any particular product. This is an example using price lining as a pricing tactic.

_____ 10. Rag fibers for paper and cotton seeds for cottonseed oil are two byproducts of the cotton textile industry. Because these products are produced together, they are complementary products.

5 Describe the role of pricing during periods of inflation and recession

_____ 11. PorterCo hand-manufactures reproduction Civil War cannons, which can take as long as two years to build. To protect themselves and cover costs, PorterCo should use delayed-quotation pricing or escalator pricing.

6 Discuss what the future might hold for pricing in the marketing mix

_____ 12. In the future, the concept of price will become more abstract as consumers handle less and less actual money.

Check your answers to these questions before proceeding to the next section.

AGREE/DISAGREE QUESTIONS

For the following statements, indicate reasons why you may agree and disagree with the statement.

1. Firms would be better off to use EDLP (everyday low price) rather than to engage in a series of complex pricing discounts, rebates, and allowances.

   Reason(s) to agree:

   Reason(s) to disagree:

2. Leader pricing should not be practiced on a regular basis by supermarkets wishing to make a profit.

   Reason(s) to agree:

   Reason(s) to disagree:

3. Psychological pricing rarely works; consumers know what is going on!

   Reason(s) to agree:

   Reason(s) to disagree:

4. The only effective pricing strategy during a recession is to lower the price of the product.

   Reason(s) to agree:

   Reason(s) to disagree:

MULTIPLE-CHOICE QUESTIONS

Select the response that best answers the question, and write the corresponding letter in the space provided.

1 Describe the procedure for setting the right price

_____ 1. The marketing manager of Techie-TV finds that the firm can gain market share and become the industry leader if it slashes prices by 50 percent. However, the vice-president of finance is committed to reporting a 25 percent return on investment at all times. This conflict illustrates:
a. need for eliminating low-profit products
b. a lack of concentration on the marketing concept
c. pricing in a mature marketplace
d. ignoring the target market
e. trade-offs in pricing objectives

_____ 2. Mona Lisa toothpaste positions its product on its ability to whiten teeth after a few weeks of use. Its price is 30 percent higher than other toothpaste brands. Mona Lisa is using a:
a. price-skimming strategy
b. penetration pricing strategy
c. status quo pricing strategy
d. flexible pricing strategy
e. leader pricing strategy

_____ 3. Southwest Airlines charges some of the lower prices in the industry. As a result, the airlines has been able to reach the mass market and to increase the incidence of air travel among those who might have chosen another means of travel. The company's current price policy would best be described as:
a. skimming
b. flexible
c. penetration
d. zone
e. absorption

_____ 4. Kmart employees regularly shop at Wal-Mart stores in order to make certain that Kmart is charging comparable prices. Kmart is engaging in:
a. leader pricing
b. status quo pricing
c. corporate espionage
d. flexible pricing
e. functional pricing

2 Identify the legal and ethical constraints on pricing decisions

_____ 5. The Specialty Surgical Practice has published a minimum fee schedule for services, and distributed this schedule throughout the medical profession. Specialty Surgical is guilty of:
a. bait pricing
b. price fixing
c. unfair trade practices
d. price discrimination
e. predatory pricing

_____ 6. The Sharp Razor Company sells its disposable razors to several large discount retailers but gives special allowances only to one retailer. Sharp Razor is practicing:
a. bait pricing
b. price fixing
c. unfair trade practices
d. price discrimination
e. predatory pricing

_____ 7. Price discrimination violates the:
a. Sherman Antitrust Act
b. Price Discrimination Act
c. Wheeler-Lea Amendment
d. Robinson-Patman Act
e. Clayton Act

3 Explain how discounts, geographic pricing, and other special pricing tactics can be used to fine-tune the base price

_____ 8. When a customer of Cona Coffee Beans chooses to pay immediately on delivery rather than wait to be billed in thirty days, the salesperson is authorized to offer that customer a 5 percent discount. This 5 percent is an example of a:
a. quantity discount
b. rebate
c. cash discount
d. functional discount
e. promotional allowance

_____ 9. When the salesperson from Ample Appliances calls on retail appliance stores, she is authorized to offer the retailers a 25 percent discount from the list price in recognition of several retailer activities, including appliance unpacking, testing, and floor display setup. This 25 percent is called a:
a. functional discount
b. promotional allowance
c. quantity discount
d. seasonal discount
e. rebate

_____ 10. A sale on water skis and swimsuits at the Wisconsin-based Sunski Store during November is an example of which of the following pricing tactics?
a. seasonal discount
b. quantity discount
c. zone pricing
d. promotional allowance
e. functional discount

_____ 11. The Audria Auto Shop has agreed to set up a special display of Lubri-car motor oils beside the store entrance and also to run an advertisement in the local newspaper. Lubri-car has agreed to supply the display case free of charge and to pay for half the cost of the advertisement. This is an example of a:
a. bundled pricing tactic
b. promotional allowance
c. functional discount
d. quantity discount
e. product rebate

_____ 12. A manufacturer of computer laser printers is offering $100 cash to consumers who buy one of their printers, produces a cash register receipt, a completed certificate, and proof of purchase. This is an example of a(n):
a. rebate
b. quantity discount
c. instant rebate
d. functional discount
e. promotional allowance

____ 13. Shipping grain to international buyers can be risky because of price changes during the time for shipment, expense incurred over long distances, and quality of product delivered. To minimize exposure, a seller would likely employ:
a. freight absorption pricing
b. FOB origin pricing
c. zone pricing
d. basing-point pricing
e. uniform delivered pricing

____ 14. The Image Gift mail-order catalog lists one dollar's worth of shipping charges for every $10 of merchandise purchased. The pricing tactic used is:
a. quantity discounting
b. uniform delivered pricing
c. zone pricing
d. freight absorption pricing
e. FOB origin pricing

____ 15. Buffington Bookcases sells specialty bookcases and office furniture accessories nationally through its catalog. The company wants to simplify pricing and reduce its risk. Buffington also desires some type of difference in prices due to distance; therefore, the company uses:
a. two-part pricing
b. uniform delivered pricing
c. freight absorption pricing
d. zone pricing
e. flexible pricing

____ 16. Ben and Jerry's, Inc., would like to expand distribution of its frozen yogurt to a new market area, but competition is intense. Ben and Jerry's should use the geographic pricing tactic of:
a. freight absorption pricing
b. zone pricing
c. FOB origin pricing
d. basing-point pricing
e. multiple unit pricing

____ 17. The Mirasha Car Parts Company has eight warehouses and has a pricing policy of charging freight from the closest warehouse to the customer, regardless of where parts are shipped. For instance, if the customer is in Houston, Texas, the closest warehouse to the customer is in Dallas. If the ordered car part actually comes from the California warehouse, the customer still pays freight from Dallas. This pricing policy is called:
a. FOB origin pricing
b. basing-point pricing
c. zone pricing
d. uniform delivered pricing
e. freight absorption pricing

____ 18. The Two-Bit Candy Store is a small retail establishment where all candies are sold for 25 cents per piece, regardless of the candy type or size. This pricing method is known as:
a. price lining
b. inflexible pricing
c. single-price tactic
d. price bundling
e. leader pricing

____ 19. Yolanda owns a yacht dealership and often will sell essentially the same type of yacht to different customers at very different prices. This policy is:
a. two-part pricing
b. flexible pricing
c. illegal
d. bait and switch
e. price lining

_____ 20. At the Sports Stop, there are tennis rackets priced at $50, $75, $90, $125, and $250. The Sports Stop has chosen this price line structure because it will:
a. reach several different target market segments
b. maintain the product line at the same stage in the product life cycle
c. result in customers determining a price-quality relationship, thus resulting in more sales of the expensive models
d. enable the store to carry a larger total inventory
e. force competitors out of the market

_____ 21. When a local grocery store runs an advertisement for 29-cent cake mixes and watermelon for 10 cents per pound, there is a good chance that it is utilizing:
a. price bundling
b. leader pricing
c. price lining
d. psychological pricing
e. variable pricing

_____ 22. Microsoft offers spreadsheet software, word processing software, and graphics software as part of its "Microsoft Office" suite of products. This is an example of:
a. price lining
b. multi-part pricing
c. psychological pricing
d. basing-point pricing
e. price bundling

_____ 23. The U-Storem facility charges a monthly warehouse fee of $25 for each ten-foot-square storage unit. In addition, each time a customer needs to enter the security-locked warehouse to add or remove products, the customer is charged $5. The U-Storem facility is using a pricing tactic known as:
a. multiple unit pricing
b. variable pricing
c. price lining
d. two-part pricing
e. price bundling

4 Discuss product line pricing

_____ 24. When deciding on prices for an entire product line, the manager should consider all of the following EXCEPT whether the:
a. products in the line could be substitutes for one another
b. products will affect demand for the other product lines
c. products share joint costs
d. buyer considers the brand or the price first
e. products in the line are complementary to one another

5 Describe the role of pricing during periods of inflation and recession

_____ 25. The government is requesting one dozen military tilt-rotor flying vehicles. It will take approximately four years for manufacturing and order filling by an aircraft company. Delayed-quotation pricing will be used because:
a. the seller will place a later date on the product invoice to help accounts receivable in recording transactions
b. flexible price-shading can then take place
c. it will prevent a competitor from submitting an earlier bid
d. submitting a bid after the closing date is possible
e. it allows the final selling price to reflect cost increases incurred between the time the order is placed and the final delivery date

_____ 26. The Tempo Textile mill is writing up a contract with NASA for a heat-resistant space fabric that will take two years to design, test, and manufacture. Tempo has added an escalator clause to the contract, which is:
a. similar to delayed-quotation pricing
b. a demand-oriented pricing tactic
c. similar to price shading
d. a form of market penetration pricing
e. also called "postage stamp" pricing

_____ 27. The salespeople at Piffle Printers routinely use discounts to increase demand for one or more products in a line, especially during times of inflation when customers are more price sensitive. This practice is called:
a. escalator pricing
b. price shading
c. bid pricing
d. delayed quotation
e. proposition specification

Check your answers to these questions before proceeding to the next section.

ESSAY QUESTIONS

1. You are the marketing manager for a new athletic shoe company that offers trendy but functional shoes. You are in the process of deciding how to price your shoes. Using the four-step process of price setting, determine how you would achieve the right price.

2. List and define the three basic methods for setting a price on a new good or service. For each method name advantages and disadvantages of using that method.

3. Many sellers sell their products to customers that are geographically dispersed, resulting in significant freight costs. Define four types of geographic pricing tactics that can be selected by a marketing manager to moderate the impact of freight costs on distant customers. For each tactic defined, specify the circumstances that would prompt the selection of that particular pricing method, and then give specific examples of products that are commonly priced in that manner.

4. Marketing managers can use a wide variety of special pricing tactics to fine-tune prices. Name and define six of these special pricing tactics. For each tactic, give an example of a company, industry, or product that would use the tactic. Then give advantages and disadvantages of using each tactic.

5. You work for a luxury car manufacturer located in the United States. Unfortunately for your company, the country has just gone into a recession, and you fear that sales of the luxury vehicles will drop dramatically as consumer confidence in the economy decreases. Using the two possible tactics for pricing during a recession, what could you do?

6. The E-Lam Corporation manufactures three types of laminating machines: portable, vending, and desktop. The joint costs of land leasing, production equipment leases, insurance, and so on are allocated on an equal basis to the three types of machines sold. Last year's sales figures and allocated joint costs follow. Should E-Lam stop selling its portable machine? Why or why not?

	Portable	**Vending**	**Desktop**
Sales	$40,000	$80,000	$90,000
Less: Cost of goods sold	50,000	50,000	50,000
Gross margin	($10,000)	$30,000	$40,000

CHAPTER NOTES

Use this space to record notes on the topics you are having the most trouble understanding.

ANSWERS TO THE END-OF-CHAPTER DISCUSSION AND WRITING QUESTIONS

Use this space to work on the questions at the end of the chapter.

ANSWERS TO THE END-OF-CHAPTER CASE

Use this space to work on the questions at the end of the chapter.

CHAPTER GUIDES and QUESTIONS

PART SEVEN

TECHNOLOGY DRIVEN MARKETING

CHAPTER 19 Marketing and the Internet

LEARNING OBJECTIVES

1 Understand the development and structure of the Internet and World Wide Web

The Internet is an international network of computers that provides people **global** communication and access to millions of information sources. Although first developed for the U.S. military, the Internet is now an international medium for commerce, entertainment, and information. The World Wide Web is one subset of the Internet, but the popularity of its multimedia and hypertext capabilities has fueled the rapid growth of and interest in the Internet by giving people access to pictures, sound tracks, and video clips, as well as interactive text-based documents.

2 Describe the changing demographics of the Internet population

Although the typical Internet user was once almost certainly a young, affluent, white American male, the demographic profile of Internet users is changing. By 1996, the average age of Internet survey respondents climbed to 33 and the percentage of females rose to 31.5. Use of the Internet is still dominated by North Americans, but European and Asian involvement is increasing. As computer usage infiltrates society, the educational level and affluence of Internet survey respondents is also falling.

3 Discuss the effects of the Internet on marketing strategy

With market projections of 250 million Internet users by the turn of the millennium, this international and technically savvy population represents a marketing segment that can support billions of dollars in annual sales. Furthermore, the international reach of the Internet expands markets beyond the geographic boundaries once imposed on various small businesses.

4 Explain how marketing research may be conducted online

Specialized online search engines can help marketers obtain business-related information, locate e-mail addresses and scan electronic business directories. Other specialized sources of business information on the Internet include electronic news services and online archives of trade magazines and journals. Conducting surveys via the Internet reduces administration and coding costs, prevents "missing data" problems, and enhances the convenience factor for respondents.

5 Describe the privacy and security issues surrounding Internet-based commerce

Broadcasting unsolicited commercial e-mail is called "spamming" and it's generally considered an invasion of privacy and possibly even a form of theft. Privacy concerns have driven some consumers to prefer a form of payment that can't be traced directly to their credit card account through some form of "electronic cash." The realization that unauthorized computers could conduct undetected searches of information located on a person's disk drive generated privacy concerns about the use of "cookie" technology. Finally, to keep corporate drives free from electronic intrusion, many companies created "firewalls" that monitored the flow of information between corporate files and external computers. However, such monitoring devices can also prevent relevant marketing information from reaching its intended corporate audience.

6 Explain how the Internet impacts the traditional marketing mix

As technological advances continue to change the way people shop for goods and services, the number and scope of business competitors increase. Increased competition drives down prices, and new technology such as "intelligent agents" help consumers locate the lowest price available on the Internet. To avoid low-margin, price-based competition, companies must employ technology to incorporate mass customization and relationship marketing as the fundamental basics of doing business via the Internet.

PRETEST

Answer the following questions to see how well you understand the material. Re-take it after you review to check yourself.

1. List and describe four ways in which companies can communicate with customers over the Internet.

2. List five ways to conduct marketing research online.

3. List and briefly describe three privacy issues for online commerce.

CHAPTER OUTLINE

1 Understand the development and structure of the Internet and World Wide Web

I. The Internet and the World Wide Web

A. Developed in the 1960s by the U.S. military, the **Internet** is a network of computers originally designed to provide uninterruptable communication in the event of sabotage or attack.

1. Eventually, network usage by research institutes and universities around the world displaced the military involvement and the system of computers became known as the Internet.

2. However, it wasn't until 1991 that commercial activity was permitted on the Internet, causing the network to expand dramatically.

3. Between January 1993 and June 1998, the number of Internet host systems grew from 1,313,000 to 36,739,000. As more and more computers were connected to the Internet, it began to double in size every year.

B. In 1996, Internet Solutions estimated that more than 78 million people had access to the Internet and recorded more than 700,000 different sites on the World Wide Web (WWW or Web), a subset of the Internet that has *multimedia* and *hypertext* capability.

1. *Multimedia* refers to any combination in a single document of text, graphics, sound, or other data formats.

2. *Hypertext* involves text-based communication links among documents; when a linked word or phrase in one document is selected, another related document is displayed.

3. The Web comprises hundreds of thousands of virtual *sites* that only exist as data in computers. Therefore, each of these sites is really nothing more than a collection of associated electronic documents called *pages,* and the first page displayed at a Web site is called the home page.

2 Describe the changing demographics of the Internet population

II. Who Is Using the Internet?

Because the computers that house the Internet's data, pages and sites are located all over the world, people living in different countries can communicate through the Internet as easily and effectively as people physically located in the same city.

By 1999, 241 countries were listed by the Internet Assigned Numbers Authority and more than 150 countries had direct access to the Internet.

A. The demographics of the people who use the Internet are also changing rapidly.

In 1994, when Georgia Tech conducted the first publicly accessible Web-based survey, the results indicated that the majority of Web users lived in North America and were between 21 and 30 years of age. At that time more than 94% of the respondents were men. A typical Internet browser, or **surfer**, tended to be young, male, and on the West Coast, like an ocean surfer.

By the time of Georgia Tech's April 1998 survey, North Americans still represented the majority of the respondents, but the average age rose to 35.1 and the percentage of females climbed to 38.7. Demographic changes indicate that the Internet is becoming more widely available to the average consumer.

3 Discuss the effects of the Internet on marketing strategy

III. The Internet's Effect on Marketing Strategy

The Internet market was estimated to include a global population of 28.8 million people in 1995. By the turn of the century that figure should exceed 327 million.

Given the demographic characteristics of these individuals, this segment represents a large and profitable market for companies to approach.

Of particular interest are the dollar figures that the Internet is generating as the pace of commercial activity matches the rise in the number of users. Online commerce is expected to increase from the 1998 level of $26.47 million to an estimated $270 billion in 2002.

Over half of the people accessing the Internet are watching less television.

Instead, businesses are developing new marketing channels that will help them access online customers. In order to communicate with these individuals via the Internet, companies have several options:

1. *Send e-mail*: Sending and receiving electronic messages remains the most frequently performed Internet activity.
2. *Create a newsgroup or discussion list*: These Internet services allow people to participate in online discussions about specific topics of interest.
3. *Launch a home page*: Companies can establish a presence on the Internet in the form of World Wide Web sites.
4. *Sponsor an established group, list, or site*: Providing sponsorship for established Internet communication groups, discussion lists, and Web sites instantly gives businesses a targeted audience.

A. Electronic Mail

1. E-mail communication is the most frequently cited reason for using the Internet. Although 9 million people browsed the World Wide Web in 1995, almost four times as many people sent and received e-mail.
2. E-mail is predicted to maintain its usage lead over the WWW so that by the year 2000, 200 million people will regularly send and receive electronic mail.
3. However, not all electronic communications are welcome. Broadcasting unsolicited commercial e-mail across the Internet is scornfully referred to as "spamming."
4. Although the Internet makes it simple to send messages to large groups of e-mail addresses, spamming is considered more than invasion of privacy because many people who get e-mail must pay for every message they receive whether requested or not, or they may have to pay for the time they spend online reading their e-mail. The situation is

similar to a post office forcing people to pay for junk mail. Companies that broadcast unrequested e-mail often become the targets of vicious responses called "flames."

5. To restrict spamming, some Internet Service Providers (ISPs) refuse to broadcast e-mail having distribution lists in excess of some maximum level-typically 10 addresses.

B. Newsgroups and Discussion Lists

Usenet is an electronic bulletin board system that allows subscribers to exchange information by forming discussion groups, generally referred to as *newsgroups*.

More than 15,000 newsgroups exist for subscribers interested in a host of topics ranging from antique cars to zoology.

C. World Wide Web Site

1. Automated programs generate highly standardized and pre-formatted Web pages-and marketing success on the WWW almost requires the use of exciting and original content and images to lure customers past the home page and deeper into the site.

2. Many businesses, even small ones, with in-house programming knowledge and available equipment capacity are launching Web sites without any additional investment. But it's easy to spend hundreds of thousands of dollars on expensive equipment and high-priced development services.

3. Once the site is available on the Web, the next step is to generate traffic by getting people to visit the site. Ways of accomplishing this objective include:

   a. Include the Web address in print advertisements and stationery-but make sure the phone number is also available for customers who aren't connected to the Internet.

   b. Register with online search engines-by linking a Web site to keywords that describe relevant goods or services, search engines such as Infoseek, Yahoo, and Lycos can send new customers directly to a company's home page.

   c. Advertise through an off-line subscription service companies like FreeLoader and PointCast automatically load changing Web site content onto subscribers' computers. Although the consumer selects which sites to save, the subscription service also generates prominently displayed links to advertisers' sites.

   d. Frequently change the Web site's content. Off-line retrieval programs like Web-Ex and Web-Compass also automatically download any information that changes on user-selected Web sites. When a company regularly changes its Web site's content, retrieval programs (and people) have a reason to frequent the site.

   e. Pay consumers to view the sit~Instead of paying for advertising space, why not pay for viewers? CyberGold and GoldMail are Web sites that pay registered surfers who have relevant demographics and interests to view interactive ads and Web sites.

D. Tracking the Traffic

1. When people begin accessing a Web site, there are several ways to determine its success. Computers called servers that host Web sites can keep track of the number of times each page is accessed.

2. If an advertisement is displayed on a page, but the person doesn't click on it to access any information it may contain, the server should also record that fact, because the ad's opportunity to be noticed is considered an "impression." Impressions can therefore be used to measure the effectiveness of each ad on a Web page.

3. Finally, when a visitor actually provides information by filling in a form or providing an e-mail address, two-way interaction occurs and the "uploaded" information is either stored on the Web site's server or sent via e-mail to the Web site's administrator.

4. If the completed form includes a request to receive electronic notification anytime the information at the site changes, the visitor then becomes a subscriber. "Subscription" represents the highest form of interactivity, because the subscriber indicates an interest in maintaining an extended relationship with the content available at a particular site.

4 Explain how marketing research may be conducted online

IV. Conducting Marketing Research Online

There are a wide variety of tools that the Internet offers to the market researcher.

A. Online information search engines allow people to find Internet-based information. Placement services help companies register their web sites with hundreds of search engines.

B. Trade magazine and journal archives provide businesses with useful marketing information.

C. Electronic news services relay late-breaking events across the globe and transmit up-to-the-minute economic data.

D. The Internet makes the marketer's job of obtaining and analyzing information easier than ever.

1. By posting a survey form at a Web site, marketers can selectively invite individuals to fill in the requested data without even leaving their home or office.

2. Survey participants can complete the form at a convenient time-day or night rather than comply with some schedule set by the researcher.

3. Another benefit of digital surveys is the elimination of coding errors that can occur when paper-based survey answers are keyed into a computer database for statistical analysis. Because each survey respondent inputs his or her own data, the information can be sent directly to the statistical program and used without further human intervention.

4. Transmitting an e-mail questionnaire or posting a Web-page survey costs much less than paying for copying, stuffing envelopes, and paying for postage.

5. The fact that respondents can participate from where they live or work eliminates the cost of renting a facility or paying for transportation.

6. Even personnel costs plummet when surveys are administered online, because fewer people are needed to assist the respondents and no coders are needed to input the data.

E. Specialized newsgroups and discussion lists are online venues for exchanging ideas, debating issues, providing intellectual support and promoting new technology.

5 Describe the privacy and security issues surrounding Internet-based commerce

V. Privacy and Security Issues for Online Commerce

Among the practical problems facing companies that seek to do business via the Internet are concerns about the safety of conducting commerce online.

A. **Electronic Cash**

The most basic problem facing online commerce involves methods of payment.

1. Currency cannot be exchanged in the virtual world.

2. Checks may be delayed, held by banks, or never sent.

3. Possible credit card number theft is a concern for consumers.

B. Firewalls

1. To keep unauthorized individuals and data-destroying software viruses from accessing or corrupting corporate information, it became common for companies to install firewalls in the mid- 1990s.

2. Instead of being built of brick and mortar, **firewalls** for data communication are special computer programs that check all incoming and outgoing information streams for proper identification and authorization.

C. Spamming

1. Broadcasting commercial messages to individuals who don't want them is nothing more than an invasion of privacy that is referred to in the Internet community as spamming.

2. To aid marketers in realizing the ramifications of their online activities, the Tenagra Corporation offers a list of Web sites that suggest acceptable standards for doing business on the Internet. This list includes guidelines from several sources that suggest online marketing tactics which don't involve spamming or any other privacy-impairing activities.

6 Explain how the Internet impacts the traditional marketing mix

VI. The Internet's Impact on the Marketing Mix

Over the past century, technological advances have significantly altered the way people buy and sell.

A. Improvements in transportation speed people and goods around the globe to enhance our appreciation for other cultures and create a unified global marketplace for products and services.

B. Enhanced learning opportunities through public education and distance teaching programs allow more people to understand and explore the knowledge-base of scientific concepts that we continue to accumulate and refine.

C. Dramatic changes in communication technology have provided us with radios, telephones, motion pictures, televisions, and the Internet.

D. The Internet also allowed merchants and manufacturers to find the best possible price for raw goods and materials so as to produce products that maximized value for their customers.

E. Enhanced communication has even facilitated other price-reducing and value-increasing practices such as just-in-time delivery to reduce warehousing expenses and electronic data interchange to minimize accounting costs and mass customization designed to deliver increased value to the customer.

VOCABULARY PRACTICE

Fill in the blank(s) with the appropriate term or phrase from the alphabetized list of chapter key terms.

cookies	Internet
download	multimedia
electronic cash	servers
firewall	spamming
hypertext	surfer
intelligent agents	World Wide Web

1 Understand the development and structure of the Internet and World Wide Web

1. The network of computers that was originally designed for military use is the ______________________________. A subset of this network that has media and hypertext capabilities is known as the ______________________________.

2. Any combination of text, graphics, sound, or other data formats in a single document is called ______________________________. Text-based links that allow different documents to communicate is known as ______________________________.

2 Describe the changing demographics of the Internet population

3. A person who explores the Internet or World Wide Web on a computer is a(n) ______________________________.

3 Discuss the effects of the Internet on marketing strategy

4. Broadcasting unsolicited commercial e-mail messages across the Internet is ______________________________. Recipients often respond with flames.

5. Computers that host web sites are ______________________________.

6. Intentional storage of information from a web site is a(n) ______________________________.

5 Describe the privacy and security issues surrounding Internet-based commerce

7. A method used to transfer a payment online for electronic shopping is ______________________________.

8. Computer codes that save small bits of information to the user's hard drive are called ______________________________. These were originally intended to customize Web sites.

9. Special computer programs that check all incoming and outgoing information streams for proper identification and authorization are ______________________________.

6 Explain how the Internet impacts the traditional marketing mix

10. Software tools that automatically comparison ship for the lowest prices on internationally available brand-name merchandise are known as ______________________________.

Check your answers to these questions before proceeding to the next section.

TRUE/FALSE QUESTIONS

Mark the statement **T** if it is true and **F** if it is false.

1 Understand the development and structure of the Internet and World Wide Web

_____ 1. The Internet was originally designed by the U.S. military and adopted next by research institutes and universities around the world.

2 Describe the changing demographics of the Internet population

_____ 2. Today's typical Internet surfer is young, male, and lives on the west coast.

_____ 3. The Internet is becoming more widely available and usage is increasing.

3 Discuss the effects of the Internet on marketing strategy

_____ 4. Browsing web sites for shopping is the number-one use of the Internet.

_____ 5. From the perspective of the marketer, newsgroups and discussion lists are attractive because they allow commercial messages to be passed to users, whether they want them or not.

4 Explain how marketing research may be conducted online

_____ 6. Although responses to electronic surveys is quicker, it is more expensive for the marketing researcher.

_____ 7. Spamming is illegal in the U.S.

5 Describe the privacy and security issues surrounding Internet-based commerce

_____ 8. Firewalls are used for data security and checking all incoming and outgoing information streams.

6 Explain how the Internet impacts the traditional marketing mix

_____ 9. Intelligent agents are private security detectives who trace data streams to their source to identify spammers and illegal web site authors.

_____ 10. Communication innovations have lowered the cost of both standardized and customized products, allowing for mass customization.

Check your answers to these questions before proceeding to the next section.

AGREE/DISAGREE QUESTIONS

For the following statements, indicate reasons why you may agree and disagree with the statement.

1. The Internet has made marketing much easier.

   Reason(s) to agree:

   Reason(s) to disagree:

2. Marketing research conducted online can be biased and unscientific.

   Reason(s) to agree:

   Reason(s) to disagree:

3. The Internet has greatly changed the meaning of the marketing mix (the 4 "Ps").

Reason(s) to agree:

Reason(s) to disagree:

MULTIPLE-CHOICE QUESTIONS

Select the response that best answers the question, and write the corresponding letter in the space provided.

1 Understand the development and structure of the Internet and World Wide Web

_____ 1. The Internet was originally developed by:
a. research institutions
b. the U.S. military
c. state-funded colleges and universities
d. private colleges and universities
e. a large corporation

_____ 2. Ellis has put a picture of a horse, a video clip of a horse jumping, a sound-byte of a whinnying horse, and some text onto his home page. Ellis is using:
a. multimedia
b. hypertext
c. Netscape
d. spamming
e. cookies

2 Describe the changing demographics of the Internet population

_____ 3. The majority of Internet users are:
a. male engineers between 18 and 35 years of age
b. college students
c. North Americans with an average age of 35
d. professionals with post graduate degrees
e. European businesspeople

3 Discuss the effects of the Internet on marketing strategy

_____ 4. Abby would like to market her hedgehog portrait services. The most effective way for her to do this, using the Internet, would be to:
a. create a home page on the Web
b. use electronic mail to reach users who have expressed an interest in hedgehog portraits
c. start a discussion list on hedgehog art
d. sponsor an existing site on hedgehogs
e. create advertising messages to show up on Web search engines

_____ 5. Phil has collected e-mail addresses from an OverEaters Anonymous discussion Web page and has just sent an unsolicited ad for his new weight-loss candies to all these addresses. Phil is ____________ this group with e-mail.

a. surfing
b. firewalling
c. spamming
d. baking cookies for
e. mass customizing

_____ 6. Trish has just started her own business selling custom-designed Victorian hatpin holders. She would like to market her products via the Internet and knows that:

a. spamming will build her business the most effectively
b. launching a Web site is always prohibitively expensive
c. she should sponsor a discussion group of grapefruit juice drinkers, because many people in Florida collect hat pins
d. it will be many years before hatpin owners get online
e. it is no longer necessary to be an experienced programmer to launch a Web site

_____ 7. The highest level of involvement with a Web site is a(n):

a. download or upload
b. impression or view
c. hit or click
d. access or search
e. subscription

4 Explain how marketing research may be conducted online

_____ 8. Electronic surveys provide all the following benefits to marketing researchers EXCEPT:

a. the demographics of Internet users is a true picture of average consumers' characteristics
b. a computer-based survey helps with completion of all items and prevents missing data from compromising data analysis
c. marketers can selectively invite individuals to fill in the requested data without leaving their home or office
d. coding errors are eliminated because data does not have to be re-keyed
e. online surveys are less expensive than other forms of research

5 Describe the privacy and security issues surrounding Internet-based commerce

_____ 9. Yvette wants to make sure that her tax clients are the only ones who can access their files on the network at the accounting firm. Yvette should make sure that her network is equipped with a:

a. surf wave crasher
b. spammer jammer
c. cookie detector
d. firewall
e. Web agent

6 Explain how the Internet impacts the traditional marketing mix

_____ 10. Technological advances have impacted the marketing mix in all the following ways EXCEPT:

a. a unified global market is being created for products and services
b. consumers can easily search for and obtain the lowest possible price
c. the cost of customizing products is lower
d. customer feedback mechanisms are enhanced with interactive web sites and e-mail
e. the economy will be based on price competition and standardization

Check your answers to these questions before proceeding to the next section

ESSAY QUESTIONS

1. What are four benefits of conducting electronic surveys?

2. You are about to launch your own business and would like to build your own home page on the World Wide Web. What are four guidelines you should follow for increasing traffic to your page? What other Internet marketing tools can you use besides a web page?

3. You are going to give a lecture to marketing students about how new technologies have influenced the four Ps of the marketing mix. What topics should you cover?

CHAPTER NOTES

Use this space to record notes on the topics you are having the most trouble understanding.

ANSWERS TO THE END-OF-CHAPTER DISCUSSION AND WRITING QUESTIONS

Use this space to work on the questions at the end of the chapter.

ANSWERS TO THE END-OF-CHAPTER CASE

Use this space to work on the questions at the end of the chapter.

CHAPTER 20 One-to-One Marketing

LEARNING OBJECTIVES

1 Define one-to-one marketing and discuss its dependence on database technology

One-to-one marketing is a customer-based, information-intensive, and long-term-oriented, individualized marketing method that focuses on share of customer rather than share of market. The focus in one-to-one marketing is to develop a customer and to try to find products for that customer rather than to develop a product and try to find customers for the product. Because of the individualized nature of one-to-one marketing, database technology can be used to sift through millions of pieces of data to target the right customers and to communicate with those customers.

2 Discuss the forces that have influenced the emergence of one-to-one marketing

The forces that have influenced one-to-one marketing include a more diverse society, more demanding and time-poor customers, a decline in brand loyalty, the explosion of new media alternatives, changing channels of distribution, and demand for marketing accountability. As mass media becomes less important in reaching more demanding customers, one-to-one marketing will increase in importance.

3 Compare the one-to-one marketing communications process to the traditional mass marketing communications process

In the traditional communications process, the marketer encodes a message through a mass media channel (such as TV) to the target audience. Noise from competing messages can affect the encoding and decoding of the message, and feedback from the audience must be gathered through marketing research. In the one-to-one marketing communications process, the messages encoded by the marketer is more personalized, there is less noise in the process, and feedback can be direct and more cost effective through the use of communications technology.

4 List the advantages and disadvantages of one-to-one marketing

Advantages to one-to-one marketing include: (1) the ability to identify the most profitable and least profitable customers, (2) the ability to create long-term relationships with customers, (3) the ability to target marketing efforts only to those people most likely to be interested, (4) the ability to offer varied messages to different consumers, and (5) increased knowledge about customers and prospects. Disadvantages include the cost and time in creating a one-to-one marketing database and increasing privacy concerns by consumers.

5 List eight common one-to-one marketing applications

Eight applications of one-to-one marketing include: (1) identifying the best customers, (2) retaining loyal customers, (3) cross-selling other products or services, (4) designing targeted marketing communications, (5) reinforcing consumer purchase decisions, (6) inducing product trial by new customers, (7) increasing the effectiveness of distribution channel marketing, and (8) maintaining control over brand equity.

6 Discuss the basics of one-to-one marketing database technology

Database technology is very important to the use of one-to-one marketing. A computerized database of customer names and profiles is used to direct marketing communications. The system relies on transactional processing systems which record the details of individual purchase transactions. This data an then be enhanced using external sources of modeled data or customer data.

7 Describe the three levels of one-to-one marketing databases

The three levels of one-to-one marketing databases (from the least sophisticated to the most sophisticated) are: (1) a customer information system (CIS), which simply captures contact information and historical transaction data; (2) a marketing intelligence system (MIS), which enhances historical data with demographic, behavioral or psychographic data; and (3) a marketing decision support system (DSS), which enables managers to obtain and manipulate information as they are making decisions.

8 Describe the future of one-to-one marketing over the Internet

The Internet is a more recent but very effective tool for one-on-one marketing. The Internet provides marketers with the ability to deliver personalized promotional messages using past customer transaction history to identify buying patterns. Marketers can also send personalized e-mail messages based on previous online transactions or other information it knows about the customer. If promotions do not get the expected response, marketers can change messages immediately.

9 Discuss privacy issues related to one-to-one marketing

Many customers are concerned about one-to-one marketing because of the potential for invasion of privacy. There is widespread misunderstanding about how personal information is collected, used, and distributed, and most consumers are unaware of the laws that protect privacy. The Internet has been a source of major concern for those consumers who fear that their privacy will be harmed.

PRETEST

Answer the following questions to see how well you understand the material. Re-take it after you review to check yourself.

1. List six forces that have influenced the emergence of one-to-one marketing.

2. How is the communications process different in one-to-one marketing than in traditional marketing?

3. List five advantages and two disadvantages of one-to-one marketing.

4. List eight common applications of marketing databases.

5. What are the three levels of one-to-one marketing databases?

CHAPTER OUTLINE

1 Discuss the evolution of one-to-one marketing

I. **What Is "One-To-One Marketing?"**

One-to-one marketing is a customer-based, information-intensive, and long-term-oriented, **individualized** marketing method that focuses on *share of customer* rather than *share of market.*

A. The Evolution of One-To-One Marketing

1. Prior to the Industrial Revolution, small merchants knew their customers and often customized goods to better suit their needs.

2. Mass production introduced the concept of "make and sell." All customers were perceived as having the same needs.

3. In the 1950's the concept of the "average customer" emerged.

4. The 1980's were characterized by fragmented markets.

5. In the 1990's customers demanded the ability to buy precisely what met their individual needs and wants.

B. Forces Shaping One-To-One Marketing

1. Increasing diversity of the family, and more importantly, the acceptance of diversity.

2. Consumers have less and less time to spend. Therefore they have became more demanding and more impatient.

3. Decreasing brand loyalty can be attributed to excessive couponing and other "deals" proliferation of brands, and increased retailer power.

4. Emergence of new media alternatives.

5. The demand for advertising accountability as reflected by sales.

C. The impact of these trends is that today's consumer wants to be recognized as an individual requiring:

1. Satisfaction of unique needs and wants

2. More direct and personal marketing efforts

3. Loyalty will have to be rewarded

4. One-to-one marketing will increase in importance

2 Compare the one-to-one marketing communications process to the traditional marketing communications process

II. A Revised Marketing Communications Process

A. The revised **one-to-one marketing communications process** flows as follows:

1. The *sender* encodes individualized messages for customers and prospects identified from the database.

2. The message is then sent through a direct communications channel.

3. The customer or prospect, *receiver*, interprets the personalized message.

4. The customer or prospect responds to the communication.

5. The one-to-one marketer captures the response, feeding it back into the database.

B. Major differences between the traditional and revised communications process include:

1. Personalization of the message

2. Use of a direct channel

3. Near elimination of noise

4. Ability to capture the individual's response

3 Explain the importance of database technology to one-to-one marketing

III. Why One-To-One Marketing Needs Database Technology

In one-to-one marketing using *database technology*, relationship building is pre-planned and consciously implemented.

A. Basics of database technology

1. One-to-one marketing using database technology is commonly referred to as **data driven marketing**.

2. A **database** is "a collection of data, especially one that can be accessed and manipulated by computer software." A **marketing database** is the compilation of names, addresses, and other pieces of pertinent information about individual customers and prospects that affects what and how marketers sell to them.

B. Database technology investments

1. 8 in 10 companies have some sort of database

2. 85 percent of manufacturers and retailers believe they will need a marketing database to be competitive.

4 List the advantages and disadvantages of one-to-one marketing

IV. While the advantages of knowing customers on a one-to-one basis seem boundless, there are also a few disadvantages.

A. Advantages of one-to-one marketing using database technology include:

1. The ability to identify the most profitable and least profitable customers.

2. The ability to create long-term relationships with customers.

3. The ability to target marketing efforts only to those people most likely to be interested.

4. The ability to offer varied messages to different consumers.

5. Increased knowledge about customers and prospects.

B. Disadvantages of one-to-one marketing using database technology include:

1. Creation of the database can be quite costly.

2. Developing the technology to practice one-to-one marketing can be slow.

3. Consumer's privacy concerns.

5 List eight common one-to-one marketing applications

V. One-To-One Database Marketing Applications

A marketing database is a *tool* that helps marketers reach customers and prospects with one-to-one marketing communications.

A. Identify the most profitable customers

Customers are most likely to purchase again because they have bought most recently, bought most frequently, or spent a specified amount of money.

B. Retain loyal customers

Loyalty programs reward loyal consumers for making multiple purchases.

C. Cross-sell other products or services

A database allows marketers to match product profiles and consumer profiles to cross-sell customers other products that match their demographic, lifestyle, or behavioral characteristics.

D. Design targeted marketing communications

Using transaction and purchase data in addition to personal or demographic information allows marketers to tailor a message.

E. Reinforce consumer purchase decisions

A database offers marketers an opportunity to reach out to customers to reinforce the purchase decision.

F. Induce product trial by new customers

Using the profile of its best customers, marketers can easily find new customers that look like its most profitable segment.

G. Increase the effectiveness of distribution channel marketing

Marketing databases enable manufacturers to advise retailers how to better meet customer needs, and makes it possible to serve customers using direct channels instead of the traditional indirect channels.

H. Maintain control over brand equity

Database marketing allows marketers to control the message that is communicated to customers.

6 Describe the three levels of one-to-one marketing databases

VI. Three Levels of One-To-One Marketing Databases

Database levels can be viewed on a continuum based upon technology sophistication and their level of integration into overall marketing and business strategy decisions.

A. Customer information system

A **customer information system** or **CIS** is used primarily to track data captured from purchase transactions and past marketing activity.

B. Marketing intelligence system

A **marketing intelligence system (MIS)** or *integrated database* builds on the basic customer and transaction information by overlaying information that provides insights into why the customer purchases. The goal is to predict the likelihood of future purchases.

C. Marketing decision support system

A marketing *decision support system* (DSS) enables managers to obtain and manipulate information as they are making decisions. The **data warehouse** is a very large corporate-wide database while a **data mart** groups information from the data warehouse pertaining to one area.

7 Discuss the steps involved in developing a one-to-one marketing database

VII. Marketing Database Development

The **database development process** incorporates strategic planning as well as technical input.

A. Identify marketing applications

Typical applications are customer retention, cross-selling, customer relationship management, sales maximization, etc.

B. Determine data requirements

After the marketing applications are identified, the data that is needed to support those applications can be determined.

C. Select the database technology

A **database management system (DBMS)** is the software program or series of programs that create, modify, and control access to the information in the database. **Hierarchal** or **structured database systems** have defined data relationships and paths. **Relational database systems** have no predetermined relationships between the data items.

D. Gather internal data

Internal data is information a company captures and hopefully stores while it conducts business. Internal data consists of customer data, transaction information, product information, salesperson information, offer information, and response data.

A **response list** includes the names and addresses of individuals who have responded to an offer of some kind.

E. Enhance the database

Database enhancement is the overlay of information to customer or prospect records for the purpose of better describing or better determining the responsiveness of customers or prospects.

1. Three reasons to use enhancements are to

    a. Learn more about customers or prospects.

    b. Increase the effectiveness of customer marketing programs.

    c. Match profiles of best customers with those of prospects.

2. Three sources of enhancement data are

    a. **Compiled data** includes names and addresses gleaned from telephone directories and membership rosters.

    b. **Modeled data** is information that has already been sorted into distinct groups or clusters of consumers or businesses based on census, household, or business-level data.

    c. Customer data could include such things as customer surveys, customer participation programs, product registration, warranty cards, or loyalty marketing programs.

F. Manipulate the data

One of the most important aspects of one-to-one marketing is the ability to manipulate the data to profile the best customers or segments of customers, analyze their lifetime value, and ultimately, to predict their purchasing behavior through statistical modeling.

1. *Customer segmentation* is the process of breaking large groups of customers into smaller, more homogeneous groups.

2. **Recency, frequency, and monetary (RFM)** values are most commonly used to define a firm's best customers. Many marketers take RFM analysis one step further by introducing *profitability* into the equation.

3. **Lifetime value analysis (LTV)** projects the future value of the customer over a period of years.

4. Through **predictive modeling**, marketers try to determine what the odds are that some other occurrence will take place in the future. The occurrence the marketer is trying to predict is described by the *dependent variable*. The *independent variables* are the things that affect the dependent variable.

G. Maintain the database

Maintenance involves keeping the information in the database up to date, and eliminating duplicate customer information through the **merge/purge** process.

8 Discuss the possibilities of one-to-one marketing over the Internet

VIII. One-To-One Marketing and the Internet

One of the most important trends in the field of one-to-one marketing is the emergence of one-to-one marketing over the Internet.

9 Discuss privacy issues related to one-to-one marketing

IX. Privacy Concerns With One-To-One Marketing

One-to-one marketing concerns many Americans because of the potential for invasion of privacy.

A. The popularity of the Internet for direct marketing, consumer data collection, and as a repository of sensitive consumer data has also alarmed privacy-minded consumers.

B. Online privacy concerns are heightened by a recent Federal Trade Commission study which revealed that relatively few online marketers have adopted comprehensive privacy guidelines.

VOCABULARY PRACTICE

Fill in the blank(s) with the appropriate term or phrase from the alphabetized list of chapter key terms.

compiled list
customer information system (CIS)
custom data
database
database enhancement
data mart
data mining
data warehouse
lifetime value (LTV) analysis
marketing database
marketing intelligence system (MIS)
modeled data
one-to-one marketing
one-to-one marketing communications process
predictive modeling
recency, frequency, and monetary (RFM) analysis
response list

1 Define one-to-one marketing and discuss its dependence on database technology

1. A customer-based, information-intensive, and long-term-oriented, individualized marketing method that focuses on share of customer rather than share of product is called ______________________.

2 Discuss the forces that have influenced the emergence of one-to-one marketing

2. The ______________________ is a revised marketing communications process that is characterized by the use of personalized communication, the lack of interfering noise, and the ability to capture the response of the customer.

6 Discuss the basics of one-to-one marketing database technology

3. The process of gathering, maintaining, and analyzing information about customers and prospects to implement more efficient and effective marketing communications is known as ______________________. At the core of this is a ______________________, which is a collection of data, especially one that can be assessed and manipulated by computer software. A ______________________is a compilation of names, addresses, and other pieces of pertinent information about individual customers and prospects that affects what and how marketers sell to them.

4. Databases can be created from a ______________________ which includes the names and addresses of individuals who have responded to an offer of some kind, such as by mail or through product rebates. Other firms create a database from a ______________________, a customer list that is developed by gathering names and addresses from telephone directories and membership rosters.

5. ______________________ is the overlay of information to customer or prospect records for the purpose of better describing or better determining the responsiveness of customers of prospects. One form of this data is called ______________________, or information that has been sorted into distinct groups or clusters of consumers or businesses based on census, household, or business-level data. Another type of data that could be used to build a database is ______________________, which is acquired by the

marketer through customer surveys, customer participation programs, product registration, warranty cards, or loyalty marketing programs.

6. By manipulating data, marketers can find answers to decision points. One manipulation technique that can be used is called ______________________________, which determines the firm's best customers by identifying those customers who have purchased most recently, most frequently, and who have spent the most money. Another manipulation technique is ______________________________, which projects the future value of the customer over a period of years. Yet another technique, called ______________________________, helps marketers determine, based on some set of past occurrences, what the odds are that some other occurrence will take place in the future. A final data manipulation technique is ______________________________, or the process of using statistical analysis to detect relevant purchasing behavior patterns in a database.

7 Describe the three levels of one-to-one marketing databases

7. There are three levels of one-to-one marketing databases. The least sophisticated is called ______________________________, which is used primarily to track data captures from purchase transactions and past marketing activity. The second level is called ______________________________ which builds on the first level by capturing a greater array of data than basic customer and transaction information through the overlay of information that provides insights into why the customer purchases. The most sophisticated database is the marketing decision support system (DSS). At the heart of a DSS is the ______________________________, a very large corporate-wide database in which the data is culled from a number of legacy systems, such as billing/accounts or order fulfillment, that are already in place within the organization. A smaller database that logically groups information from the data warehouse pertaining t one area, such as for market segmentation or campaign management, is the ______________________________.

Check your answers to these questions before proceeding to the next section.

TRUE/FALSE QUESTIONS

Mark the statement **T** if it is true and **F** if it is false.

1 Define one-to-one marketing and discuss its dependence on database technology

_____ 1. The key to successful one-to-one marketing is the effective use of database technology.

2 Discuss the forces that have influenced the emergence of one-to-one marketing

_____ 2. An increase in brand loyalty has helped the emergence of one-to-one marketing.

_____ 3. Sydney's Sleighs has turned to one-on-one marketing in order to help track consumer response to promotional monies being spent. The company is demanding accountability for funds.

3 Compare the one-to-one marketing communications process to the traditional mass marketing communications process

_____ 4. One of the key differences in the communications process of one-to-one marketing is that it generally involves less noise.

4 List the advantages and disadvantages of one-to-one marketing

_____ 5. Tina's Tax Service would like to utilize more one-to-one marketing. One of the advantages that Tina should enjoy is the relatively small up-front cost of setting up a database.

_____ 6. Privacy issues are not a primary concern of customers who are targeted through one-on-one marketing.

5 List eight common one-to-one marketing applications

_____ 7. When consumers log onto the web site of amazon.com, they are greeted by name and provided a list of books that they might be interested in purchasing based on past purchases. This is an example of retaining loyal customers.

_____ 8. An Internet marketer of fresh flowers sends a message to all new customers that they will have the satisfaction of knowing that the flowers will arrive to the recipient's house on time and that they recipient will be pleased. This is an example of reinforcing consumer purchase decisions.

6 Discuss the basics of one-to-one marketing database technology

_____ 9. Ed's Electronics store has gathered the names and addressed of individuals who responded to the store's latest advertisement through the purchase of an electronics product. The store plans to use this list in subsequent promotions. This list is an example of a response list.

7 Describe the three levels of one-to-one marketing databases

_____ 10. Vortex Corporation has developed a database that contains the names, addresses, and other vital information of key customers along with behavioral and lifestyle enhancement information. Vortex has developed a marketing decision support system (DSS).

Check your answers to these questions before proceeding to the next section.

AGREE/DISAGREE QUESTIONS

For the following statements, indicate reasons why you may agree and disagree with the statement.

1. One-to-one marketing will eventually take over mass marketing.

   Reason(s) to agree:

   Reason(s) to disagree:

2. One-to-one marketing is more conducive to business marketing than to consumer marketing.

Reason(s) to agree:

Reason(s) to disagree:

3. Without database technology, one-to-one marketing would not be possible.

Reason(s) to agree:

Reason(s) to disagree:

4. Consumer concerns over privacy will make one-to-one marketing more difficult in the future.

Reason(s) to agree:

Reason(s) to disagree:

MULTIPLE-CHOICE QUESTIONS

Select the response that best answers the question, and write the corresponding letter in the space provided.

1 Define one-to-one marketing and discuss its dependence on database technology

_____ 1. Which of the following is NOT a characteristic of one-to-one marketing?
a. it is long-term-oriented
b. it focuses on market share
c. it is individualized
d. it is customer-based
e. it is information-intensive

2 Discuss the forces that have influenced the emergence of one-to-one marketing

_____ 2. All of the following have influenced the emergence of one-to-one marketing EXCEPT:
- a. new technology to lower production costs
- b. the increasing diversity of the U.S. population
- c. the demand for accountability
- d. decreasing brand loyalty
- e. new media alternatives

_____ 3. The decline in brand loyalty can be attributed to:
- a. excessive sales promotions
- b. the increasing power of retailers
- c. the proliferation of brands available
- d. none of the above
- e. all of the above

3 Compare the one-to-one marketing communications process to the traditional mass marketing communications process

_____ 4. The communications process for one-to-one marketing differs from that of traditional marketing in that the former:
- a. does not involve message encoding
- b. contains less "noise"
- c. utilizes the Internet
- d. has not require translation of the message
- e. requires no marketing research

_____ 5. Andrea's favorite web site is a music retailer that offers the largest variety of music in the world. Each time she logs onto the web site, she receives a personalized message that suggests that she consider buying CDs by new jazz artists, based on her past jazz purchases. Andrea reads and interprets the message before deciding what to do. Andrea is in the act of:
- a. encoding
- b. providing feedback
- c. decoding
- d. channeling
- e. receiving

4 List the advantages and disadvantages of one-to-one marketing

_____ 6. Which of the following is NOT an advantage of one-to-one marketing?
- a. the ability to identify the most profitable and least profitable customers
- b. the ability to create long-term relationships with customers
- c. the ability to target marketing efforts only to those people most likely to be interested
- d. the ability to offer varied messages to different consumers
- e. the relatively low cost of creating a marketing database

_____ 7. A catalog company selling premium-priced and durable children's clothing would like to find parents who are willing to spend top dollar for good quality clothing. The company should consider one-to-one marketing because it has the ability to:
- a. target marketing efforts only to those people most likely to be interested
- b. offer varied messages to different customers
- c. create long-term relationships with customers
- d. provide in-depth knowledge about customers and prospects
- e. create a database that can be used for the long term

_____ 8. Amazon.com, the Internet's leading bookseller, boasts that more than 64 percent of its sales in 1998 came from repeat customers. This demonstrates which of the following advantages of one-to-one marketing?
a. the ability to identify the most profitable and least profitable customers
b. the ability to create long-term relationships with customers
c. the ability to target marketing efforts only to those people most likely to be interested
d. the ability to offer varied messages to different consumers
e. increased knowledge about customers and prospects

_____ 9. All of the following are disadvantages to one-to-one marketing EXCEPT:
a. the high cost of creating a marketing database
b. mounting privacy concerns by consumers
c. the limited number of market segments that can be reached
d. the relatively high failure rate of building a database
e. none of the above

5 List eight common one-to-one marketing applications

_____ 10. Which of the following is NOT a common application of one-to-one marketing?
a. retaining loyal customers
b. identifying the best customers
c. maintaining control over brand equity
d. providing sales leads to telemarketers
e. inducing product trial by new customers

_____ 11. Many high-end hotel chains award "points" for every dollar spent in one of their hotels. Customers who earn a high number of points are given special privileges that may include upgraded hotel rooms or several free nights. This application of one-to-one marketing is an example of:
a. retaining loyal customers
b. reinforcing customer purchase decisions
c. maintaining control over brand equity
d. designing targeted marketing communications
e. inducing product trial by new customers

_____ 12. When a customer buys one of its cars, a Neptune Motor Company dealership sends out the entire team that was responsible for handling the customer's business. The team congratulates the customer for his/her purchase in a small celebration. This is an example of:
a. increasing the effectiveness of distribution channel marketing
b. reinforcing customer purchase decisions
c. identifying the best customers
d. designing targeted marketing communications
e. inducing product trial by new customers

_____ 13. An Internet bookseller has noticed through your past purchases that you are interested in science fiction novels. As a result, each time you log into the bookseller's web site, you are greeted with a list of suggested best sellers in the area of science fiction. This is done in an attempt to:
a. increase the effectiveness of distribution channel marketing
b. reinforce customer purchase decisions
c. cross-sell other products
d. maintain control over brand equity
e. induce product trial by new customers

6 Discuss the basics of one-to-one marketing database technology

_____ 14. The compilation of names, addresses, and other pieces of pertinent information about individual customers and prospects that affects what and how marketers sell to them is called a:
a. marketing intelligence system
b. marketing database
c. predictive model
d. data mart
e. data mine

_____ 15. Evan recently purchased a new CD player with a $100 rebate offer. He completed the rebate certificate and sent it to the manufacturer. The rebate certificate contained Evan's full name, his phone number, his address, and some purchase information. Evan has most likely become part of the manufacturer's:
   a. data mine
   b. demographic list
   c. compiled list
   d. data mart
   e. response list

_____ 16. As a member of a professional women's association, Sarah receives special offers from women's clothing catalog companies that specialize in business clothing. Sarah is most likely part of a:
   a. data mine
   b. demographic list
   c. compiled list
   d. data mart
   e. response list

_____ 17. The PRIZM database breaks down U.S. Census data by zip codes and analyzes each code for social rank, mobility, ethnicity, and family life cycle, among other criteria. This form of data enhancement is an example of:
   a. predictive modeling
   b. custom data
   c. lifetime value analysis
   d. modeled data
   e. recency-frequency-monetary analysis

_____ 18. A greeting card company gave away a free card to any customer who purchased two cards and completed a small questionnaire on the back of the coupon. The coupon captured the customer's name, address, phone number, birthday, and several pieces of information about purchasing behavior. The data received by the card company is an example of:
   a. predictive modeling
   b. custom data
   c. lifetime value analysis
   d. modeled data
   e. recency-frequency-monetary analysis

_____ 19. An Internet marketer of fresh flowers would like to build a database of customers who purchased flowers in the last twelve months and who spent over $100. The company should consider conducting:
   a. predictive modeling
   b. a custom data analysis
   c. a lifetime value analysis
   d. a modeled data analysis
   e. a recency-frequency-monetary analysis

_____ 20. A luxury car company has enjoyed repeat purchases from its most loyal customers and calculates the "value" of its most loyal customers at over $300,000. The car company has conducted:
   a. predictive modeling
   b. a custom data analysis
   c. a lifetime value analysis
   d. a modeled data analysis
   e. a recency-frequency-monetary analysis

7 Describe the three levels of one-to-one marketing databases

_____ 21. A large retail chain of pet supplies and food has created a database that captures customer information, including names, addresses, past transactions, and promotions used by customers. The company has created:
a. marketing decision support system (DSS)
b. sales lead system
c. customer information system (CIS)
d. marketing intelligence system (MIS)
e. management information system (MIS)

_____ 22. A gourmet food store has built a database that captures customer data, such as transactions and demographic data, and integrates it with customer profiles. The store then uses this integrated data to predict the future purchases of customers and can send out special offers. The store has created a:
a. marketing decision support system (DSS)
b. sales lead system
c. customer information system (CIS)
d. marketing intelligence system (MIS)
e. management information system (MIS)

_____ 23. The most sophisticated marketing database system is a:
a. marketing decision support system (DSS)
b. sales lead system
c. customer information system (CIS)
d. marketing intelligence system (MIS)
e. management information system (MIS)

_____ 24. A very large, corporate-wide database culled from a number of legacy systems, such as billing/accounting, order fulfillment, distribution, customer service, and marketing and sales, is called a:
a. data mart
b. data mine
c. customer information system (CIS)
d. a decision support system
e. a data warehouse

Check your answers to these questions before proceeding to the next section

---

ESSAY QUESTIONS

1. You are a marketing consultant. One of your major clients is a retailer of rugged clothing for hiking, camping, and other sports. Lately, your client has felt that it has lost sales due to its rather unfocused marketing efforts and is intrigued by one-to-one marketing. As the consultant, list some advantages that your client might enjoy from conducting one-to-one marketing.

2. You are a marketing manager for a vacation resort in Jamaica called "Island Breeze." You have been asked by the Vice-President of Marketing to look into building a marketing database to conduct one-to-one marketing. List eight common uses for this database, and indicate one specific example as it applies to your company.

3. As the general manager for a luxury car dealership, you have built a database that includes all kinds of information about past and current customers. List five data manipulation techniques and indicate by example how each one could be used.

CHAPTER NOTES

Use this space to record notes on the topics you are having the most trouble understanding.

ANSWERS TO THE END-OF-CHAPTER DISCUSSION AND WRITING QUESTIONS

Use this space to work on the questions at the end of the chapter.

ANSWERS TO THE END-OF-CHAPTER CASE

Use this space to work on the questions at the end of the chapter.

SOLUTIONS

PART ONE	Chapters 1 - 4
PART TWO	Chapters 5 - 9
PART THREE	Chapters 10 - 12
PART FOUR	Chapters 13 - 14
PART FIVE	Chapters 15 - 16
PART SIX	Chapters 17 - 18
PART SEVEN	Chapters 19 - 20

Solutions

CHAPTER 1 An Overview of Marketing

PRETEST SOLUTIONS

1. The process of planning and executing the conception, pricing, promotion, and distribution of ideas, goods, and services to create exchanges that satisfy individual and organizational goals.

   (text p. 6)

2. The five conditions are:

   - There must be at least two parties
   - Each party must have something the other party values
   - Each party must be able to communicate with the other party and deliver the goods or services ought by the other trading party
   - Each party must be free to accept or reject the other's offer
   - Each party must want to deal with the other party

   (text p. 6)

3. Product, price, promotion, and distribution

4. The production orientation, the sales orientation, the marketing orientation, and the societal orientation.

   (text p. 7)

5. **Sales orientation**:

   - Inward focus based on organization's needs
   - Focus on selling goods and services
   - Product targeted at everyone
   - Profit is gained through maximum sales volume
   - Goals achieved through intensive promotion

   Marketing orientation:

   - Outward focus, based on wants and preferences of customers
   - Focus on satisfying customer wants and needs and delivering superior value
   - Product targeted at specific groups of people
   - Profit is gained through customer satisfaction
   - Goals achieved through coordinated marketing and interfunctional activities

   (text p. 10)

6. The seven steps in the marketing process are: (1) Understanding the organization's mission and the role marketing plays in fulfilling that mission; (2) Setting marketing objectives; (3) Gathering, analyzing, and interpreting information about the organization's situation; (4) Developing a marketing strategy by deciding exactly which wants and whose wants the organization will try to satisfy and by developing appropriate marketing activities to satisfy these wants; (5) Implementing the marketing strategies; (6) Designing performance measures; and (7) Evaluating marketing efforts and making changes, if needed.

   (text p. 17)

7. Four reasons for studying marketing are: (1) Marketing plays an important role in society; (2) Marketing is important to business; (3) Marketing offers outstanding career opportunities; and (4) Marketing affects your life every day.

   (text pp. 17-18)

VOCABULARY PRACTICE SOLUTIONS

1. marketing
2. production orientation, sales orientation, marketing orientation, societal marketing concept
3. marketing concept
4. customer value
5. customer satisfaction
6. relationship marketing
7. teamwork
8. empowerment

TRUE/FALSE SOLUTIONS

(question number / correct answer / text page reference / answer rationale)

1. T 6
2. T 7 In a typical marketing exchange, one party offers goods and services in exchange for a price.
3. F 7 A sales orientation would be a more appropriate description for a firm that does not research consumer needs and wants but rather relies on a strong sales effort.
4. F 7 A sales orientation is based on "pushing products" to customers, while a market orientation is based on satisfying customer needs while meeting organizational objectives.
5. F 10 The customer needs to determine the benefits and the sacrifices; these perceptions cannot be defined by the marketer.
6. T 17

AGREE/DISAGREE SOLUTIONS

(question number / sample answers)

1. Reason(s) to agree: In order for a firm to be marketing oriented, it must operate efficiently (production orientation) and use aggressive selling techniques to push products through distribution channels.

   Reason(s) to disagree: Maintaining a production orientation or a sales orientation can actually hurt the firm's marketing efforts. If too much focus is given to production or sales, the firm will lost focus on customer satisfaction, the ultimate goal of marketing.

2. Reason(s) to agree: An organization with a marketing orientation must ensure that all employees—and especially those who are in direct contact with customers—understand that the firm's goal is to satisfy customer needs. Organizations such as Nordstrom's, Disney, and Southwest Airlines understand this concept well, and all employees are trained in delivering good customer service.

   Reason(s) to disagree: Though this concept sounds good, it is not practical. By making marketing "everyone's job," no one in the organization has accountability for marketing. While customer service can be delivered by everyone, there are many other aspects of marketing--such as promotion, planning, and marketing research--that require the expertise of a trained marketing department.

3. Reason(s) to agree: Just as engineering majors or journalism majors do not need to take marketing, neither should finance, management, accounting, or other majors. Marketing is a separate field that requires a level of expertise that should be reserved for those who plan to make a career out of it.

   Reason(s) to disagree: Any student majoring in a business area—and even those in related non-business areas, such as communications or public relations—should be required to take at least one marketing course. A business professional needs to know marketing in order to accept and work toward the organization's long-term goals, which are likely to be marketing-driven.

MULTIPLE-CHOICE SOLUTIONS

(question number / correct answer / text page reference / answer rationale)

1. c 6 Marketing must involve at least two parties in order for an exchange to occur.
2. b 6 Exchange involves the trade of items of value but does not necessarily involve formal organizations, profit, or money/legal tender.
3. e 6 As long as there is an organization and a client/user/customer/consumer group willing to engage in the exchange act, then marketing activities are relevant.
4. d 6 Marketing and exchange are not limited to profit seeking transactions, and there can be many types of costs other than direct costs and monetary payment.
5. d 7 The production orientation is a philosophy that focuses on the internal capabilities of the firm rather than on the desires and needs of the of the marketplace.
6. e 7 The production orientation guides a company to build whatever it builds best; that is, whatever it has the experience and expertise in doing.
7. a 7 Only the sales orientation assumes that aggressive sales techniques will sell more products, regardless of customer desires and needs.
8. d 7 Beth is concerned with meeting the needs and wants of the marketplace and, therefore, has a marketing orientation.
9. d 8 Family Shelter is exemplifying the marketing concept by concentrating on the needs of a specific group of customers.
10. e 9 Organizations with a societal marketing orientation seek the long-term best interests of society. The donation of earnings to an environmental cause is illustrative of this orientation.
11. e 10 A sales orientation has the short-term goal of increasing sales, which can be easily done through intensive sales promotions, such as discount pricing. A marketing orientation involves coordination among many organizational functions, such as production, research and development, finance, and marketing.
12. b 17 Appraising marketing personnel is generally a human resources activity (management) rather than a marketing activity.
13. e 17 Developing a new advertising campaign is part of promotion, or communicating with the target market.
14. e 17 Marketing is an important conceptual base that will help assess the needs and wants of the various business contacts and customers. Marketing is a key component of every business.
15. d 17 The marketing concept stresses the commitment to satisfying customer needs and wants with an entire range of marketing tools, not just selling or advertising.

ESSAY QUESTION SOLUTIONS

1. The five conditions of exchange are:
   - There must be at least two parties.
   - Each party must have something the other party values.
   - Each party must be capable of communicating with the other party and deliver the good or services sought by the other trading party.
   - Each party must be free to accept or reject the other's offer.
   - Each party must believe that it is appropriate or desirable to deal with the other party.

   Marketing can occur even if an exchange does not take place. Many of the activities of marketing (distribution, promotion, pricing, product development, and so on) can take place without a final exchange.

   (text p. 6)

2. The **production orientation** focuses firms on their internal production capabilities rather than the desires and needs of the marketplace.

   The **sales orientation** assumes that buyers resist purchasing items that are not essential, and that buyers will purchase more of any item if aggressive selling techniques are used. Again, this orientation does not address the needs and wants of the marketplace.

   The **marketing orientation** is dependent on the customer's decision to purchase a product and provides increased responsiveness to customer needs and wants.

   The **societal marketing orientation** refines the marketing orientation by stating that the social and economic justification for an organization's existence is the satisfaction of customer wants and needs while meeting the organization's objectives and preserving or enhancing both the individual's and society's long-term best interests.

   (text pp. 6-9)

3. Five key areas in which a market orientation differs from a sales orientation are as follows:

   - **The organization's focus**: A market orientation has an "outward" focus based on the wants and preferences of customers, while a sales orientation has an "inward" focus based on the organization's needs.
   - **The firm's business**: A market orientation defines business as satisfying customer needs and delivering value, while a sales orientation defines business as selling goods and services.
   - **Those to whom the product is directed**: A market orientation targets specific groups of people who have needs for products, while a sales orientation targets everybody in order to maximize short-term sales.
   - **The firm's primary goal**: A market orientation has the goal of gaining profit through customer satisfaction, while a sales orientation has the goal of gaining profit through maximum sales volume.
   - **Tools the organization uses to achieve goals**: A market orientation achieves goals through coordinated marketing and interfunctional activities, while a sales orientation achieves goals through intensive promotion.

   (text p. 10)

4. Marketing process activities include:

   - Understanding the organization's mission
   - Setting marketing objectives
   - Performing a situation analysis, including strengths, weaknesses, opportunities, and threats
   - Developing a marketing strategy, including a target market specification, and marketing mix, including product, place, promotion, and price
   - Implementing the marketing strategy
   - Designing performance measures
   - Periodically evaluating marketing efforts, and making changes if needed

   (text p. 17)

CHAPTER 2 Strategic Planning

PRETEST SOLUTIONS

1. The six major elements of a marketing plan are:

   - **Business mission**: The firm's long-term vision.
   - **Objectives**: Statements of what is to be accomplished through marketing activities.

- **Situation analysis**: An analysis of the firm's or product's strengths, weaknesses, opportunities, and threats
- **Target market selection**: Selection of the group of people to whom marketing efforts will be targeted
- **Marketing mix**: Strategies for the 4 Ps (product, price, promotion, and place)
- **Implementation**: Action plan for marketing activities

(text p. 32)

2. Four strategic alternatives are:
   - **Market penetration**: Sell more products to the present market
   - **Product development**: Sell new products to the present market
   - **New market**: Sell the present products to a new market
   - **Diversification**: Sell new products to new markets

   (text pp. 37-38)

3. Two tools that can be used to select a strategic alternative are:
   - **Portfolio matrix**: Allocates resources among product or SBUs on the basis of relative market share and market growth rate. Categories are:
     - Stars: High growth, high share products that need high investment to continue firm's growth;
     - Cash cows: Low growth, high share products that are "milked" for their excess profits;
     - Problem children: High growth, low share products that either need investment or they will turn into dogs;
     - Dogs: Low growth, low share products that should not receive many resources.
   - **Market attractiveness/company strength matrix**: Allocates resources among SBUs on the basis of how attractive a market is and how well the firm is positioned to take advantage of opportunities in that market.

   (text pp. 39-42)

4. A marketing strategy is the activity of selecting and describing one or more target markets and developing and maintaining a marketing mix that will produce mutually satisfying exchanges with target markets.

   (text p. 43)

5. The four Ps are:
   - **Product**: The product offering, its packaging, warranty, brand name, company image, etc.
   - **Place (distribution)**: Making product available when and where customers want them.
   - **Promotion**: Communication of the product to the target market; includes personal selling, advertising, public relations, and sales promotion.
   - **Price**: What a buyer must give up to obtain a product.

   (text p. 44)

VOCABULARY PRACTICE SOLUTIONS

1. strategic planning
2. planning, marketing planning
3. marketing plan
4. mission statement
5. marketing myopia
6. strategic business unit (SBU)
7. marketing objective
8. SWOT analysis
9. environmental scanning
10. strategic window

11. market penetration, market development, product development, diversification
12. portfolio matrix, star, cash cow, problem children (or question marks), dog, market attractiveness/company strength matrix
13. differential advantage, sustainable competitive advantage
14. marketing strategy
15. market opportunity analysis
16. marketing mix, 4 Ps
17. implementation, evaluation, control
18. marketing audit

TRUE/FALSE SOLUTIONS

(question number / correct answer / text page reference / answer rationale)

1. T 28
2. F 33 The first step is to create, or to review, the business mission statement.
3. F 33 Because Mike's Motos defines its business in terms of the benefits customers seek rather than in terms of specific goods or services, it does not suffer from marketing myopia.
4. F 34 This objective is neither measurable nor time specific.
5. T 35
6. T 36
7. F 36 This firm is a prospector.
8. T 37
9. T 37
10. T 44
11. F 45 The use of theaters to communicate a motion picture release is an example of promotion. Using theaters to show the film would be an example of place.
12. T 46
13. F 47 A marketing audit is not preoccupied with past performance but instead looks to the future allocation of marketing resources. Additionally, both small and large firms use marketing audits.

AGREE/DISAGREE SOLUTIONS

(question number / sample answers)

1. Reason(s) to agree: The basic outline for a marketing plan takes into account a variety of organizational types, including non-profit organizations. All organizations should have a mission statement, objectives, a situation analysis, a target market, a marketing mix, and implementation and evaluation and control.

   Reason(s) to disagree: Some small firms may not find it necessary to go through all the steps of a basic marketing plan. Doing a complete analysis of a small firm's situation may be too tedious, as the owners probably already know what the situation is. Also, non-profit organizations cannot put together strategies for the "4 Ps" since there may not be a product or a price.

2. Reason(s) to agree: Once a situation analysis is complete, the situation can change immediately. Though a firm's strengths and weaknesses do not change quickly, opportunities and threats can change in a matter of minutes. Documenting the analysis seems silly when the business environment is so fluid.

   Reason(s) to disagree: It is never a waste of time to thoroughly analyze a firm's situation. Though the environment can change quickly, it is important for a firm to understand what its current resources are, what its weaknesses are, and to be responsive to both threats and opportunities. Unless these are stated explicitly, there may not be acceptance from all managerial levels.

3. Reason(s) to agree: The tools assume that an organization has either separate strategic business units (SBUs) or many different products. If a small firm had only one product to sell, it is simple to decide what resources the product should get: all of them!

   Reason(s) to disagree: Even small firms can benefit from these tools. By understanding where the firm's product(s) fit in the marketplace, it can find a positioning strategy that will help it compete.

4. Reason(s) to agree: If a strategy is sound, then the implementation should follow naturally. It is more important to do the right thing than to do something right.

   Reason(s) to disagree: A strategy will fall apart if it is not implemented well. If a new product is launched with a sound strategy (good positioning, good advertising, etc.), and the sales force has structural problems that prevent it from selling the product to the customer, then the new product may fail. The failure will be due to the poor implementation, not strategy. Implementation is at least as important as strategy.

MULTIPLE-CHOICE SOLUTIONS

(question number / correct answer / text page reference / answer rationale)

1. d 28 Only designing a marketing information system is not part of the six-step process in strategic planning.
2. c 28 All the listed activities are part of the strategic planning process.
3. b 33 A mission statement answers the question, "What business are we in and where are we going?"
4. e 33 The mission statement dictates the firm's business based on a careful analysis of benefits sought by current and potential customers, as well as existing and anticipated environmental conditions.
5. b 34 The concept of running department stores is a much narrower focus than the broad range of opportunities found in "providing a range of products and services."
6. b 34 This mission statement is too broad and does not state the business that the firm is in.
7. d 34 A broad mission statement for a telephone company would recognize that the firm's business is total communications service.
8. c 34 An SBU usually has its own mission statement, target markets, and separate functional departments.
9. a 34 Marketing objectives should consistent with organization objectives, should be measurable, should be realistic, and should specify a timeframe. The first objective fits these criteria the best.
10. d 35 A situation analysis attempts to ascertain the present situation and forecast trends.
11. d 36 Corporate culture is the pattern of basic assumptions held by an organization's members in order to cope with its environments.
12. e 36 This firm is a prospector since it focuses on identifying and capitalizing on emerging market opportunities.
13. b 37 Market development involves finding new uses for a product to stimulate sales among new and existing customers.
14. d 37 Product development involves selling a new product to existing markets.
15. e 38 Diversification is defined as selling a new product to a new market.
16. a 40 The portfolio matrix designates the SBUs as dogs, cash cows, stars, or problem children.
17. b 40 Cash cows generate cash, have dominant market share, and are in low-growth industries.
18. c 40 Problem children are SBUs with large cash demands because of their rapid growth and poor profit margins.
19. c 40 This program is a cash cow. Though enrollment has been declining, the program is still large and "pays the bills." The best strategy is to hold or preserve the program.
20. a 41 The company appears to have medium attractiveness, which calls for a strategy of maintenance.
21. b 41 Superior skills are the unique capabilities of managers and workers that distinguish them from the personnel of competing firms.
22. c 43 Target markets are the chosen market segments that have a need for the firm's product offerings.

23. d 44 The product is the starting point for any marketing mix. Without it, pricing, distribution, and promotion are irrelevant. The production capacity can be changed to fit the proposed product.

24. e 45 Distribution strategies are concerned with making products available when and where customers want them.

25. c 45 Promotion covers a wide range of communication vehicles. Publicity is generally a nonpaid form of communication and is a subset of promotion.

26. c 45 The only element that is often subject to quick and easy change is price.

27. d 45 Implementation involves all of the steps listed.

28. a 46 Control involves that mechanism for correcting actions in the planning and implementation phases.

29. c 46 A marketing audit is a thorough, systematic, periodic evaluation of the goals, strategies, organization, and performance of the marketing organization. A marketing audit will evaluate the past, present, and future performance of all aspects of the marketing department.

30. d 47 Planners need to stretch their imaginations and search for creative solutions to problems. Planning is a constant process and should include top management.

31. e 47 Creating a special department where none has been before may make management feel as though something has been accomplished, but it may also alienate other employees and even slow down the progress toward a marketing orientation unless the other four steps are taken.

ESSAY QUESTION SOLUTIONS

1. The elements of the marketing plan are:

   1. Define the business mission
   2. Set marketing objectives
   3. Conduct a situation analysis (SWOT)
   4. Select target market(s)
   5. Establish the marketing mix (product, place, promotion, price)
   6. Implement, evaluate, and control the plan

   (text, p. 31)

2. A marketing objective is a statement of goals—of what is to be achieved—through marketing activities. The criteria for a good objective are to be: (1) realistic; (2) measurable; (3) time specific, and (4) consistent. A good example of an objective is given in the textbook (for a retail pet food company): "To achieve 10% dollar market share in the specialty pet food market within 12 months of product introduction."

   (text, pp. 34-35)

3 A target market is a group of individuals or organizations toward whom marketing activities are focused. The three general strategies for selecting target markets are:

   - Appealing to the entire market with one marketing mix; an example is Coca-Cola targeting the mass market for its Coke Classic;
   - Concentrating on one segment; an example is Topol toothpaste targeting smokers who are concerned about white teeth; and
   - Using multiple marketing mixes; an example is the Gap clothing store targeting a young adult market for its casual clothing outlets and the children's market with an even younger clothing line.

   (text, pp. 43-44)

4.

Market Growth Rate (In constant dollars)	High	Low	
	Star	Problem Child, or Question Mark	High
	Cash Cow	Dog	Low

High Low
Market Share Dominance
(Share relative to largest competitor)

Stars: Stars are market leaders that are growing quickly. Star SBUs often have large profits but require much cash to finance the rate of growth. The tactic for marketing management is to protect existing market share and to obtain a majority of new users entering the market.

Cash cows: These SBUs usually generate more cash than is required to maintain market share. The basic strategy is to maintain market dominance with price leadership and technological product improvements. Excess cash can be allocated to other areas where growth prospects are the greatest.

Problem children: Rapid growth coupled with poor profit margins causes cash demands for this class of SBUs. Three alternative strategies can be enacted: (1) invest heavily to obtain better market share, (2) acquire competitors to get the necessary market share, or (3) drop the SBU. If cash is not provided, these SBUs will become dogs.

Dogs: Because dogs have low-growth potential and a small market share, they usually end up leaving the marketplace. Mature markets, no new users, stiff competition, and market leader dominance characterize SBUs in this category. Tactics include resegmenting markets, harvesting, or dropping the SBU.

(text, pp. 40-41)

5. **Product:** This includes development of the new cereal, production assistance, packaging and labeling, branding, and other components.

   Place (or distribution): The cereal must move through a channel of distribution to get from the manufacturer to the final consumer. This channel will probably involve wholesalers and retailers (grocery stores). Physical distribution (stocking and transportation logistics) is also part of the place or distribution component.

   Promotion: Promotion of the cereal might include any or all of the following to inform, educate, persuade, and remind target markets about the cereal's benefits: personal selling, advertising (TV, radio, magazines, billboards, and so on), sales promotion (such as coupons and rebates), and public relations.

   Price: Pricing is an important component of the marketing mix because it is flexible and therefore can be used as a powerful strategic tool against competitive cereals.

   (text pp. 44-45)

CHAPTER 3 The Marketing Environment

PRETEST SOLUTIONS

1. A target market is a defined group of people most likely to buy a firm's product.

   (text, p. 60)

2. Social factors that affect marketing are: the values held by various consumers (such as "cultural creatives" or "traditionalists;" dual-career families and their "poverty of time;" the growth of component lifestyles; the changing role of families; and the growth of women in the workplace.

   (text pp. 61-62)

3. Demography is the study of people's vital statistics, such as their age, race and ethnicity, and location.

   (text, p. 64)

4. Four generations that are often used as target markets are:

   - Generation Y: Born between 1980 and the present, Generation Y is composed of children of younger Baby Boomers. They are "born to shop" and have become savvy consumers at a young age. This generation is culturally, ethnically, and economically diverse.
   - Generation X: Born between 1964 and 1980, Generation X is composed of children of older Baby Boomers. Many Generation Xers were latchkey children who were left to fend for themselves as their dual-career parents worked. Members of this generation do not mind indulging themselves and are cynical, savvy, and technology-driven.
   - Baby Boomers: Born between 1945 and 1964, Baby Boomers are the largest generation in U.S. history. At 78 million strong, Baby Boomers are America's mass market. Baby Boomers are entering their 50s but still cling to their youth. They were taught to think for themselves and were more independent than their own children, the Generations X and Y.
   - Older Consumers: Today, many older consumers, who are the parents of the Baby Boomers, are healthy and robust compared to the older people of past generations. They have driven marketers to develop innovations in order to help them in advancing years—such as snap-on lids for detergent. Older consumers spend on their grandchildren, causing retailers such as F.A.O. Schwartz to create special sections for these consumers.

   (text pp. 65-68)

5. Legislation that impacts marketing include:

   - Sherman Antitrust Act: Makes trusts and conspiracies in restraint of trade illegal.
   - Clayton Act: Outlaws discrimination in prices to different buyers; prohibits tying contracts; and makes illegal the combining of two or more competing corporations by pooling ownership of stock.
   - Federal Trade Commission Act: Creates the FTC; outlaws unfair methods of competition.
   - Robinson-Patman Act: Prohibits price discrimination to different buyers; requires sellers to make available supplementary services and allowances to all purchasers on a proportionately equal basis.
   - Wheeler-Lea Amendment: Outlaws false and deceptive advertising; broadens FTC's power to prohibit practices that might injure the public.
   - Lanham Act: Establishes protection for trademarks.
   - Celler-Kefauver Antimerger Act: Strengthens the Clayton Act to prevent corporate acquisitions that reduce competition.
   - Hart-Scott-Rodino Act: Requires large companies to notify the government of their intent to merge.

   (text p. 76)

VOCABULARY PRACTICE SOLUTIONS

1. target market
2. environmental management
3. poverty of time, component lifestyles
4. demography, discretionary income
5. Generation X, baby boomers, personalized economy
6. multiculturalism
7. stitching niches
8. inflation, recession
9. basic research, applied research
10. Consumer Product Safety Commission (CPSC), Federal Trade Commission (FTC), Food and Drug Administration (FDA)
11. ethics, morals
12. code of ethics
13. corporate social responsibility
14. pyramid of corporate social responsibility

TRUE/FALSE SOLUTIONS

(question number / correct answer / text page reference / answer rationale)

1.	F	60	A firm's target market is a defined group of people most likely to buy the firm's product. The target market may be *based* on a geographic location but is not actually the location.
2.	F	61	Marketers can control the four P's, but they have no direct control over external environmental elements.
3.	F	63	More Americans actually have modernist values (47%). Traditionalists represent only about 29% of the population.
4.	T	64	
5.	F	66	Generation X followed the Baby Boomers.
6.	F	66	Baby boomers are defined by their year of birth. Marital status, employment, and number of children are not factors.
7.	T	71	
8.	F	71	U.S. incomes are on the rise.
9.	T	77	
10.	F	78	A purely competitive market does not exist in the real world.
11.	F	83	Business ethics are actually a subset of the values held by society.
12.	T	84	
13.	F	85	If firms act in a socially responsible manner, additional government regulations may not be necessary, and this is a powerful argument on behalf of firms' social responsibility. Avoiding existing regulations is not an example of socially responsible behavior, and it's also illegal.

AGREE/DISAGREE SOLUTIONS

(question number / sample answers)

1. Reason(s) to agree: Generation X—composed of people aged 18 to 30—are rebellious, distrustful of authority, and cynical. This is directly a function of age, not necessarily of the times. When Baby Boomers were younger, they had some of the same characteristics. Marketing to Generation X, then, should be similar to the way firms marketed to Baby Boomers when they were aged 18 to 30.

Reason(s) to disagree: Despite the fact that Baby Boomers were also once young, rebellious, and distrustful of authority, Generation X is different because they grew up under different circumstances. They were bombarded even more with media messages, and many were latchkey children. They came into the workforce during tough economic times and were unable to find jobs as easily. Marketers cannot treat the two generations the same way because they have different values.

2. Reason(s) to agree: The U.S. is composed of many different cultures, and these cultures blend together to become what is known as "American" culture. It is to this collective group that firms should market their products.

   Reason(s) to disagree: The U.S. is not a melting pot; rather, it is a "mosaic" of cultures, each being distinct and separate from the others but working together to form what is known as "American" culture. Because they are distinct and separate, firms should adapt their marketing programs to each culture.

3. Reason(s) to agree: Though the demographic, technological, political, legal, and ethical environments are important, it is the economic environment that has the greatest impact on marketing efforts. It does not matter what demographic trend is occurring; consumer spending habits are directly influenced by what is happening in the economy. When the economy is in prosperity, consumers spend more; when the economy slows, consumers confidence decreases and they spend less. This has the most significant impact on business.

   Reason(s) to disagree: The demographic, technological, political, legal, ethical, and economic environments all have equal importance to most firms. Though the economic environment may have the most significant short-term impact, the other factors can have a significant long-term impact.

4. Reason(s) to agree: Let's face it: the for-profit firm is in business in make profits. Anything that is done—philanthropic actions, social responsibility—has the ultimate goal of making money in the long run. The upside is that society can benefit from the actions, regardless of the ulterior motive.

   Reason(s) to disagree: Organizations are made up of people, and people have a conscience. They make decisions based on many levels of morality—preconventional, conventional, or postconventional—but the person who makes decisions based on postconventional morality is doing what he or she thinks is the right thing to do, regardless of the profit or other impact on the firm.

MULTIPLE-CHOICE SOLUTIONS

(question number / correct answer / text page reference / answer rationale)

1.	c	60	The Baby Boomers would be considered the firm's target market: the group of people to whom its marketing efforts are targeted.
2.	e	60	Attempting to influence external environmental factors is environmental management.
3.	d	62	The factors of the other choices are demographic, economic, and technological; not social.
4.	a	61	Managers have the least amount of information available about social trends, which are the most important factor for estimating the success of clothing styles in the future.
5.	a	61	A trend to view anything disposable as wasteful and harmful to the environment is a threat for this firm.
6.	b	62	A component lifestyle pieces together products and services that fit a variety of interests and needs and does not conform to a certain stereotype.
7.	d	64	Each of the demographic factors stated are true and have influence on the sales of consumer packaged goods. The slowing of population growth has caused Kraft General Foods to niche-market to meet the needs of smaller groups of homogenous customers. Campbell Soup Company responded by offering single serving canned soups to the fast growing single market. Markets are selling packaged goods products by featuring famous older celebrities such as Lena Horne and Steve Allen.
8.	b	65	Generation Y is the cause of the increase of children's product offerings in the 1990s.
9.	d	66	Generation X is most likely to be unemployed or underemployed.
10.	d	66	The number one trait was "to think for themselves." This created a need for marketers to target the individualistic characteristics of this important but large target group.
11.	b	70	The Asian-American population outpaced Hispanics in the growth during the 1980s. However, Hispanics are projected to be the fastest growing ethnic group in the U.S. in the future.

12.	d	70	Although fast growth is projected for all minority segments, by the year 2023, the white segment will still be a majority, accounting for approximately two-thirds of the U.S. population.
13.	b	71	Stitching niches means seeking common interests, motivations or needs across ethnic groups. Mass marketing does not take the needs of various distinct cultures into account, and creating unique marketing mixes for individual mixes does not try to find a common basis among niches.
14.	b	72	In times of inflation, consumers are more price conscious and less brand-loyal.
15.	a	73	Reduced income, production, and employment result in reduced demand. These conditions define a recession.
16.	e	73	Applied research attempts to develop new or improved products.
17.	e	77	The sole purpose of the Consumer Product Safety Commission is to protect the health and safety of consumers in and around their homes. The commission has the power to prescribe mandatory safety standards for almost all products consumers use.
18.	b	77	Running advertisements that state that the former advertisements were inaccurate or incorrect is an example of corrective advertising.
19.	b	78	The nature of the competitive environment dictates the amount of flexibility a firm will have with its pricing.
20.	b	81	Preconventional morality is described as calculating, self-centered, and even selfish, and based on what will be immediately punished or rewarded. The firm will be immediately rewarded by higher profits if it sells the potentially harmful meat.
21.	b	82	Conventional morality moves from an egocentric viewpoint toward the expectations of society. The beer company is trying to operate by the standards set by society.
22.	c	82	Postconventional morality represents the morality of the mature adult. People are less concerned about how others might see them and more concerned about how they judge themselves over the long run. The chemical company is concerned with the long-term good of society.
23.	e	84	Social responsibility is the obligation that business feels for the welfare of society. By hiring the blind, the firm is helping to meet the needs of society.
24.	e	84	Social responsibility is a long-term commitment that business makes to the betterment of society.
25.	b	84	A national survey indicated that nearly one-third of employees felt pressure to engage in misconduct.

ESSAY QUESTION SOLUTIONS

1. There are six environmental factors discussed in the text. Here are some possibilities of how these factors might affect cigarette marketing:

   Social Change: As U.S. citizens become more health conscious, cigarette smoking is less socially acceptable. Smoking is now often banned in public and workplaces.

   Demographics: Because the older baby boomers tend to smoke, markets may remain solid. However, the number of new smokers will be declining because the number of young people is declining. Additionally, people are becoming more educated, and higher education levels mean lower smoking levels. Finally there is an increase in multicultural markets; different ethnic groups have varying demand levels for cigarettes.

   Economic Conditions: Changing economic conditions may cause people to spend their income on more important items than cigarettes. Penalties from car and health insurance companies for smokers may also prompt less smoking.

   Technology: New technologies in tobacco crop production and cigarette manufacturing may reduce costs for manufacturers.

   Competition: Competition in the mature cigarette industry is heavy, especially from new low-cost entrants. On a broader level, cigarettes may compete against other substitutes such as nicotine chewing gums, smokeless cigarettes, and chewing tobacco. Global competition is also a factor.

Political and Legal Factors: Taxes imposed on cigarettes, local laws banning smoking in public places, and laws requiring health warnings on packaging and advertising all serve to make the marketing of cigarettes more difficult.

(text pp. 61-80)

2. Several trends are apparent in the 1990s:

   The United States is growing older, or approaching middle-age. Between 1980 and 1994, the number of 24- to 44-year-olds increased by 32 percent, the population aged 65 or older grew by 24 percent, and the number of people over 85 expanded by 44 percent. This has led to marketing for products that cater to the elderly: easy-open packages, insurance, easy-to-read labels, items designed for those with arthritic hands, food products for people with dentures, and health-care programs.

   The United States is growing at a slower rate. The annual growth rate is less than 1 percent, and populations of younger U.S. citizens are shrinking. Marketers should shift their emphasis to groups with a higher growth rate, such as the elderly, or focus on the specialized needs of market niches.

   The singles market is growing, creating needs for new restaurants, convenience foods, and travel.

   There are shifts in population to Florida, Georgia, North Carolina, Virginia, Washington, and Arizona. Marketers may have to shift distribution and promotion emphasis to southern or western states.

   Rapidly growing minority markets--Blacks, Hispanics, and Asian-Americans--are all becoming significant market segments. Marketers have to recognize the specialized needs of all these groups.

   (text pp. 64-69)

3. **Preconventional morality** is childlike in nature. It is calculating, self-centered, and even selfish, and based on what will be immediately punished or rewarded. The Bolt salesperson might say, "It does not matter to whom I sell my products or how the products are used in the long run. That is not my problem. All that matters is the great profit I can get from every sale."

   Conventional morality moves from an egocentric viewpoint toward the expectations of society. Loyalty and obedience to the organization (or society) become paramount. The Bolt salesperson might say, "I need to make sure that I sell my products in such a way that the company remains profitable, but I should only sell to carefully screened, responsible, adult customers."

   Postconventional morality represents the morality of the mature adult. At this level people are less concerned about how others might see them and more concerned about how they see and judge themselves over the long run. The Bolt salesperson might say, "Even though the sale of my product is legal and will increase company profits, is it right in the long run? Might it do more harm than good in the end? Maybe I should quit this job and go back to school."

   (text, pp. 81-82)

4. The pyramid of corporate social responsibility portrays four kinds of responsibility: economic, legal, ethical, and philanthropic. Economic performance is the foundation for the structure, because if the company does not make a profit, then the other three responsibilities are moot. While maintaining a profit, business is expected to obey the law, do what is ethically right, and be a good corporate citizen.

   - Philanthropic responsibilities are to be a good corporate citizen, contribute resources to the community, and improve the quality of life.
   - Ethical responsibilities are to be ethical, do what is right, just, and fair, and to avoid harm
   - Legal responsibilities are to obey the law, which is society's codification of right and wrong, and play by the rules of the game.
   - Economic responsibilities are to be profitable, because profit is the foundation on which all other responsibilities rest.

   (text, p. 85)

CHAPTER 4 Developing a Global Vision

PRETEST SOLUTIONS

1. Global marketing is marketing to target markets throughout the world.

   (text, p. 96)

2. Five advantages that multinational firms have over domestic firms are:

   - They can overcome trade problems;
   - They can often sidestep regulatory problems.
   - They can shift production from one plant to another as market conditions change.
   - They can tap new technologies from around the world.
   - They can often save a lot in labor costs.

   (text p. 100)

3. Five external environments that global marketers face are:

   - Culture: the values, beliefs, and lifestyles shared by citizens of that culture.
   - Economic and technological development: whether the society is traditional, preindustrial, take-off, industrializing, or full industrialized.
   - Political: government policies, political systems, regulations, and legal structures.
   - Demographic makeup: population density, age distribution, urban versus rural distribution, and income levels.
   - Natural resources: such as petroleum or precious metals.

   (text pp. 101-114)

4. Five methods of entering the global marketplace are:

   - Exporting: selling domestically produced products to buyers in another country. The risk is low.
   - Licensing: selling the rights to use property, such as manufacturing processes, trade secrets, patents, trademarks, or other proprietary knowledge. Risk factors are higher than exporting but still relatively low.
   - Contract manufacturing: private label manufacturing by a foreign company. Risks are moderate.
   - Joint venture: ownership of an entity between two or more firms. Risk is relatively high.
   - Foreign direct investment: active ownership of a foreign company or of overseas manufacturing or marketing facilities. The risk is high.

   (text p. 114)

5. Four product/promotion options are:

   - One product/one message: a firm can use the same product and same basic promotional message throughout the world. An example is Kodak.
   - Product invention: creating a new product for a market or drastically changing an existing product. An example is that Nabisco had to change the taste of Oreo cookies to be less sweet when marketing to Japanese children.
   - Message adaptation: maintain the same basic product but alter the promotional message. Harley-Davidson sells the same product but changes its advertising message in Japan to show American motorcycle riders speeding through the Japanese countryside.
   - Product adaptation: maintain the same basic message but alter the product. An example is Domino Pizza's line extensions in Japan, which include toppings such as squid, corn, curry, and spinach.

   (text pp. 120-124)

VOCABULARY PRACTICE SOLUTIONS

1. global marketing, global vision
2. multinational corporation (MNC)
3. global marketing standardization
4. traditional society, preindustrial society, takeoff economy, industrializing society, fully industrialized society
5. Uruguay Round, World Trade Organization (WTO), General Agreement on Trades and Tariffs (GATT)
6. keiretsu
7. North American Free Trade Agreement (NAFTA), Mercosur, Maastricht treaty
8. exporting, buyer for export, export broker, export agent
9. licensing
10. contract manufacturing, joint venture, direct foreign investment
11. dumping
12. countertrade

TRUE/FALSE SOLUTIONS

(question number / correct answer / text page reference / answer rationale)

1.	F	100	For a firm to have the resources to be multinational, it must usually be an enormous company.
2.	F	100	Global marketing standardization makes this presumption.
3.	F	104	A society in the earliest stage of economic development is called a "traditional society." The preindustrial society is the second stage of economic development.
4.	F	107	This is an example of a boycott.
5.	F	110	GATT and NAFTA are trade agreements, EC is a market grouping.
6.	T	113	
7.	T	115	
8.	F	118	Athena is engaged in contract manufacturing, not licensing. Licensing would allow the manufacturer to sell the product and provide royalties to Athena.
9.	T	119	
10.	T	124	

AGREE/DISAGREE SOLUTIONS

(question number / sample answers)

1. Reason(s) to agree: Even domestic firms may want to expand their operations eventually, and one of the most obvious ways is to target new geographic markets outside one's home country. Whether a firm intends to become international or not, it will be faced with foreign competition entering its own home country, and the lack of a global vision will allow the foreign competitors to steal market share.

   Reason(s) to disagree: Very small firms—such as small clothing boutiques or family businesses—need to develop a global vision. Doing so would take focus away from its more immediate goals.

2. Reason(s) to agree: Culture is the foundation for the behavior of customers, the most important people in marketing.

   Reason(s) to disagree: Cultural factors as well as economic factors, technological factors, political and legal factors, and demographic factors are all equally important. Ignoring any one of these factors may lead to a fatal mistake.

3. Reason(s) to agree: By adapting both the product and the promotion to the local market, the marketer will always appeal to the market.

   Reason(s) to disagree: Adapting both product and promotion may not be necessary—and may even be undesirable—to the local market. In addition, adaptation is generally more expensive, since new advertising must be created, new packaging must be created, and a modified product with higher per unit production costs may be required.

4. Reason(s) to agree: Exporting is generally the least risky and least expensive form of international expansion. By exporting, firms can "test the waters" of a foreign market before committing more resources.

   Reason(s) to disagree: Foreign market conditions may make it more desirable to use another method of expansion, such as licensing or foreign direct investment. If a foreign government provides financial incentives—such as tax credits or subsidies—then foreign direct investment may be more desirable.

MULTIPLE-CHOICE SOLUTIONS

(question number / correct answer / text page reference / answer rationale)

1.	e	96	The reasons for thinking globally are summarized in all answers.
2.	a	98	A company heavily engaged in global trade moves resources, goods, services, and skills across national boundaries and is called a multinational corporation.
3.	b	100	Global marketing standardization is the product of uniform products that can be sold the same way throughout the world. McDonald's Big Mac is a good example of this.
4.	b	101	The first step is research--examining the external environment for opportunities and threats.
5.	b	101	Central to any country is its culture--a common set of values shared by its citizens that determine what is socially acceptable. Culture may dictate which colors are good or appropriate.
6.	a	104	Traditional societies are largely agricultural with a social structure and a value system that provide little opportunity for upward mobility.
7.	e	105	The takeoff economy is the period of transition from a developing to a developed nation, with new industries rising and a healthy social and political climate.
8.	c	105	In the industrializing period, technology spreads from leading sectors of the economy to the remainder of the nation, producing both capital goods and consumer durable products. Secondary markets and a large middle class begin to emerge.
9.	c	105	A preindustrial society has an emerging middle class with an entrepreneurial spirit but still lacks modern distribution and communication systems.
10.	b	107	A limit on the amount of a specific product that can enter a country is called a quota.
11.	c	114	Important factors for marketers to consider about a global market are the distribution of the wealth, the average household income, and the division of the population between rural and urban areas. Demography is the study of statistics of populations and their changes in age, income, distribution, and so on. In Indonesia, much of the population is poor and cannot afford luxuries such as blenders.
12.	d	115	An export broker matches buyers and sellers across international borders.
13.	e	115	Given the limited resources of the firm, and given that the firm's efforts are limited to exporting, the best choice is to build upon what they do best, creation of fruit juices.
14.	e	115	By continuing to control a vital part of the process and not train the licensee in 100 percent of the procedures, Blazer Lazer should be able to retain control.
15.	a	118	Marketing is usually handled by the domestic firm, as the contract manufacturer is involved in production only.
16.	d	117	Selling the rights to another company to market one's property in exchange for royalties is known as licensing.
17.	c	119	For the highest control and highest return, the firm will have to build a plant in South America.

18.	e	120	Though the system of marketing research remains the same its implementation, range of research choices, and even overall accuracy are impacted by different environments (especially so in a developing nation).
19.	e	122	Offering the same product in a foreign market but changing the advertising is an example of message adaptation.
20.	d	122	Product invention, in a global context, refers to a new market or use without changing the product.
21.	c	124	Developing nations often lack economic purchasing power.
22.	e	125	When multinational corporations are trying to gain fast market share, they may dump their products in the global market.
23.	a	125	Barter is the trading of a good for a good rather than a good for money.

ESSAY QUESTION SOLUTIONS

1. **Culture:** A global marketer must look at (1) assumptions and attitudes, (2) personal beliefs and aspirations, (3) interpersonal relationships, and (4) the social structure in each country. Culture affects the way products are used and could have a significant impact on the purchase and use of stereo systems.

   Level of economic and technical development: Techtronix should find out which economic stage describes the countries under consideration. These include the traditional society, preindustrial society, takeoff economy, industrializing society, and fully industrialized society. It is unlikely that countries at the lowest levels would be willing or able to purchase a digital tapedeck.

   Political structure: Techtronix should keep in mind the market-orientation of the economic system in each country. It should also be aware of the growth of nationalist sentiments in many nations. Finally, the company must become aware of international legal structures designed to either encourage or limit trade.

   Demographic makeup: Whether the population is urban or rural is a significant demographic consideration because marketers may not have access to many rural consumers. Another important factor is the amount of income within a country. In addition to the per capita income, the distribution of this income should be considered because often wealth is not evenly distributed.

   Natural resources: An important factor in the global external environment that has become more evident in the past decade is the shortage of natural resources. Shortages of raw materials in some countries may make it difficult to export the digital tapedecks.

   (text pp. 101-114)

2. **Traditional society:** These countries are in the earliest stage of economic growth and technical development. They are agricultural and custom-bound, and the economy operates at the subsistence level because of backward or no technology. The social structure and value system provide little opportunity for upward mobility. The culture may be highly stable and require a powerful disruptive force to initiate economic growth.

   Preindustrial society: This second stage of economic growth and technical development involves economic and social change along with the emergence of a rising middle class with an entrepreneurial spirit. Nationalism may rise, with resulting restrictive policies toward multinational corporations. Like traditional societies, marketing is difficult because of the lack of distribution and communication systems.

   Takeoff economy: The third stage is the period of transition from a developing to a developed nation. New industries arise, and a healthy social and political climate emerges.

   Industrializing society: In the fourth stage, technology spreads from the leading sectors of the economy that powered the takeoff to the remainder of the nation. At this point, the nation begins to produce capital goods, consumer durable products, and component parts. As a result, economic growth continues.

   Fully industrialized society: This society is the fifth and last stage of economic growth and technical development. In this stage the nation is an exporter of manufactured products, many of which are based on high technology. Great Britain, Japan, Germany, and the United States fall in this classification. The wealth of these nations creates market potential for extensive trading among industrialized countries. These nations may also trade manufactured goods for raw materials with developing countries.

   (text pp. 104-105)

3. **Tariff:** A tax levied on the goods entering a country. For example, trucks imported into the United States face a 25 percent tariff. Many products coming from developing nations receive tariff reductions or exemptions under the Generalized System of Preferences. Since the 1930s tariffs have decreased in use.

   Quota: A limit on the amount of a specific product that can enter a country. Quotas are usually sought as a means of protection from foreign competition. The United States has strict quotas on the importation of textiles, sugar, and dairy products. Harley-Davidson sought quotas on large motorcycle imports so it could remain competitive against Japanese firms.

   Boycott: An exclusion of all products from certain countries or companies. Governments employ boycotts to exclude companies or countries with whom they have a political difference. Several Arab nations boycotted Coca-Cola because it maintained distributors in Israel.

   Exchange control: A law compelling a company earning foreign exchange from its exports to sell it to a control agency such as a central bank. A company wishing to buy goods abroad must first obtain foreign exchange from the control agency. Generally, exchange controls limit the importation of luxuries by way of rationing exchange. Avon Products cut back production in the Philippines because of exchange controls.

   Market grouping: Also known as common trade alliances, market groupings such as the EC include several countries that agree to work together to form a common trade area that enhances trade opportunities.

   Trade agreements: These are legal agreements to stimulate global trade such as the Uruguay Round, which created the World Trade Organization, replacing the Agreement on Trade and Tariffs (GATT). These WTO is an organization whose objective is to increase global trade by lowering tariffs and reducing other trade barriers. Other countries (such as the United States and Canada) have signed bilateral trade agreements to increase global trade.

   (text pp. 107-108)

4. Japanese keiretsu, NAFTA, and the European Union are all types of alliances that provide benefits to participating members while sometimes creating barriers to outside firms.

   Keiretsu: Keiretsu is a Japanese society of business in which various domestic firms form an alliance in order to benefit each other. Keiretsu take on one of two forms: bank-centered keiretsu are massive industrial combines of 20 to 45 core companies centered around a bank. A good example of this is Sumitomo Bank, which has whole or partial ownership of many different companies, such as Sumitomo Construction, NEC Corporation (electronics), and Nippon Sheet Glass. A supply keiretsu is a group of companies dominated by a major manufacturer such as Matsushita, which not only manufactures electronics products but also controls the supply chain (retail stores). Keiretsu are sometimes considered to be trade barriers to foreign firms, which are blocked from distributing goods and services to Japan.

   NAFTA: The North American Free Trade Agreement is an agreement among the US, Canada, and Mexico that created the world's largest free trade zone. Though many trade barriers between the US and Canada had already been eliminated prior to NAFTA's 1993 ratification, NAFTA continued to eliminate trade barriers between these two countries and Mexico. Many tariffs, quotas, and licensing requirements have been lifted.

   The European Union: The EU is an alliance among 15 western European countries that calls for economic, monetary, and political union. The EU was officially formed with the ratification of the Maastricht Treaty in 1993. Most trade barriers such as tariffs and quotas have been completely eliminated among EU members. In 1999, eleven of the fifteen EU members formed a monetary union under a European central bank and a common currency called the "Euro."

   (text pp. 109-111)

5. **Direct foreign investment** is complete ownership of manufacturing and marketing subsidiaries in foreign countries. This option offers the greatest potential rewards but carries the highest risk.

   Joint ventures are similar to licensing agreements, except that the domestic firm assumes an equity position (partial ownership) in the foreign company. This is more risky but gives management a voice in company affairs that it might not have under licensing.

   Contract manufacturing is private-label manufacturing by a foreign company. The foreign company produces a certain volume of products to specification, with the domestic firm's brand name on the goods. The domestic company usually handles the marketing. This method carries medium levels of risk and return.

Licensing is a legal process whereby a licensor agrees to let another firm use its manufacturing process, trademarks, patents, trade secrets, or other proprietary knowledge. The licensee then agrees to pay the licensor a royalty or fee. This method has lower risk than direct or contract manufacturing. Franchising is one form of licensing that has grown rapidly in global markets in recent years.

Exporting is selling domestically produced products in another country. It is the least complicated and least risky alternative for entering the global market. A company deciding to export can sell directly to foreign importers or buyers, or it may sell to independent exporting intermediaries located in its domestic market. The most common intermediary is the export merchant, or buyer for export. This intermediary is treated like a domestic customer; the buyer assumes all risks and sells globally for its own account. Export brokers operate primarily in agriculture and raw materials. Export agents include foreign sales agent-distributors and purchasing agents of foreign customers.

(text pp. 115-120)

6. A joint venture is an alliance between two or more firms. An international joint venture is created when a domestic firm buys part of a foreign company or when it joins with a foreign company to create a new entity. Advantages of joint ventures include gaining quick market access, sharing the expenses of foreign expansion, and gaining valuable skills from your partner. Disadvantages include disagreement on management strategies, profit sharing, and high risk.

   (text p. 118)

7. **Product:** A firm could choose a one-product strategy and attempt to sell the same product, without major modifications, to all countries. However, minor modifications should be expected for a one-product strategy, including changes in packaging, sizes, and labeling. A company may not be able to follow a one-product strategy and may elect to make more drastic modifications to adapt products to the needs and wants of people in different countries. Finally, a firm may have to use the product invention strategy, which means developing a new product or drastically changing an existing one for a global market. Products in the global market may also need to meet ISO 9000 standards.

   Promotion: A global marketing standardization concept indicates that the global promotions strategy should be "one product, one message." In this case, marketers should attempt to standardize as much of their advertisements as possible. Advertising elements (such as a model or picture) should have universal appeal so they can appear in many different countries. Commercials with no voice-over or dialogue will not need translation, while print ads could share the same picture of the product and just use translated copy. Care should be taken that advertising elements are not culturally offensive in the different countries. Alternatively, marketers can develop individual promotional campaigns for each country, keeping in mind that media availability differs from country to country. Additionally, some cultures consider hard-sell promotions taboo and view advertising as something that actually devalues the product.

   Price: The pricing component of the marketing mix presents some unique problems and challenges in global marketing. Exporters must not only cover production costs but also consider transportation costs, insurance, taxes, and tariffs. Additionally, marketers should consider level of economic development. Because developing nations lack mass-purchasing power, prices may have to be lowered substantially. Firms should not assume that these countries will be willing to accept lower quality in exchange for lower price. Some companies that overproduce items end up dumping them in the global market. These products are sold either below cost or below their sale price in their domestic market. This can cause unfair competition. Dumping of products by foreign producers is illegal in the United States and Europe but is not controlled in other nations. Government tariffs and decrees further confuse pricing, In addition, fluctuation in global monetary exchange rates can make a product's price change drastically. Finally, pricing may not involve cash exchanges; about 30 percent of global trade involves barter, or the exchange of one product for another. Companies must determine the value of goods they are trading.

   Place: Often it is difficult for a firm to obtain adequate global distribution. For example, the Japanese distribution system may be the most complicated in the world. Imported goods must travel through many layers of intermediaries. The channels are based on historical and traditional patterns of socially arranged tradeoffs that are difficult for the government to change. In addition, retail establishments are very different. Japanese supermarkets are large multistory buildings that sell many nonfood items. Alternatively, China does not have the advantages of fully industrialized Japan. Transportation is primitive, with most products being transported by humans, wheelbarrows, carts, and bicycles.

   (text pp. 120-126)

CHAPTER 5 Consumer Decision Making

PRETEST SOLUTIONS

1. The five steps of the consumer decision-making process are:
   - Need recognition: An imbalance between actual and desired states.
   - Information search: Consumer seeks information about various alternatives to satisfy the need.
   - Evaluation of alternatives: Consumer evaluates the different products that will satisfy the need.
   - Purchase: Consumer purchases the product(s).
   - Postpurchase behavior: Consumer evaluates the purchase decision and may be satisfied or unhappy with the decision.

   (text p. 143)

2. Five factors that could influence the level of consumer involvement in decision-making are previous experience with the purchase or product, interest in the purchase, perceived risk of negative consequences, the situation around the purchase, and social visibility of the product.

   (text pp. 149-150)

3. Three cultural factors influencing consumer decision-making are:
   - Culture and values: A consumer's own culture and the values that it holds will influence decisions.
   - Subculture: A homogeneous group of people who share values with the overall culture but also have values unique to its own smaller group.
   - Social class: A group of people who are considered nearly equal in status or community esteem, who socialize among themselves, and who share behavioral norms.

   (text 153-160)

4. Three social factors influencing consumer decision-making are:
   - Reference groups: All the formal and informal groups that influence a consumer's decision making.
   - Opinion leaders: People within specific reference groups who influence others.
   - Family: Has a large influence on consumer decision-making.

   (text pp. 160-165)

5. Three individual factors influencing consumer decision-making are:
   - Gender: Men and women make purchase decisions differently.
   - Age and family life cycle: Consumers may make decisions based on either their age or their stage in the family life cycle (such as being "married with children" or "single").
   - Personality, self-concept, and lifestyle: An individual's personality traits, self-concept (how he or she sees themselves), and lifestyle (a mode of living) influence decision-making.

   (text pp. 166-168)

6. Four psychological factors influencing consumer decision-making are:
   - Perception: The process by which we select, organize, and interpret stimuli into meaningful and coherent pictures.
   - Motivation: The driving force that causes a person to take action.
   - Learning: The process that creates changes in behavior through experience and practice.
   - Beliefs and attitudes: Beliefs are an organized pattern of knowledge that an individual holds as true about his or her world. Attitudes are learned tendencies to respond consistently toward a given object.

(text pp. 168-176)

VOCABULARY PRACTICE SOLUTIONS

1. consumer behavior
2. consumer decision-making process, stimulus, need recognition
3. want
4. evoked set, internal information search, external information search, marketing-controlled information source, nonmarketing-controlled information source
5. cognitive dissonance
6. involvement
7. routine response behavior, limited decision making, extensive decision making
8. culture, subculture
9. reference groups, primary membership group, secondary membership group, aspirational group, nonaspirational reference group, norms, opinion leaders
10. socialization process
11. social class
12. personality, self-concept, ideal self-image, real self-image
13. lifestyle
14. perception, selective exposure, selective distortion, selective retention
15. motive, Maslow's hierarchy of needs
16. learning, stimulus generation, stimulus discrimination
17. value, belief, attitude

TRUE/FALSE SOLUTIONS

(question number / correct answer / text page reference / answer rationale)

1.	T	143	
2.	T	146	
3.	F	148	This describes a low-involvement buying situation.
4.	F	149	Acquiring information about an unfamiliar brand in a familiar product category is called limited decision making.
5.	T	149	
6.	T	150	
7.	F	153	The entire United States is probably a culture. The U.S. can, however, be divided into a number of subcultures, such as age, region, or ethnicity.
8.	F	158	The United States' social class system may not be similar to those in other countries. Additionally, the class delineations are not clearly defined.
9.	T	161	
10.	F	169	This describes selective retention because the information was not remembered. If selective distortion had occurred, Jon would have changed or distorted the information he read.
11.	T	171	
12.	F	174	Attitudes are formed toward brands, while beliefs are formed about specific attributes.

AGREE/DISAGREE SOLUTIONS

(question number / sample answers)

1. Reason(s) to agree: Even in making trivial decisions, such as deciding which restaurant to go to, consumers go through every step of the decision making process. Consumers recognize a need ("we're hungry"), search for information ("which restaurant is near us?"), evaluate alternatives ("there are three restaurants, which one sounds the best to you?"), purchase (go to the restaurant and buy lunch), and engage in postpurchase behavior (evaluate the food).

   Reason(s) to disagree: Certain types of purchases, such as impulse purchases, do not require all the steps in the decision making process. Consumers may not even think about the purchase until they are at the checkout counter of a grocery store, for instance.

2. Reason(s) to agree: Culture is beliefs, values, and lifestyles that are shared and learned by members of a society. As such, cultural values cut across all social groups that belong to that culture.

   Reason(s) to disagree: Different social classes are different subcultures. While some basic values are shared, subcultures have distinctions that set them apart from each other. For example, the higher classes may place a high value on higher education, while lower social classes may place a higher value on economic independence.

3. Reason(s) to agree: Manufacturers of cosmetics, clothing, and even sports cars portray aspirational reference groups in their advertisements in order to appeal to a specific target market. Cosmetics companies show beautiful women with flawless skin, while sports car manufacturers may show a savvy man who "has everything." Marketers want to show specific target markets what they could become if they bought the product.

   Reason(s) to disagree: Some of the most effective advertisements show "real people," not beautiful people. By showing real people, marketers are trying to get target markets to identify more directly with the product as "something for me."

4. Reason(s) to agree: In marketing, it does not matter which product is better, it only matters what the consumer perceives is the better product.

   Reason(s) to disagree: Reality (scientific studies) can influence how consumers perceive things.

MULTIPLE-CHOICE SOLUTIONS

(question number / correct answer / text page reference / answer rationale)

1.	c	143	The TV commercial is an external stimulus.
2.	d	144	A want is often brand specific, whereas a need is something an individual depends on in order to function efficiently. A person may need clothing but want specific brands.
3.	b	144	Public sources of information such as magazines, consumer rating organizations or even professionals not associated with the marketing company are known as nonmarketing-controlled information sources.
4.	e	147	Each option offered would help to reduce the postpurchase anxiety a consumer might experience.
5.	a	149	The buying of frequently purchased, low-cost goods is typically routine response behavior.
6.	e	149	Because of the situational factors, this low-cost item is a high-involvement product in this case.
7.	d	149	This is a low-involvement purchase that requires little search. The best strategy is to put the candy at the point-of-purchase, where impulse decisions are often made.
8.	e	154	Asians do not value freedom of expression very highly, as this promote the importance of the individual above the group.
9.	b	158	As a college-educated professional, Joe is part of the upper middle class.
10.	e	161	Aspirational groups are those groups that someone would like to join but currently is not a member.
11.	b	161	A nonaspirational or dissociative group is one that the consumer attempts to maintain distance from and does not want to imitate in purchase behavior.
12.	c	161	Frozen corn is a low-cost, low-involvement good. Reference group influence is more strongly felt on high-risk (beer or cigarettes), socially visible (clothing, beer, or cigarettes), or high-cost (car) products.

13.	a	161	The family is the most important social institution for many consumers.
14.	b	161	Consumer is used to denote the user of the good or service. This question emphasizes how wide a difference there can be between consumer and purchaser.
15.	d	167	Ideal self-image represents the way an individual would like to be.
16.	a	166	Lifestyle is defined by one's activities, interests, and opinions.
17.	c	169	Selective retention is the process whereby a consumer remembers only that information which supports personal feelings or beliefs.
18.	b	172	Safety needs are one of the first needs that individuals seek to satisfy, and the alarm system's selling point is the safety and security it provides.
19.	b	173	Lever Brothers' intention is to encourage trial of the product in order to engage learning (process that creates a change in behavior).
20.	e	174	Stimulus generalization occurs when one response (positive attitude for a product) is extended to a second, similar stimulus (new product, same brand).
21.	c	174	A belief is often developed about the attributes of a product. Attitudes, however, are more complex and encompass values.
22.	d	175	The text offers the first three choices as methods to employ when attempting an attitude change.
23.	a	176	The cereal company is attempting to change beliefs about taste, a product attribute, from negative to positive.

ESSAY QUESTION SOLUTIONS

1. For this high-involvement decision process, you would go through these steps:

    1. **Need recognition:** You want to go to graduate business school.
    2. **Information search:** You check both internal and external sources of information. Internal sources can include your own skills, knowledge, and feelings about geographic location of the schools. External sources can include college catalogs, college visits and interviews, and college guides that rank and describe various business programs.
    3. **Evaluation of alternatives:** You consider attributes of various schools in an evoked set. These attributes might include overall reputation of the school, tuition rates, availability of scholarships or other financial aid, and geographic location.
    4. **Purchase:** You apply to the schools you select. Upon hearing which schools will accept you, you decide which school you'll enter and accept the offer. You start your studies at the school.
    5. **Postpurchase behavior:** You are satisfied with your purchase, which was the result of extensive decision making. Alternatively, you are dissatisfied with your purchase, leave the school, and begin the process again.

    (text pp. 143-148)

2. The factors that affect involvement level are previous experience, interest, perceived risk of negative consequences, situation, and social visibility.

    Previous experience. Because there may be no previous experience or familiarity with the product, level of involvement will be high.

    Interest. Areas of interest vary by individual. You may or may not be interested in athletic shoes. However, purchasing the shoes indicates an interest in the social group and probably a high level of involvement.

    Perceived risk of negative consequences. Several types of risks are involved in the purchase. With expensive shoes, loss of purchasing power and opportunity costs result in financial risk. A social risk is taken, because wearing these shoes may cause a positive or negative reaction from other peer groups. For example, some might view the purchase as frivolous. Finally, there is a psychological risk involved in the form of anxiety or concern about whether the "right" shoes have been purchased and are acceptable to other members of the social club.

    Situation. The circumstances of the social club make the shoes a high-involvement purchase.

Social visibility. Because these shoes are on social and public display, wearing the shoes makes a statement about the individual. This would also make the purchase a high-involvement one.

(text pp. 149-150)

3. There are five levels in Maslow's hierarchy of needs:

   Physiological needs are the most basic level of human needs and include food, water, and shelter.

   Safety needs include security and freedom from pain and discomfort.

   Social needs involve a sense of belonging and love.

   Self-esteem needs include self-respect, feelings of accomplishment, prestige, fame, and recognition.

   Self-actualization is the highest human need. It refers to self-fulfillment and self-expression.

   (text pp. 171-172)

4. A reference group is a formal or informal group of people that influence the buying behavior of an individual. Four reference groups are:

   Primary membership groups are all people with whom an individual interacts regularly, such as family, friends, and coworkers. Friends and coworkers probably have the greatest influence on your wardrobe.

   Secondary membership groups are people with whom an individual interacts more formally, such as clubs, professional groups, and religious groups. A church group may influence how an individual dresses at church functions.

   Aspirational reference groups are those that a person would like to join. An individual may see an attractive model in a magazine advertisement and want to dress like that model.

   Nonaspirational reference group are those from whom an individual tries to dissociate him- or herself. The individual may avoid dressing like others at school in order to present a more professional image.

   (text p. 161)

5. The three methods are listed with some possible examples.

   Changing beliefs about attributes: The company could work to promote the image of a family brand by changing consumers' beliefs about adult versus children's toothpastes. Any negative beliefs or misconceptions should also be changed. For example, consumers may believe the toothpaste wears enamel or stains teeth when in fact it does not.

   Changing the importance of beliefs: The company could start emphasizing certain attributes that already exist. These might include environmental concerns (a package made of 100 percent recycled materials) or consumer preferences (the favorite choice of all consumers).

   Adding new beliefs: The company could try to expand the consumption habits of consumers by encouraging them to brush more than three times a day. The company could also emphasize additional attributes to the ones already in use such as more cavity fighting ingredients, enamel builders, or patriotism (a red, white, and blue package with a U.S. flag on it).

   (text pp. 175-176)

CHAPTER 6 Business Marketing

PRETEST SOLUTIONS

1. Business marketing is the marketing of goods and services to individuals and organizations for purposes other than personal consumption.

   (text p. 186)

2. A strategic alliance is a cooperative arrangement between business firms.

   (text p. 187)

3. Four major categories of business market customers are:
   - **Producers**: organizations that produce finished goods or services.
   - **Resellers**: retailers and wholesalers that buy finished goods and resell them for a profit.
   - **Governments**: federal, state, or local governments.
   - **Institutions**: organizations that seek to achieve goals other than the standard business goals of profit, market share, and return on investment. These could include schools, hospitals, or charities.

   (text pp. 190-191)

4. Seven types of business goods and services are:
   - **Major equipment**: capital goods such as large or expensive machines, mainframe computers, furnaces, generators, airplanes, and buildings.
   - **Accessory equipment**: such as portable tools and office equipment that are less expensive and shorter-lived than major equipment.
   - **Raw materials**: unprocessed products such as mineral, ore, lumber, wheat, corn, fruit, vegetables, and fish.
   - **Component parts**: either finished items for assembly or products that need very little processing before becoming part of some other product.
   - **Processed materials**: products used directly in manufacturing other products.
   - **Supplies**: consumable items that do not become part of the final product.
   - **Business services**: expense items that do not become part of a final product.

   (text pp. 197-198)

5. Five important aspects of business buying behavior are:
   - **Buying center**: all those persons who become involved in the purchase decision.
   - **Evaluative criteria**: criteria that are used to evaluate a purchase decision, such as price, service, and quality.
   - **Buying situations**: whether the product being considered is a "new buy," a "modified rebuy," or a "straight rebuy."
   - **Purchasing ethics**: ethical decisions involved in purchasing.
   - **Customer service**: service given by potential suppliers.

   (text p. 200)

VOCABULARY PRACTICE SOLUTIONS

1. business marketing
2. strategic alliance (strategic partnership)
3. North American Industry Classification System (NAICS)
4. derived demand, joint demand, multiplier effect (or accelerator principle)
5. reciprocity

6. major equipment, accessory equipment, raw materials, component parts, processed materials, supplies, business services

7. buying center

8. new buy, modified rebuy, straight rebuy

TRUE/FALSE SOLUTIONS

(question number / correct answer / text page reference / answer rationale)

1.	F	186	Since the end user of the product is the consumer, Tokia is engaged in consumer marketing, not business marketing. Just because a manufacturer uses retailers in its distribution does not mean that it is engaged in business marketing.
2.	F	186	Because of the Internet, GE's business market sales are larger than its consumer market sales.
3.	T	187	
4.	F	190	Institutions, one type of business organization, have these primary goals. Other business firms, such as resellers, share the same goals as many consumer firms.
5.	F	192	As its name indicates, NAICS (North American Industry Classification System) has been adopted only by Mexico, Canada, and the United States.
6.	T	192	
7.	F	195	This represents inelastic demand.
8.	T	196	
9.	F	197	Personal selling plays a very important role throughout the business purchasing process. Advertising has less emphasis.
10.	T	200	
11.	F	202	Adding features is an example of a modified rebuy.

AGREE/DISAGREE SOLUTIONS

(question number / sample answers)

1. Reason(s) to agree: Because business customers are fewer and far more important on an individual basis, it is more important for firms to develop strategic alliances in order to provide strong products and services to important business customers.

   Reason(s) to disagree: Strategic alliances are no more important to business marketers than to consumer marketers. Each type of organization can use alliances in any aspect of their business: from R&D to distribution.

2. Reason(s) to agree: Because business customers are fewer, marketers can get to know them well. Products can be targeted—and even tailored—to these customers to increase the chance for a sale.

   Reason(s) to disagree: Business customers are much more sophisticated buyers than consumers. Business marketing requires a thorough understanding of the buying situation, the people involved in the buying decision, and the customer's overall goals. Business buying can be highly complex and even political.

3. Reason(s) to agree: Just like consumers, business customers start with need recognition, seek information, evaluate alternative solutions, purchase, and engage in postpurchase behavior.

   Reason(s) to disagree: Business buying behavior is much more complex and includes several other steps in the process, such as establishing evaluative criteria, setting up buying situations, and establishing purchasing ethical guidelines.

4. Reason(s) to agree: A lot of hard work would go into a single sale to a business market, but the sale would be a big one.

   Reason(s) to disagree: There are rewards to selling to consumers markets as well. Since consumer goods tend to be better known than most business goods, the salesperson could enjoy the prestige of a well known product.

MULTIPLE-CHOICE SOLUTIONS

(question number / correct answer / text page reference / answer rationale)

1.	e	186	All the offered transaction types define business exchanges.
2.	d	186	Business markets will continue to be larger and more powerful than consumers markets with the use of the Internet.
3.	c	187	Strategic alliances are difficult to maintain over time.
4.	e	190	All of these businesses are producers.
5.	d	190	Reseller is the best answer because Sysco purchases finished goods and resells them. Sysco does not produce the goods or change their form and is not an institution. Because it actually purchases the goods, it is not just a transportation company. Inventory carrier is an unrelated term that does not explain why Sysco would break down quantities into cases.
6.	b	191	The most common process involves sealed bids, with selection going to the lowest price.
7.	a	190	The only example of a for-profit organization in the list is restaurants, and they would not be customers of an institutional-only supplier.
8.	a	192	NAICS is only used to classify firms within North America.
9.	c	193	NAICS currently has 20 economic sectors.
10.	c	192	The demand for soft drink containers is driven by the consumer demand for soft drinks; therefore, the containers have a derived demand.
11.	a	195	A product is price inelastic if a change in price leads to a small change in quantity demanded or does not significantly affect demand for the product. This often happens with utilities or other services that are highly demanded.
12.	e	195	When two or more items are used in combination in the final product, they have a joint demand.
13.	d	196	Businesses tend to be more formal in purchasing procedures and perform more paperwork and more analysis.
14.	a	196	Reciprocity is the normal business practice of using customers as suppliers of goods or services.
15.	e	198	Fishing is an extractive industry, and the fish are sent with little or no alteration to Star-Kist. This fits the definition of raw materials.
16.	b	198	The engines are finished items, ready for assembly into another item.
17.	e	198	As the manufacturer of the finished software that gets stored on the computer diskettes, Microsoft is a good example of an OEM (original equipment manufacturer) market.
18.	a	200	A gatekeeper is anyone who regulates the flow of information in a purchasing decision situation.
19.	d	202	Make or buy decisions are always a tradeoff with no clear solution.
20.	c	202	Because this is a new and complicated purchase, it will require a thorough analysis by the industrial buyer.
21.	a	202	When a previously purchased item needs to be reordered, but with changes or additions, it is a modified rebuy.
22.	c	204	Ordering the product again without consideration of competitive products represents a straight rebuy.

ESSAY QUESTION SOLUTIONS

1. A strategic alliance is a cooperative arrangement between business firms. Strategic alliances may include licensing agreements, joint ventures, distribution agreements, research and development consortia, and other general forms of partnership. Strategic alliances are becoming very important in business marketing because firms are realizing that they can become more competitive by taking on a partner that may have more expertise

in a certain area or that has access to certain markets. Successful strategic alliances allow all partners to gain substantial benefits from the alliance.

(text p. 187)

2. **Producers** include individuals and organizations that purchase goods and services for the purpose of making a profit by using them to produce other goods, to become part of other goods, or to facilitate the daily operations of a firm. Text examples include General Motors, AT&T, and IBM.

   Resellers include those wholesale and retail businesses that buy finished goods and resell them for a profit. Examples could include any grocery store or retail clothing store.

   Government organizations include a large number of buying units that purchase goods and services. The federal government, and state, county, and city governments are all examples.

   Institutions are nonprofit organizations that have different primary goals from ordinary businesses. This category includes schools, churches, hospitals, civic clubs, foundations, and labor unions.

   (text pp. 190-191)

3. Eleven differences are:

   Demand: There are several differences between organizational and consumer demand. Organizational demand is derived from the demand of consumer products, tends to be price inelastic, has joint demand with related products used in combination with the final product, and tends to fluctuate more than consumer demand.

   Purchase volume: Business customers buy in much larger quantities (both in single orders and in total annual volume) than do consumers.

   Number of customers: Business marketers tend to have far fewer customers than consumer marketers.

   Location of buyers: Unlike consumer markets, business customers tend to be geographically concentrated.

   Distribution structure: Channels of distribution tend to be much shorter in business marketing. Direct channels are also more common.

   Nature of buying: Business buying is usually more formalized with responsibility assigned to buying centers or purchasing agents.

   Nature of buying influence: More people are involved in business purchasing decisions than in consumer purchases because many levels and departments of the firm are involved in the purchase.

   Type of negotiations: Bargaining and price negotiation are more common in business marketing, including lengthy stipulations of final contracts.

   Use of reciprocity: Business purchasers often buy from their customers, and vice versa.

   Use of leasing: Businesses often lease equipment, unlike consumers who more often purchase products.

   Primary promotional method: Personal selling is often emphasized in business marketing, while advertising is emphasized in consumer marketing.

   There are several similarities between business and consumer buying behavior. First, both types of buyers use a decision process to make choices, although the steps can be different. Additionally, the personal makeup of individual buyers in a business purchasing situation continues to influence the purchase. Finally, both types of buyers react to environmental and situational factors when making a purchasing decision, including the influences of other role players (household members versus other organization members), influence of culture (either subcultures or organizational culture), or other environmental conditions (lifestyle or work style).

   (text pp. 192-197)

4. The seven types of business goods and services are:

   Major equipment such as large or expensive machines, computer systems, or generators.

   Accessory equipment, or shorter-lived and less expensive equipment, such as fax machines, power tools or microcomputers.

   Raw materials, or unprocessed extractive or agricultural products, such as mineral ore, lumber, and wheat.

Component parts such as spark plugs, tires, or electric motors for cars.

Processed materials, or materials used in manufacturing other products, such as sheet metal, chemicals, and plastics.

Supplies such as business stationery, cleansers, or paper towels.

Business services such as advertising, marketing research, or air freight.

(text pp. 197-199)

5. The **initiator** of the buying decision could be identified as the salesperson who identified the need but more likely would be the sales force manager who suggested the purchase be made.

   Influencers/evaluators might include the finance office (which would control the amount of money available for the purchase), members of the sales force (who might provide information about phones that competitors are using), and the purchasing department (which would have a good knowledge of alternative suppliers).

   Gatekeepers could include management (which would only approve of certain phone models that are compatible with existing phone systems), the finance office (which may only approve a limited budget), and the purchasing department (which would recommend matches with likely vendors).

   The **decider** might be the president of the company, the president of marketing, or the sales force manager; the decider is the person with the power to approve the brand of phone.

   The **purchaser** will be the purchasing agent in the purchasing department who will negotiate the terms of the sale.

   Users will include all sales force members who will use the cellular phone.

   (text p. 200)

6. **New buy:** A new buy is a situation requiring the purchase of a product or service for the first time. In this case, Pike's would have no experience buying color copiers or has not established any relationship with a vendor of color copiers. Pike's may be a new or small company that currently does not have any color copiers at all. Alternatively, Pike's may be involved in value engineering and searching for less expensive alternatives than buying color copiers. These could include photography or hand reproduction of color materials.

   Modified rebuy: In this case Pike's would have experience with copiers in general and an established relationship with office equipment vendors. The focus would be on the new need of added color capabilities.

   Straight rebuy: In this case the purchase of color copiers would be a routine purchasing decision or a reorder of previously ordered color copiers from the same vendor. Perhaps Pike's is a reseller of color copiers.

   (text p. 202)

CHAPTER 7 Segmenting and Targeting Markets

PRETEST SOLUTIONS

1. A market is people or organizations with needs ro wants and with the ability and the willingness to buy. A market segment is a subgroup of the market who share one or more characteristics that cause them to have similar product needs.

   (text p. 212)

2. The four basic criteria for segmenting consumer markets are: 1) substantiality; 2) identifiability and measurability; 3) accessibility; and 4) responsiveness.

   (text p. 214)

3. Five bases for segmenting consumer markets are:

   - Geographic segmentation: Segmenting by region of the country or world, market size, market density, or climate.

- Demographic segmentation: Segmenting by age, gender, income, ethnicity, or family life cycle.
- Psychographic segmentation: Segmenting by personality, motives, lifestyles, or geodemographics (neighborhood lifestyle characteristics such as segmenting by zip codes).
- Benefit segmentation: Segmenting by benefits that consumers seek from products.
- Usage-rate segmentation: Segmenting by the amount of product bought or consumed.

(text pp. 214-225)

4. Two bases for segmenting business markets are:
   - Macrosegmentation: Segmenting according to location, customer type, customer size, or product use.
   - Microsegmentation: Segmenting according to key purchasing criteria, purchasing strategies, importance of purchase, or personal characteristics.

   (text p. 225)

5. The six steps in segmenting a market are:
   - Select a market or category for study.
   - Choose a basis or bases for segmenting the market.
   - Select segmentation descriptors.
   - Profile and evaluate segments.
   - Select target markets.
   - Design, implement, and maintain appropriate marketing mixes for each segment.

   (text pp. 226-227)

6. Three strategies for selecting target markets are:
   - Undifferentiated targeting: targeting the entire mass market.
   - Concentrated targeting: focusing on a market niche, or a single segment.
   - Multisegment targeting: targeting two or more well-defined segments.

   (text pp. 227-231)

6. Positioning is developing a specific marketing mix to influence potential customers' overall perception of a brand, product line, or organization in general.

   (text p. 231)

VOCABULARY PRACTICE SOLUTIONS

1. market
2. market segment, market segmentation
3. segmentation bases (variables)
4. geographic segmentation
5. demographic segmentation, family life cycle (FLC)
7. psychographic segmentation, geodemographic segmentation
8. benefit segmentation, usage rate segmentation, 80/20 principle
8. macrosegmentation, microsegmentation, satisficer, optimizer
9. target market
10. undifferentiated targeting strategy, concentrated targeting strategy, niche, multisegment targeting, cannibalization
11. positioning, position, product differentiation, perceptual mapping, repositioning

TRUE/FALSE SOLUTIONS

(question number / correct answer / text page reference / answer rationale)

1.	F	212	Animals are not able to make purchases, nor do they have the authority to buy, and, therefore, they cannot be called a market.
2.	F	213	Even large companies like Coca-Cola practice market segmentation, as evidenced by the large variety of products that are offered at specific market segments.
3.	T	215	
4.	F	220	Targeting baby boomers indicates age as a segmentation variable; no reference is made to marital status and parental status, which are considered in the family life cycle.
5.	F	221	Psychographic segmentation specifically using variables of personality and lifestyle would better suit the Sharper Image.
6.	T	223	
7.	T	223	
8.	T	227	
9.	T	230	
10.	F	232	Perceptual maps indicate how consumers perceive a product's positioning. The perception may or may not be the firm's intended positioning.

AGREE/DISAGREE SOLUTIONS

(question number / sample answers)

1. Reason(s) to agree: Market segmentation helps firms to define customer needs and wants more precisely. By doing this, firms can allocate resources more efficiently and more accurately define marketing objectives.

   Reason(s) to disagree: Market segmentation does not necessarily have to happen. It could be argued that undifferentiated targeting is actually the absence of market segmentation in that it views the market as one mass market with no individual segments. Some firms use this strategy very effectively.

2. Reason(s) to agree: Substantiality indicates whether a market is large enough to sustain sales and profit over the long term. The other criteria—identifiability and measurability, accessibility, and responsiveness—are secondary after determining the market size.

   Reason(s) to disagree: All the criteria are equally important. If any of the criteria is lacking, the market segmentation will not be successful.

3. Reason(s) to agree: All firms decide where to distribute their products. Retail outlets practice geographic segmentation by locating themselves near their geographic segments. Manufacturers practice geographic segmentation by selecting which distribution channels and retail outlets will offer their product.

   Reason(s) to disagree: Global markets are, by definition, not geographically segmented. A firm can sell products through the Internet with no intention to target one geographic market versus another.

4. Reason(s) to agree: Psychographic segmentation is very difficult and requires a lot of primary research into consumer personalities, lifestyles, and motives. Firms cannot pick up secondary sources of research and "find" their psychographic segments easily. Besides, the psychographic elements of consumers (personalities, etc.) are self-reported and may not be as accurate as pure demographic or geographic data.

   Reason(s) to disagree: Regardless of the inaccuracy of psychographic data, elements of consumer personalities, lifestyles, or motives are very real to the consumer. Though most firms choose demographic and geographic segmentation because of the ease of finding accurate data, psychographic segmentation may allow a firm to differentiate itself in the minds of consumers.

5. Reason(s) to agree: Repositioning rarely works. Once a brand is established and well known (whether for good or bad reasons), it is very difficult and expensive to change the image that the consumer has of the brand. one example is Oldsmobile's failed attempt to change its image to be a young adult's car. Another example is Coke's attempt to target a younger market with a sweeter product ("New Coke"). It is better to launch a new product with a clean slate than to change old attitudes and views about a product.

Reason(s) to disagree: It is more expensive to launch a new product than to reposition an old one. A new product must build awareness before it builds a large consumer franchise. An old product can be repositioned, as long as the repositioning is believable.

MULTIPLE-CHOICE SOLUTIONS

(question number / correct answer / text page reference / answer rationale)

1.	e	212	A market is a group of people or organizations that has wants and needs that can be satisfied by particular product categories, has the ability to purchase these products, is willing to exchange resources for the products, and has the authority to do so.
2.	d	212	Dividing one large market into groups is called market segmentation.
3.	c	213	The purpose of segmentation is to group similar consumers and to serve their needs with a specialized marketing mix.
4.	a	213	The rapidly changing nature of most markets (business and consumer) dictates a frequent and regular reexamination of the segmentation process.
5.	e	213	Like any marketing institution, each of the factors mentioned are important criteria to consider when employing a segmentation strategy.
6.	e	214	Complexity is not a criterion; simplicity would be preferred.
7.	d	214	Substantiality means that the selected segment will be large enough to warrant developing and maintaining a special marketing mix.
8.	a	214	Serving the specific needs of a segment must be commercially viable, even if the number of potential customers is small.
9.	a	214	Identifiability and measurability is the first problem: How does one know who is shy and timid and how many shy and timid people are out there? The company would have to know that before it can gauge if there are enough of them. If there were enough timid people, they would probably be responsive to a special product and accessible because of mass media.
10.	b	214	Accessibility is the ability to communicate information about the product offering to the segment.
11.	c	214	Responsiveness is in force when a target segment responds differently (hopefully more positively) to the marketing mix than other segments.
12.	c	214	A segmentation base is the characteristic used to segment the market.
13.	b	215	Targeting by terrain is an example of geographic segmentation.
14.	d	215	Generation X is an age variable and is therefore demographic.
15.	a	215	The only demographic variable is age.
16.	b	219	By targeting marital status and the number of children in a household, the minivan is using family life cycle as its segmentation base.
17.	d	220	Adventurous and fun-loving describe personality, which is a psychographic base.
18.	e	224	By targeting heavy smokers, Wrigley's is using usage-rate segmentation (targeting "heavy ;users of cigarettes).
19.	a	223	The pens are designed to offer different benefits and uses to different target groups.
20.	c	225	The 80/20 principle proposes that a minority of a firm's customers will purchase a majority of the volume of the product.
21.	b	226	Satisficers use a simple, quick purchasing strategy of looking for the first available adequate supplier.
22.	e	226	Firms that use an optimizer purchase strategy consider many suppliers and carefully analyze options.
23.	d	227	An undifferentiated strategy would put Industry-Quip up against entrenched competitors, and the firm would have no product advantage.

24.	a	227	Because it is the only outlet in this small community and is likely to serve a variety of needs, it would have a substantial marketing cost savings if it employed an undifferentiated segmentation strategy.
25.	a	229	Left-Out is concentrating on one specific target market, which is a concentrated (or niche) targeting strategy.
26.	e	231	Cannibalization occurs when sales of a new product cause a decline in sales of a firm's existing products.
27.	d	232	The graphical display shows a perceptual map of consumers' perception of tea brands.
28.	e	233	Oldsmobile tried to reposition itself toward a younger target market, which may or may not have perceived the car as being for them.
29.	b	235	VALS is applicable to United States values and lifestyles and may be inappropriate to apply to people in other countries.

ESSAY QUESTION SOLUTIONS

1. There are five possible ways to segment this market:

   - One homogeneous market consisting of ten people (one segment)
   - A market consisting of ten individual segments
   - A market composed of two segments based upon age group (five adults and five children)
   - A market composed of three segments based upon region (four South, two Northeast, four West)
   - A market composed of five segments based on age group and region (one Adult/South, two Adult/Northeast, two Adult/West, three Child/South, two Child/West)

   (text pp. 212-213)

2. **Substantiality:** A selected segment must be large enough to warrant developing and maintaining a special marketing mix. (This means that the segment is commercially viable.) One could assume that many people have a dry scalp problem, especially in dry climates, after a sunburn, or when participating in outdoor activities with high winds (skiing, sailing, and so on).

   Identifiability and Measurability: The segments must be identifiable and their size measurable. Descriptive data regarding demographic, geographic, and other relevant characteristics of segment members must be available. It may be difficult to identify people who have dry scalps and describe them in terms of relevant characteristics. People may not realize that they have dry scalps and therefore may not respond to marketing research probes.

   Accessibility: The firm must be able to reach members of targeted segments with customized marketing mixes. If the segment is not measurable and cannot be described, it may be difficult to precisely know how to reach the dry scalp segment. Where do people with dry scalps shop? What media do they watch or read? What are their buying habits?

   Responsiveness: The market segment must respond differently from other segments to a marketing mix; otherwise, there is no need to treat that segment separately. People with dry scalps may not consider their problem to be important enough to respond differently and may continue to buy their current shampoo, which seems to meet needs.

   (text p. 214)

3. [Many different examples can be used. Make sure that each variable was used correctly.]

   Geographic variables: Target neighborhoods near college campuses.

   Demographics variables: Target young women (aged 15-30).

   Psychographics variables: Target "New Age" followers.

   Benefits-sought variables: Target people who need a way to relax.

Rate of product usage: Target people who are heavy users of aromatherapy products, such as shampoo or cologne.

(text pp. 214-225)

4. Possible benefits for toothpaste could include:

   Cavity/decay prevention--Crest

   White/bright teeth--Ultra Brite, Gleem, MacLeans, Plus White, Rembrandt

   Fresh breath--Close-Up, Pepsodent

   Tartar control--Tartar Control Crest, Tartar Control Colgate

   Plaque reduction--Dental Care, Viadent, Dentagard, Peak

   Stain remover--Pearl Drops, Topol, Caffree, Zact, Clinomyn, Rembrandt

   Gingivitis/gum disease prevention--Crest, Colgate, Metadent

   Flavor/great taste--Aim, Colgate, Stripe

   Fun for kids--Crest Sparkle, Oral-B Sesame Street, Colgate Jr.

   No mess--any pump or squeeze bottle toothpaste

   Sensitive teeth--Sensodyne, Denquel, Promise

   All-in-one--Aquafresh

   Denture cleaning--Dentu-Creme, Dentu-Gel, Complete

   Baking soda--Arm & Hammer, Metadent

   Low price--Arm & Hammer, store brands, brands on sale

   (text p. 223)

5. Undifferentiated strategy: Target everyone within a five-mile radius of the store. (Your thinking is that everyone should eat organic foods!)

   Concentrated strategy: Target higher income families that are extremely health conscious.

   Multisegment strategy: Target higher income professional women who are health conscious and elderly people who are health conscious.

   (text pp. 227-228)

6. Although exact locations will vary by student, likely quadrant locations are shown below. The positioning should indicate that there is room in the marketplace for a low-priced sporty car.

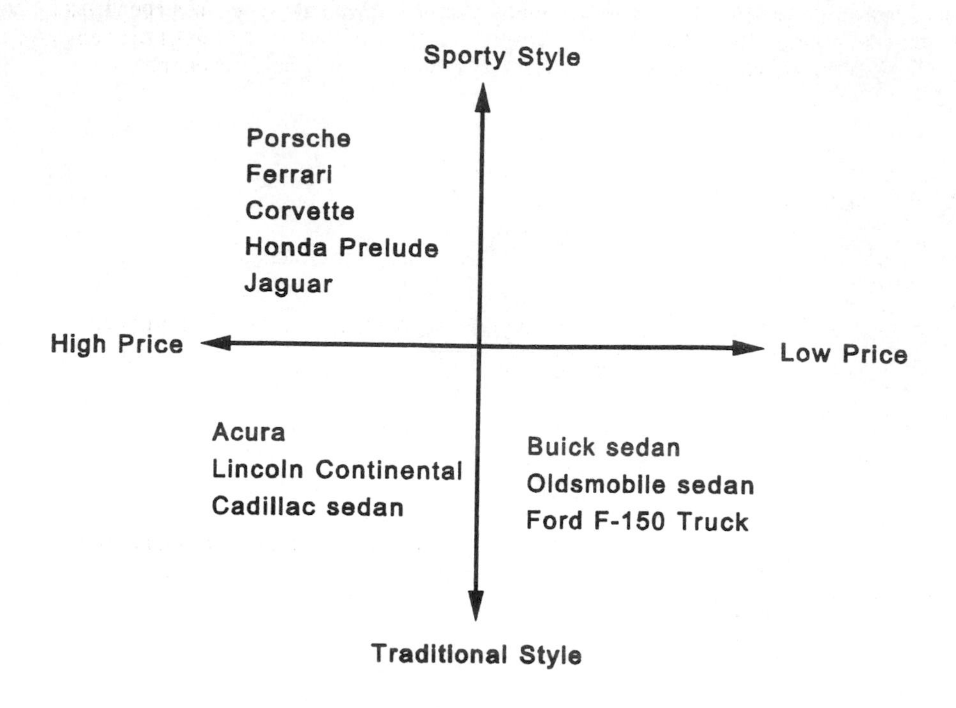

(text pp. 232-234)

CHAPTER 8 Decision Support Systems and Marketing Research

PRETEST SOLUTIONS

1. DSS systems should be interactive, flexible, discovery-oriented, and accessible.

   (text p. 244)

2. Marketing research is the process of planning, collecting, and analyzing data relevant to a marketing decision. Three roles are:

   - Descriptive: Gathering and presenting factual statements.
   - Diagnostic: Explaining data.
   - Predictive: Addressing "what if" questions.

   (text p. 245)

3. Steps in the marketing research process are: 1) Identifying and formulating the problem or opportunity; 2) Planning the research design and gathering primary data; 3) Specifying the sampling procedures; 4) Collecting data; 5) Analyzing the data; 6) Preparing and presenting the report; and 7) Following up on recommendations.

   (text p. 248)

4. Nine types of survey research are: 1) in-home personal interviews; 2) mall intercept interviews; 3) telephone interviews from the interviewer's home; 4) central location telephone interviews; 5) self-administered and one-time mail surveys; 6) mail panel surveys; 7) Internet interviews; 8) computer disk by mail; and 9) focus groups.

   (text pp. 257-267)

5. Scanner-based research is a system for gathering information from a single group of respondents by continuously monitoring the advertising, promotion and pricing they are exposed to and the things they buy. An example is BehaviorScan research, which tracks purchases of households through store scanners.

   (text p. 271)

VOCABULARY PRACTICE SOLUTIONS

1. marketing intelligence, decision support system (DSS)
2. marketing research
3. marketing research problem, marketing research objectives, management decision problem
4. secondary data, Internet, World Wide Web, Uniform Reference Locator (URL)
5. primary data, research design
6. survey research, mall intercept interview
7. computer-assisted personal interviewing, computer-assisted self-interviewing
8. central-location telephone (CLT) facility, in-bound telephone surveys, integrated interviewing
9. computer disk by mail survey
10. unrestricted Internet sample, screened Internet sample, recruited Internet sample
11. focus group, group dynamics
12. open-ended question, closed-ended question, scaled-response question
13. observation research, audit
14. experiment
15. universe, sample
16. probability sample, random sample, nonprobability sample, convenience sample
17. measurement error, sampling error, frame error, random error
18. field service firms, cross-tabulation
19. scanner-based research, BehaviorScan, InfoScan

TRUE/FALSE SOLUTIONS

(question number / correct answer / text page reference / answer rationale)

1.	F	244	Marketing intelligence is everyday information about the marketing environment.
2.	T	245	
3.	F	248	Marketing research focuses on a specific problem or opportunity that has arisen, not on general environmental information.
4.	F	248	The next step in the marketing research process is to determine marketing research objectives.
5.	F	249	Because the results from this survey were collected at another time, for another (although similar) purpose, the survey would represent secondary research data.
6.	T	256	
7.	T	251	
8.	T	267	
9.	F	268	This describes a convenience sample.
10.	T	273	

AGREE/DISAGREE SOLUTIONS

(question number / sample answers)

1. Reason(s) to agree: Since secondary data are not designed for specific marketing problems faced by a specific firm, they will most likely not help the firm solve its own specific problem.

   Reason(s) to disagree: Some secondary data may provide enough information for general problems or opportunities, such as discovering how many people comprise a demographic target market.

2. Reason(s) to agree: Because focus groups are derived from small sample sizes and because data coming from these groups are qualitative, they should only be used as a first step to any important decisions made by a firm.

   Reason(s) to disagree: Focus groups can be used to gauge certain situations, such as consumer feelings about new product ideas. By conducting focus groups across different geographic markets, a firm can "unbias" the information coming from the groups.

3. Reason(s) to agree: Information gathered from marketing research will always be flawed because of many different issues, such as sampling errors or the inaccuracy of attitudinal data (consumers' self reports may not be accurate).

   Reason(s) to disagree: Whether information is perfect or not is in the eyes of the marketer. If research uncovers information that is good enough for certain decision making, then it can be deemed to be perfect.

MULTIPLE-CHOICE SOLUTIONS

(question number / correct answer / text page reference / answer rationale)

1.	d	244	Decision support systems are easy to learn and use, even by people with little computer knowledge.
2.	b	244	Managers' ability to give instructions and see results illustrates the interactivity of a DSS.
3.	e	244	Marketing decision support systems are easy to learn and use, even by people with little computer knowledge.
4.	e	245	One of the roles of market research is to be diagnostic and to explain what happened.
5.	a	243	Researching consumer attitudes without researching why they have these attitudes is considered to be descriptive research.
6.	e	245	Predictive research tells managers the impact (decrease in sales) of a marketing decision (pricing).
7.	d	248	The first step must be to recognize the marketing problem.
8.	e	248	The key to conducting marketing research is systematically conducting research activities by first defining the problem.
9.	c	249	A marketing research objective should identify the specific information that the research will find out.
10.	c	249	Secondary data have been previously collected for some other purpose and may not fit the current research problem.
11.	b	249	Because the data is immediately available, internal company data can help create a profile quickly.
12	b	251	By typing the competitor's URL, you can often find annual reports that contain sales information available to the public.
13.	d	251	.com is used by for-profit businesses.
14.	a	254	Infoscan is a service that provides retail sales, market share, and other types of data by scanning UPC codes at retail outlets, such as supermarkets.
15.	a	256	Collecting primary data is more expensive than collecting secondary data.
16.	a	257	Piggyback studies gather data on two different projects using one questionnaire to save money.
17.	a	258	A mall intercept allows demonstration of the product, and the others do not. A laboratory experiment is not a survey.

18.	a	258	Telephone interviews are of moderate cost, when compared to in-home or mall interviews, yet are more expensive than mail surveys. Telephone interviews are also a fast survey method.
19.	c	260	The major disadvantage to using the Internet is that the sample will be biased; Internet users are heavily skewed toward well-educated, technically-oriented males.
20.	a	261	By calling potential respondents and asking them to participate in a survey, the company is using a recruited Internet sample.
21.	c	266	A type of research that does not rely on direct interaction with people is observation research.
22.	a	266	Observation research is the only data collection process in which there is no interaction and no possibility of influencing the behavior of the subjects. Response rates are a factor of survey research only.
23.	b	268	A probability sample is characterized by every element in the population having a known nonzero probability of being selected, allowing an estimate of the accuracy of the sample.
24.	b	268	A random probability sample occurs when every population member has an equal chance of being selected.
25.	c	268	A probability sample allows an estimate of the accuracy of the sample. In nonprobability samples there is no way to estimate the sampling error.
26.	a	268	A convenience sample is based on the use of respondents who are readily accessible.
27.	d	268	Measurement error occurs when there is a difference between the information desired by the researcher and the information provided by the measurement process.
28.	b	271	Scanner-based research systems gather information over time from a single panel of respondents. This attempts to develop an accurate picture of the direct causal relationship between marketing efforts and actual sales.
29.	a	273	The future benefits of the research must outweigh the cost of performing the research.

ESSAY QUESTION SOLUTIONS

1. Marketing research entails planning, collecting, and analyzing data relevant to marketing decision making, and communicating results of this analysis to management. Marketing research provides decision makers with data on the effectiveness of current marketing strategies. Marketing research is the primary data source for the DSS. The three functional roles of marketing research are descriptive, diagnostic, and predictive.

   The descriptive role of marketing research includes gathering and presenting factual statements. The diagnostic role of research explains and assigns meaning to data. The predictive role allows the researcher to use the descriptive and diagnostic research to predict the results of a planned marketing decision.

   Marketing research improves the quality of marketing decision making by shedding light on the desirability of various marketing alternatives. It can also help managers trace problems. Managers might use research to find out why something did not work out as planned. Marketing research can identify incorrect decisions, changes in the external environment, and strategic errors. Finally, marketing research provides insight into questions about the marketplace. Marketing research can help managers develop the marketing mix by providing insights into lifestyles, preferences, and purchasing habits of target consumers. Finally, marketing research can help foster customer value and quality.

   (text p. 244)

2.
   1. Identify and formulate the problem or opportunity
   2. Plan the research design and gather primary data
   3. Specify the sampling procedures
   4. Collect the data
   5. Analyze the data
   6. Prepare and present the report
   7. Follow up

(text p. 248)

3. Six advantages to conducting surveys on the Internet are:

- The speed with which a survey can be created, distributed to respondents, and received back from respondents
- Low cost
- The ability to track attitudes, behavior, and perceptions over time
- The convenience of asking two or three "quick" questions
- The ability to reach large numbers of people
- The ability to create visually pleasing surveys

Three disadvantages to conducting surveys on the Internet are:

- The biased sampling of the Internet (Internet users are not representative of the population as a whole)
- Problems with security on the Internet
- Unrestricted Internet samples (anyone who desires can complete a survey; this results in biased sampling)

Researchers can avoid getting unrestricted Internet samples two ways: by using screened Internet samples (whereby respondents are selected based on certain criteria, such as age or income or geographic region) and by suing recruited Internet samples (whereby respondents are recruited by telephone or other methods before the administration of the Internet survey).

(text pp. 250-255)

4. C Closed-ended multiple choice

B Closed-ended dichotomous

A Open-ended

F Two questions in one (Nikes? Reeboks? Both at once?)

G Biased/leading (Excellence is assumed.)

D Scaled-response

E Ambiguous (What does "soon" mean?)

(text p. 264)

CHAPTER 9 Creating a Competitive Advantage Using Competitive Intelligence

PRETEST SOLUTIONS

1. A competitive advantage is the set of unique features of a company and its products that are perceived by the target market as significant and superior to the competition. Three types of competitive advantage are:

- Cost competitive advantage: Being the low cost competitor in an industry while maintaining satisfactory profit margins.

- Differential competitive advantage: When a firm provides something that is unique that is valuable to buyers beyond simply offering a low price.
- Niche competitive advantage: When a firm seeks to target and effectively serve a small segment of the market.

(text pp. 284-288)

2. Competitive intelligence is the creation of an intelligence system that helps mangers assess their competition and their vendors in order to become more efficient and effective competitors. Benefits of competitive intelligence include: allows managers to predict changes in business relationships, identify marketplace opportunities, guard against threats, forecast a competitor's strategy, discover new or potential competitors, learn from the success or failures of others, learn about new technologies that can affect the company, and learn about how government regulations are impacting the competition.

   (text p. 288)

3. Five sources of internal competitive intelligence data include company employees, such as salespeople; warranty cards; returned merchandise forms; repair records; customer order files; back order reports; trade show inquiries; accounts receivable histories; and literature requests based upon company advertisements.

   (text p. 290)

4. Sources of non-computer-based external competitive intelligence are: experts, CI consultants, government agencies, UCC filings, suppliers, photographs, newspapers or other publications, yellow pages, trade shows, speeches by competitors, neighbors of competitors, and advisory boards.

   (text pp. 290-296)

5. Industrial espionage is an attempt to learn about competitors' trade secrets by illegal or unethical or both means.

   (text p. 298)

VOCABULARY PRACTICE SOLUTIONS

1. competitive advantage, differential competitive advantage, value impression, augmented product
2. cost competitive advantage, experience curve
3. niche competitive advantage
4. competitive intelligence
5. CI audit, CI directory
6. experts, Uniform Commercial Code (UCC)
7. industrial espionage

TRUE/FALSE SOLUTIONS

(question number / correct answer / text page reference / answer rationale)

1.	F	284	A cost competitive advantage is difficult to sustain over the long-term because technology can be transferred and because other competitors can cause price competition at any time.
2.	T	284	
3.	T	285	
4.	F	288	Competitive intelligence is part of an entire marketing intelligence system.
5.	T	290	
6.	F	290	The description is a CI directory, not a CI audit.
7.	F	296	Advisory boards are not an internal source but an external source of intelligence.
8.	F	293	UCC filings are open to the public, so accessing them is completely legal and ethical.
9.	F	299	Though France has been accused of some industrial espionage, it is China that poses the biggest threat to the U.S. today.
10.	T	299	

AGREE/DISAGREE SOLUTIONS

(question number / sample answers)

1. Reason(s) to agree: The free enterprise system endorses pure competition, in which firms have access to all kinds of information. As such, maintaining a competitive advantage (either in low costs, differentiated product, or a market niche) is difficult because a competitor can easily discover a firm's strategy and copy it if it is working.

   Reason(s) to disagree: Firms can maintain a competitive advantage in the long-term by creating barriers to entry. An example would be a firm that has created an innovative new product and that has patented the product, thus blocking other firms from creating the same product. This firm will be able to maintain a differential competitive advantage as long as the patent lasts.

2. Reason(s) to agree: If a firm makes a concerted effort toward maintaining an excellent competitive intelligence system, employees may be encouraged to do anything in their power to gather information. Since firms cannot control everything that an employee does, this effort will most likely lead to industrial espionage.

   Reason(s) to disagree: In its attempt at maintaining a good competitive intelligence system, management of a firm can state quite clearly that, though employees are highly encouraged to find legal and ethical means of uncovering competitive activity, all efforts to engage in illegal or "shady" activity will be punished.

3. Reason(s) to agree: Since salespeople usually work away from company headquarters, they are subject to less scrutiny from management. They may gather competitive intelligence from the field, but they may do so at a cost. Vendors and salespeople from competitive companies may trade information in a "tit-for-tat" deal, thus divulging company secrets from both sides.

   Reason(s) to disagree: Salespeople are still the best way to gather competitive intelligence. Though information may not be accurate (i.e., it may be based on rumors), this information may make management aware of what is going on in the industry. Even rumors have some value: it warns management of potential dangers, and the firm can be ready to respond if necessary.

4. Reason(s) to agree: External sources tend to be unbiased and based on factual information filed to government agencies, published by journalists, and so forth.

   Reason(s) to disagree: The analysis of factual information may be biased. Also, published sources can be as inaccurate as internal, subjective sources.

5. Reason(s) to agree: As wrong as it might be, industrial espionage is a reality among firms in certain countries. Countries such as China, France, Japan, Israel, the U.K., and even Canada have made economic espionage a top priority of their foreign intelligence services. Whether industrial espionage is right or wrong may be a cultural difference between the U.S. and the rest of the world.

   Reason(s) to disagree: No matter how many countries practice economic espionage, the U.S. should maintain its integrity and continue to punish firms that engage in it. As the economic leader of the world, the U.S. should set an example for other countries.

MULTIPLE-CHOICE SOLUTIONS

(question number / correct answer / text page reference / answer rationale)

1.	b	284	Nordstrom's customer service is an example of having a competitive advantage.
2.	a	284	By running efficient operations with a no frills service, Southwest Airlines enjoys a cost competitive advantage.
3.	e	284	Firms often gain a cost competitive advantage by having "experience" at production so that per unit costs decline over time.
4.	c	287	Augmented products do not usually provide a cost competitive advantage.
5.	c	287	Federal Express uses a differential competitive advantage to distinguish its service from that of its competitors.
6.	d	287	Cosmetics companies use packaging to give their product a higher value impression; otherwise, cosmetic brands would be perceived as not being different from each other.
7.	e	287	By targeting such a narrow group of students, Hudson's is practicing a niche competitive strategy.
8.	b	289	The description fits that of competitive intelligence.

9.	a	289	Having an good competitive intelligence system does not guarantee success; it only helps.
10.	e	290	What is described is a CI (competitive intelligence) audit.
11.	c	290	What is described is a CI (competitive intelligence) directory.
12.	d	296	Advisory boards are considered to be an external source of intelligence since board members are not employees of the firm.
13.	e	291	Though she has expertise in her field, Geena is a competitive intelligence (CI) consultant.
14.	a	294	The yellow pages are a good first source for competitive intelligence, especially for a local marketplace.
15.	d	291	FOIA is not a government agency. It stands for the Freedom of Information Act.
16.	a	291	The SEC (Securities and Exchange Commission) is a good source for financial information about publicly traded firms.
17.	b	293	They should use UCC (Uniform Commercial Code) filings to find this information.
18.	c	294	The yellow pages cannot divulge product manufacturing costs. This is generally considered to be highly private information.
19.	e	295	Trade shows are the quickest way to gather sales literature in one stop.
20.	d	296	A company's web site not only has general information about the company but may contain important recent press releases about new products introductions, etc.
21.	b	296	Lexis-Nexis is a good on-line database of recent lawsuits.
22.	c	298	The "janitor" is using unethical means of obtaining competitive information, or industrial espionage.
23.	e	299	China currently poses the biggest threat to U.S. firms.
24.	b	299	The act that makes it illegal to engage in industrial espionage is the Economic Espionage Act.
25.	c	299	Both individuals (Greg and Keith) would be guilty under the Economic Espionage Act.

ESSAY QUESTION SOLUTIONS

1. Three competitive advantage strategies and examples that would fit under each are:

   - Cost competitive advantage: Being the low cost competitor in an industry while maintaining satisfactory profit margins. You could target lower income families who would not otherwise be able to afford child care. You could receive government subsidies to run your center and operate under a "no frills" approach to keep costs low.
   - Differential competitive advantage: When a firm provides something that is unique that is valuable to buyers beyond simply offering a low price. You could focus your positioning on high quality customer service, on good relationships from parents, or on extremely high academic achievement.
   - Niche competitive advantage: When a firm seeks to target and effectively serve a small segment of the market. You could target small market niches, such as to open an "international school" for children who want to learn different languages or an "entrepreneurial school" for children who want to learn to run businesses.

   (text pp. 284-287)

2. Five internal sources that would not cost you anything are: your company's sales force, your company's technical staff, back order reports, trade show inquiries, and accounts receivables records. Five external sources that would not cost you anything are: government agency reports (some of them), advisory boards, newspapers and other periodicals search, UCC filings, and most Internet sites (especially competitor's web sites).

   (text pp. 288-289)

3. Industrial espionage is an attempt to learn competitors' trade secrets by illegal or unethical or both means. Examples include: outright theft of trade secrets, concealment, fraud artifice, deception, copying without authorization, duplication, sketches, drawings, photographs, downloads, uploads, alterations, destruction,

photocopies, transmissions, deliveries, mail, communications, or other transfers of conveyances of trade secrets without authorization.

(text p. 298)

CHAPTER 10 Product Concepts

PRETEST SOLUTIONS

1. A product is everything, both favorable and unfavorable, that a person receives in an exchange. Four types of consumer products are:

   - Convenience products: A relatively inexpensive item that merits little shopping effort.
   - Shopping products: A more expensive product that is found in fewer stores.
   - Specialty products: A product for which consumers shop extensively.
   - Unsought products: A product unknown to the buyer or a known product that the buyer does not actively seek.

   (text pp. 318-320)

2. Product item: A specific version of a product that can be designated as a distinct offering among an organization's products. Example: Gillette's MACH 3 razor.

   Product line: A group of closely related products. Example: Gillette's entire offering of blades and razors.

   Product mix: All the products an organization sells. Example: Gillette sells blades, razors, toiletries, writing instruments, and lighters.

   (text p. 320)

3. A brand is a name, term, symbol, design, or combination thereof that identifies a seller's products and differentiates them from competitors' products. Three branding strategies are:

   - Generic products versus branded products: Whether or not to brand your product.
   - Manufacturer's brand versus private brands: Whether a retailer should use a manufacturer's brand name or its own private label name.
   - Individual brands versus family brands: Whether to use individual brands names (such as Tide detergent) or family brand names (such as Sony's entire line of electronics products).

   (text pp. 323-330)

4. The three most important functions of packaging are: to contain and protect the product, to promote products, and to facilitate the storage, use, and convenience of products.

   (text pp. 332-333)

5. Three different branding strategies for entering international markets are:

   - To use one brand name everywhere (such as Coca-Cola using the same brand name everywhere);
   - To adapt or modify the brand to the local market ;
   - To use different brand names in different markets (such as Gillette changing its Silkience hair care name to "Soyance" in France and "Sientel" in Italy.

   (text pp. 334-335)

6. Warranties not only protect the buyer from defective products but also gives essential information about a product.

   (text p. 335)

VOCABULARY PRACTICE SOLUTIONS

1. product
2. business (industrial) product, consumer product
3. convenience product, shopping product, specialty product, unsought product
4. product item, product line, product mix
5. product mix width, product line depth
6. product modification, planned obsolescence
7. product line extension
8. brand, brand name, brand mark
9. brand equity, master brand, brand loyalty
10. generic product, manufacturer's brand, private brand, individual branding, family brand, co-branding
11. trademark, service mark, generic product name
12. persuasive labeling, informational labeling, universal product code (UPC)
13. warranty, express warranty, implied warranty

TRUE/FALSE SOLUTIONS

(question number / correct answer / text page reference / answer rationale)

1.	F	318	Consumers do not buy toothpaste based on its chemical makeup; they buy products based on the benefits they deliver.
2.	F	319	Because Tammy puts so much effort into an information search, makeup is better classified as a shopping product.
3.	T	320	
4.	F	320	What is described is actually part of Sony's product mix, not its product line. A product line is a group of related products.
5.	F	322	The cereal already had an existing position. Changing consumer perceptions, as in this example, is called repositioning.
6.	T	323	
7.	F	323	The Nike "swoosh" is a brand mark, not a brand name. The word "Nike" is a brand name.
8.	T	327	
9.	T	329	

AGREE/DISAGREE SOLUTIONS

(question number / sample answers)

1. Reason(s) to agree: For most products, price is the number one feature for consumers. When buying a car, for instance, consumers select cars based on a price range first, then on other features. In some industries, brand names almost do not matter any more (such as long distance service, which has become heavily price driven).

   Reason(s) to disagree: Branding is still very important. A brand name distinguishes one company's product from that of its competitors and is associated with certain features and imagery. One good example is Coca-Cola, the best known brand in the world. The equity that has been developed throughout the decades on Coke is more important than the actual taste of the product, as Coca-Cola saw when it tried to introduce New Coke in 1985.

2. Reason(s) to agree: Private brands—or brands that are owned by retailers or wholesalers (rather than manufacturers)—are usually always priced lower than the manufacturers' brands. People cannot afford to pay the prices of major brands will naturally go to the private brands, which may be perceived as "good enough" since it has the same ingredients.

   Reason(s) to disagree: It is not only lower income consumers who purchase private brands. Many higher income people, who do not see the merit of paying a premium for manufacturers' brands, also purchase private brands.

Whether a higher income consumer chooses to purchase a manufacturer's brand or a private brand has more to do with the value impression that he/she has rather than the income.

3. Reason(s) to agree: By adapting products, including brands, to local markets, the global marketer has a better chance of acceptance from the local market. Brand names often do not translate well into other languages, and products may not sell if some adaptations are not made.

   Reason(s) to disagree: Products and brands do not have to be adapted to local markets. Sometimes the image of the product's home country is just as important as the product itself. Levi's is an example of a U.S.-based company that markets "Americana" to other cultures and does not adapt the product on purpose. McDonald's is another example of a successful global brand that does not adapt its products (it adapts some of its product offerings, however) and markets American culture. It does adapt its promotions and its environment to match that of its market, however.

4. Reason(s) to agree: Family branding helps to develop overall brand equity and to tie together products that otherwise would not be related. Using a strong brand name developed on a strong product on other products that are related will help boost sales of the newer products. Gillette has been very successful at family branding, as well as Sony, because they use strong company brand names to communicate certain quality features across their entire product line or product mix.

   Reason(s) to disagree: Family branding is not necessary for strong brand equity development. Products such as Tide detergent have existed on their own for many years and may target very different markets versus other detergents made by the same company, Procter & Gamble. When targeting different market segments, individual branding may be more appropriate.

MULTIPLE-CHOICE SOLUTIONS

(question number / correct answer / text page reference / answer rationale)

1.	e	318	The product is the first decision around which the others are based.
2.	a	319	The fact that David did not consider another brand shows that the car was a specialty product.
3.	b	319	Shopping products are usually more expensive than convenience products and are found in fewer stores. Consumers usually compare items across brands or stores.
4.	c	318	The lack of forethought that went into this purchase made it a convenience product.
5.	a	320	An organization's product mix includes all of the products that it sells.
6.	e	321	Product line depth is the number of product items in a product line.
7.	a	322	Quality modifications entail changing a product's dependability or durability.
8.	b	322	Functional modifications are changes in a product's versatility, effectiveness, convenience, or safety.
9.	d	322	Planned obsolescence can take place as a result of any type of product modification.
10.	c	323	Adding products to a product line entails line extension.
11.	e	324	Coca-Cola is enjoying a strong brand equity: the value of the brand name to the company.
12.	c	324	When consumers immediately think of a specific brand when a product category, use situation, product attribute, or customer benefit is mentioned, the brand is a master brand.
13.	e	327	A private brand is one owned by the retailer.
14.	b	327	The dealer will get a higher margin on a private brand.
15.	b	328	Individual branding is the policy of using different brand names for different products. Individual brands are used when products differ greatly in use, performance, quality, or targeted segment.
16.	e	329	When two separate brands (in this case, from two separate companies) use their brands on a product or package, this is known as co-branding.
17.	d	331	A generic product name identifies a product by class of type and cannot be trademarked.
18.	a	332	Guarantees are not a function of packaging; they are a function of warranties.
19.	d	332	Currency and pricing considerations are not the prime considerations when considering branding and packaging strategies.

20.	b	335	An express warranty is any written guarantee.
21.	e	335	This is the definition of a warranty.
22.	a	335	Under the Uniform Commercial Code, all sales carry an unwritten guarantee that the good or service is fit for the purpose for which it was sold.

ESSAY QUESTION SOLUTIONS

1. **Convenience products** are relatively inexpensive items that require little shopping effort. The products are bought regularly, usually without significant planning. Convenience products may include candy, soft drinks, combs, aspirin, small hardware items, dry cleaning, car wash services, and so on.

   Shopping products are more expensive than convenience products and found in fewer stores. Consumers spend some effort comparing brands and stores. Shopping products may include washers, dryers, refrigerators, televisions, furniture, clothing, housing, choice of university, and so on. Shopping products can be homogeneous (consumers perceive the products as being essentially the same) or heterogeneous (consumers perceive the products to differ in features, style, quality, and so on).

   Specialty products are those items for which consumers are willing to search extensively. Consumers are extremely reluctant to accept substitutes for specialty products. Brand names and service quality are important. Fine watches, luxury cars, expensive stereo equipment, gourmet restaurants, and specialized medical services could all be considered specialty products.

   Unsought products are those that the buyer does not know about or does not actively seek to buy. These products include insurance, burial plots, encyclopedias, and so on.

   (text pp. 318-320)

2. Product mix width refers to the number of product lines that an organization offers. In this case the width of the product mix is three: coffees, appliances, and desserts. Product line depth is the number of product items in a product line. There are five items in the coffee line, three items in the appliance line, and two items in the dessert line.

   The company could adjust its portfolio with modifications (quality, functional, or style), additions, repositioning of products, or extending or contracting product lines.

   (text pp.320-321)

3. Strategies could include:

   - Generic branding
     - Advantages: low pricing and high volume
     - Disadvantages: no brand equity for a well known company.
   - Branded product
     - Advantages: to build brand equity over time and to gain higher profit
     - Disadvantages: none.
   - Individual branding
     - Advantages: can target other market segments that other Gillette products do not currently target
     - Disadvantages: does not build on company's good name
   - Family branding
     - Advantages: build on company's positive image and name
     - Disadvantages: may cannibalize other products

   Note: In this case, it does not make sense to co-brand.

   (text, pp. 323-329)

4. **One Brand Name Everywhere.** This strategy is useful when the company markets mainly one product and the brand name does not have negative connotations in any local market. Advantages of a one-brand strategy are greater identification of the product from market to market and ease of coordinating promotion from market to market. This strategy may be difficult for Q-T-Pie because a variety of products are sold. Additionally, it is not likely that "Q-T-Pie" carries the same meaning in all languages or communicates the benefits of the product.

   Adaptations and Modifications. If the brand name is not pronounceable in the local language, the brand name is owned by someone else, or the brand has a negative connotation in the local language, minor modifications can make the brand name more suitable. This could be a viable alternative for Q-T-Pie.

   Different Brand Names in Different Markets. Local brand names are often used when translation or pronunciation problems occur, when the marketer wants the brand to appear to be a local brand, or when regulations require localization. This could also be a viable alterative for Q-T-Pie.

   (text, pp. 334-335)

CHAPTER 11 Developing and Managing Products

PRETEST SOLUTIONS

1. The six categories of new products are:
   - New-to-the-world products
   - New product lines
   - Additions to existing product lines
   - Improvements or revisions of existing products
   - Repositioned products
   - Lower priced products

   (text pp. 344-345)

2. The seven stages of new product development are:
   - Develop a new product strategy
   - Generate new product ideas
   - Screen ideas for best ones
   - Conduct a business analysis
   - Develop the product prototypes
   - Test the product in laboratory and test markets
   - Commercialize (introduce) the product to the market

   (text pp. 345-353)

3. Reasons why new products may fail are:
   - No discernible benefit versus existing products
   - Poor match between product features and customer desires
   - Overestimation of the market size
   - Incorrect positioning
   - Price is too high or too low
   - Inadequate distribution

- Poor promotion
- Inferior compared to competitors

(text p. 354)

4. Diffusion is the process by which the adoption of an innovation spreads. Five categories of adopters are:
   - **Innovators**: The first 2 ½ percent of all those who adopt the product. They are risk-taking and like to try new ideas.
   - **Early adopters**: The next 13 ½ percent of all those who adopt the product. They are opinion leaders.
   - **Early majority**: The next 34 percent of all those who adopt the product. They are more likely to weigh pros and cons before adopting a product and rely on the group for information.
   - **Late majority**: The next 34 percent of all those who adopt the product. They are conservative and tend to adopt the product because most of their friends already have it.
   - **Laggards**: The last 16 percent of all those who adopt the product. They are very conservative and are tied to tradition.

   (text pp. 356-357)

5. The four stages of the product life cycle are:
   - **Introductory stage**: Full scale launch of the product. Sales grow slowly, and marketing costs are high.
   - **Growth stage**: Sales grow at an increasing rate as awareness of the product grows. Advertising and other promotional costs are high, and competitors enter the market.
   - **Maturity stage**: Sales start to flatten out. The market approaches saturation as new users are more difficult to identify. Profits and sales are at an all-time high.
   - **Decline stage**: A long-run drop in sales occurs. Product may eventually die or be pulled before it does.

   (text pp. 359-361)

VOCABULARY PRACTICE SOLUTIONS

1. new product
2. new product strategy
3. product development
4. brainstorming
5. screening, concept test
6. business analysis, development
7. test marketing, simulated (laboratory) test marketing, commercialization
8. new-product committee, new-product department, venture team
9. simultaneous product development
10. adopter, innovation, diffusion
11. product category, introductory stage, growth stage, maturity stage, decline stage

TRUE/FALSE SOLUTIONS

(question number / correct answer / text page reference / answer rationale)

1.	F	344	Line extensions, such as new ice cream flavors, are considered to be a type of new product.
2.	T	345	
3.	F	346	Brainstorming does not involve evaluation of the ideas as they are generated; criticism of any kind is avoided.
4.	T	346	
5.	T	347	

6.	F	351	Test markets are generally very expensive to conduct.
7.	T	353	
8.	F	355	Venture teams, not venture capitalists, would be such an organizational form.
9.	F	356	Simultaneous product development describes an organization form in which all departments work together throughout the development process, not in competition with each other.
10.	F	356	Sales of any product do not follow precise bell-shaped curves over a set period of time. The length of time a product spends in any one stage of the product life cycle may vary dramatically.
11.	F	356	Carol would be considered an innovator since she is among the first to see the new movies.
12.	T	358	
13.	T	360	

AGREE/DISAGREE SOLUTIONS

(question number / sample answers)

1. Reason(s) to agree: In today's competitive marketplace, too much testing and too much time spent on generating ideas can cause a product failure. Another company may be thinking about the same idea and may launch the product before you. Also, when a product is tested—especially in the market—competitors can find out about the new product.

   Reason(s) to disagree: Following each step methodically will help increase the success rate of the new product. Each step is necessary, though each step should be accomplished efficiently. Too much time should not be spent on idea generation, for instance, while enough time should be given to product testing to ensure that all strategies are sound.

2. Reason(s) to agree: The most important quality of a new product idea is the positioning that it holds. If the positioning is good—that is, if the new product delivers unique benefits that the target market wants—then it will certainly succeed.

   Reason(s) to disagree: Even products with a sound, unique, and desirable positioning can fail in their execution. If the advertising is incorrect or inadequate, if the price is too high or too low, or if distribution does not occur quickly enough, the product can fail, despite its original strategy.

3. Reason(s) to agree: A new products committee can only generate a limited number of ideas. Employees from around the organization should be encouraged to join in the new products idea generation, as the volume of ideas is necessary.

   Reason(s) to disagree: While getting ideas from all employees is good, the lack of a new products committee will de-emphasize the necessity of new products development. If no committee is called upon to be the leader of new products development, no one will have accountability for the effort.

4. Reason(s) to agree: Products such as fads, fashions, styles, or even some high-tech products do not follow the traditional product life cycle theory. Fads go from introduction to decline very quickly and jump over some stages. Styles (a recurring product) come and go every several years and can be "reborn" at any time. High-tech products become obsolete very quickly, and new technology take their place before the old products even die.

   Reason(s) to disagree: Regardless of the type of product, every product undergoes some kind of life cycle. The product may not go through a traditional life cycle, but it does grow, mature, and eventually decline.

MULTIPLE-CHOICE SOLUTIONS

(question number / correct answer / text page reference / answer rationale)

1.	a	344	An addition to the existing product line is considered a new product by marketers.
2.	b	344	New-to-the-world products, where the product category itself is new, are also called discontinuous innovations.
3.	c	344	Repositioning an existing product and targeting it toward new market segments (microwave users) is another type of new product because it is new to that segment.
4.	a	344	Klean's "new and improved" is an example of revising an old product.

5.	c	346	The least likely invitee would be the lawyers, who may not understand the market as well as all the other team members.
6.	d	346	The first step in the process is the development of a new products strategy that will set the stage for the other steps (even idea generation).
7.	c	346	Brainstorming is a process in which a group thinks of as many ways to vary a product or solve a problem as possible--without considering the practicality of the idea.
8.	a	348	Screening is the first filter stage of the new product development process and is used to eliminate inappropriate product ideas.
9.	c	349	Concept testing allows consumers to express opinions about a product before a prototype is developed.
10.	e	349	In the business analysis stage of new product development, preliminary demand, cost, sales, and profitability estimates are made.
11.	b	349	Many products that test well in the laboratory are next subjected to use tests; they are placed in consumers' homes or businesses for trial.
12.	d	351	Test marketing is a limited introduction of a product and a marketing program to determine the reactions of potential customers in a market situation.
13.	a	351	A good test market does not have to be large.
14.	b	353	Simulated (laboratory) market tests usually entail showing members of the target market advertising for a variety of products; then purchase behavior, in a mock or real store, is monitored.
15.	b	353	During commercialization, production materials and equipment are ordered.
16.	e	354	A poor match to customer needs is a violation of the marketing concept.
17.	e	355	The product-review committee makes decisions concerning failing or declining products.
18.	a	355	Because Blanko bases its corporate philosophy on the marketing concept, the marketing department would be the logical location.
19.	c	355	A venture team is an entrepreneurial, market-oriented group, staffed by a small number of representatives from marketing, research and development, finance, and other areas, focused on a single objective--planning their company's profitable entry into a new business.
20.	b	356	There are five categories: innovator, early adopter, early majority, late majority, and laggard.
21.	d	356	It is the product characteristics that affect the adoption rate most heavily. The "buy American" movement is external to the acceptance of the product.
22.	e	361	If a firm changes a feature of a product, it could move into another growth stage.
23.	a	360	The introductory stage of the product life cycle is affected by how customers perceive the characteristics of the product and how much better the new product is compared to the old.
24.	d	361	Since promotional spending is high and growth is still occurring, the tent is in the growth stage of the life cycle.
25.	a	361	Given the high competition and spending focused on short-term promotions, the soft drinks are battling in the maturity stage.

ESSAY QUESTION SOLUTIONS

1. **New-to-the-world products (discontinuous innovations)** are products that are introduced in an original form. Text examples include digital audiotapes and recordable compact discs.

   New product lines are products that the firm has not offered in the past but will introduce into an established market. The text example is Williamson-Dickie Manufacturing introducing a line of fashion clothing to supplement its original line of work clothes.

   Additions to existing product lines are new products that supplement a firm's established line. Complementary products can also supplement a firm's offerings. The text example is American Express introducing a Platinum Card.

Improvements or revisions of existing products are usually minor changes that may entail the addition or deletion of ingredients. Many of these products are labeled "new and improved." Text examples include L'Eggs Classics pantyhose and Pert Plus shampoo.

Repositioning means that existing products are targeted at new markets or market segments. The text example is Johnson & Johnson's repositioning of its baby shampoo as a product for everyone.

Lower-cost products are those that provide similar performance to competing brands at a lower cost. Lower cost may result from technological advantages, economies of scale in production, or lower marketing costs. Text examples include Wilkinson Sword's lower-cost high-performance disposable razors and Kimberly-Clark's Kleenex Premium toilet tissue.

(text pp. 344-345)

2. The seven steps of the new product development process are:

- New product strategy
- Idea generation
- Screening
- Business analysis
- Development
- Testing
- Commercialization

(text pp. 345-353)

3. Criteria for choosing a test market include:

- Demographics and purchasing habits that mirror the overall market
- Good distribution in test cities
- Media isolation
- Similarity to planned distribution outlets
- Relative isolation from other cities
- Availability of advertising media that will cooperate
- Diversified cross section of ages, religion, cultural-societal preferences, etc.
- No atypical purchasing habits
- Representative as to population size
- Typical per capita income
- Good record as a test city, but not overly used
- Not easily "jammed" by competitors
- Stability of year-round sales
- No dominant television station; multiple newspapers, magazines, and radio stations
- Availability of retailers that will cooperate
- Availability of research and audit services
- Free from unusual influences

An advantage of test marketing is that it gives management an opportunity to evaluate alternative strategies and to see how well the various aspects of the marketing mix fit together. Test marketing may reduce risk by allowing modification of a marketing mix before national introduction or withdrawal of a product with failure characteristics. However, test markets have several disadvantages: costs of test markets are high, and a product's success in a test market does not guarantee it will be a nationwide hit.

(text pp. 351-352)

4.

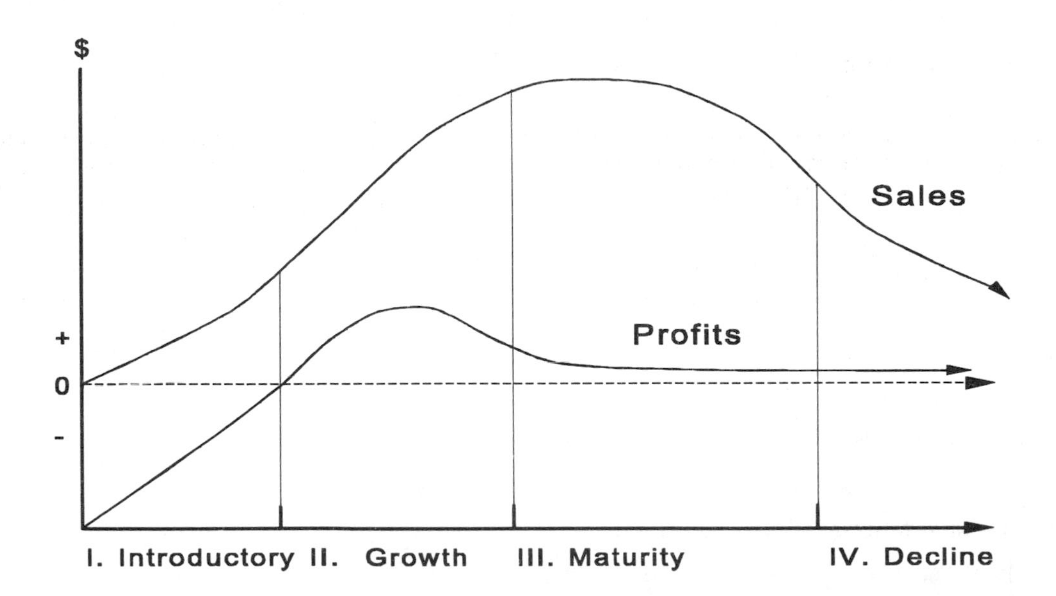

The sales line should start at zero sales at the beginning of the introductory stage, accelerate during the growth stage, peak in the maturity stage, and decrease during the decline stage.

The profit line should start in the negative range during the introductory stage, break even at the start of the growth stage, peak during the growth stage, fall during the maturity stage, and reach a near-zero asymptote during the decline stage. At no point should the profit line be above the sales line.

The introductory stage is characterized by a high failure rate, little competition, frequent product modification, limited distribution, slow sales, high marketing and production costs, negative profits, and promotion that stimulates primary demand.

The growth stage is characterized by increased sales, new competitors, healthy profits that peak, aggressive brand promotion, expanded distribution, price reductions, and possible acquisitions.

The maturity stage is characterized by a peak in sales, lengthened product lines, style modifications, price reductions, falling profits, competitor turnover, heavy promotion, and brand "wars."

The decline stage is characterized by a long-run drop in sales and profits, less demand, widespread competitor failure, reduction of advertising costs, and possible elimination of the product.

(text pp. 359-361)

5. **Innovators** represent the first 2 1/2 percent of adopters. They are venturesome and eager to try new products. They have higher incomes, better education, self-confidence, and less reliance on group norms than noninnovators. Moreover, they are active outside of their community. Innovators obtain information from scientific sources and experts. They may see a new movie in a special preview showing.

Early adopters represent the next 13 1/2 percent of adopters. They are reliant on group norms and values, oriented to the local community, and likely to be opinion leaders. They may see a new movie during the first week after it is launched

The early majority is the next 34 percent to adopt. They are deliberate in their information collection and are likely to be the friends and neighbors of opinion leaders. They may see a new movie during the season (summer or holidays) that it is launched.

The late majority is the next 34 percent to adopt. They adopt with skepticism to conform to social pressure. They tend to be older and below average in income and education. They rely on word-of-mouth communication rather than the mass media. They will see the movie after half the population sees it (maybe after it wins some awards, such as an Academy Award).

Laggards are the final 16 percent to adopt. They are tradition-bound and do not rely on group norms. Laggards have the lowest socioeconomic status, are suspicious of new products, and are alienated from a rapidly advancing society. Marketers typically ignore laggards. They may wait until the movie is in video form.

(text pp. 356-357)

CHAPTER 12 Services and Nonprofit Organization Marketing

PRETEST SOLUTIONS

1. Four characteristics of service products are:
   - **Intangibility**: services cannot be touched, seen, heard, tasted, or felt in the same manner as goods.
   - **Inseparability**: services are often sold, produced, and consumed at the same time.
   - **Heterogeneity**: services tend to be less standardized and uniform than goods.
   - **Perishability**: services cannot be stored, warehoused, or inventoried.

   (text pp. 373-374)
2. Five components of service quality are:
   - **Reliability**: ability to perform the service dependably, accurately, and consistently.
   - **Responsiveness**: ability to provide prompt service.
   - **Assurance**: knowledge and courtesy of employees and their ability to convey trust.
   - **Empathy**: caring, individualized attention to customers.
   - **Tangibles**: the physical evidence of the service.

   (text p. 375)
3. Five "gaps" that can occur in service quality are:
   - **Gap 1**: gap between what customers want and what management thinks customers want.
   - **Gap 2**: gap between what management thinks customers want and the quality specifications that management develops to provide the service.
   - **Gap 3**: gap between the service quality specifications and the service that is actually provided.
   - **Gap 4**: gap between what the company provides and what the customer is told it provides.
   - **Gap 5**: gap between the service that customers receive and the service they want.

   (text pp. 375-377)
4. Internal marketing is treating employees like customers and developing systems and benefits that satisfy their needs.

   (text p. 383)
5. Unique aspects of nonprofit organizations for the following:
   - **Objectives**: focus not on profit but on generating enough funds to cover expenses and to provide certain services to its constituencies.
   - **Target markets**: are not "buying" the product and may be apathetic or strongly opposed to the organization. Organizations may feel pressured to maximize their limited funding and to go after undifferentiated segments and may complement the positioning of for-profit organizations.
   - **Products**: offer benefits that are often complex, indirect, or weak. Many products elicit very low involvement ("don't litter").

- **Distribution**: can be difficult, but determining where the target market is located and how to deliver the services to the market is critical.
- **Promotion**: may be prohibited, or funding for promotion may be limited. Nonprofit organizations must rely on volunteers for promotion, They must also make use of existing services for sales promotion and rely on public service announcements (PSAs). Nonprofit organizations often license their names and/or images to communicate to a larger audience.

(text pp. 386-388)

VOCABULARY PRACTICE SOLUTIONS

1. service
2. intangibility
3. search quality, experience quality, credence quality
4. inseparability
5. heterogeneity
6. perishability
7. reliability, responsiveness, assurance, empathy, tangibles
8. gap model
9. core services, supplementary services
10. mass customization
11. internal marketing
12. nonprofit organization, nonprofit organization marketing
13. public service advertisement (PSA)

TRUE/FALSE SOLUTIONS

(question number / correct answer / text page reference / answer rationale)

1.	T	373	
2.	F	375	This is an example of responsiveness, not reliability.
3.	T	381	
4.	F	382	IBM is practicing internal marketing. Relationship marketing is focused on customers, not employees.
5.	F	384	Nonprofit organizations may also include pricing in their marketing mix by charging admission or charging a fee for participation in an event.
6.	F	386	Target markets are often more difficult to persuade because the benefits are less direct.
7.	T	387	
8.	F	388	Advertisements are PSAs only if there is no charge to the sponsor of the message.
9.	T	388	

AGREE/DISAGREE SOLUTIONS

(question number / sample answers)

1. Reason(s) to agree: The main objective of service providers is to deliver services to customers, and much more direct human contact is needed to accomplish this. In service organizations, almost all employees are trained well in customer service.

   Reason(s) to disagree: Whether a firm is selling a service or whether it is selling a good, good customer service is needed. Even when customers buy goods (such computers), they expect representatives from the manufacturer to answer questions and to provide assistance in operating the product.

2. Reason(s) to agree: All the gaps that can occur with services can also occur with goods. Gaps can occur when goods are not designed to fulfill the needs of customers, when management does not understand what customers want, or when the goods are not promoted accurately.

   Reason(s) to disagree: The gap model does not exist for goods as it does for services. The reason is that service quality is much more intangible and difficult to assess; every time a service is delivered, it is different. A good is tangible and more easier evaluated to determine if gaps can be fixed.

3. Reason(s) to agree: A strong organizational image will give customers confidence in the firm and its people—its most important resource for providing service.

   Reason(s) to disagree: In service marketing, it is difficult to divide products (services) from their providers. Thus, it is important to focus on promoting both the products and the organization.

4. Reason(s) to agree: Many nonprofit organizations exist to serve the needs of a specific target market, such as the nonsmoking campaign (targeted at smokers) or psychological counseling (targeted at people with specific types of psychological problems). Though the target markets may be apathetic or hostile toward the organization, they are easily identified.

   Reason(s) to disagree: Some nonprofit organizations do not have clear target markets. Political campaigns could target many different constituencies by focusing on different issues.

MULTIPLE-CHOICE SOLUTIONS

(question number / correct answer / text page reference / answer rationale)

1.	a	373	Services are described as intangible.
2.	e	373	Many service situations are complicated, and customers do not have the knowledge to judge the quality of the actual service itself.
3.	a	374	Inseparability is when services/goods are produced and consumed at the same time.
4.	e	374	Heterogeneity causes inconsistency and lack of standardization. Training helps to alleviate these conditions.
5.	a	374	Perishability is the inability to store, warehouse, or inventory a service.
6.	d	375	This illustrates a gap between what management thinks customers want (freshly cooked chicken) and the quality specifications that management develops to provide the service (to minimize chicken waste).
7.	a	376	This illustrates a gap between what customers want (low teacher to child ratio) and what management thinks customers want (lower tuition).
8.	c	376	This illustrates a gap between what the company provides (no respect for passengers) and what the customer is told it provides (respect for passengers).
9.	e	378	Gateway is likely providing services to accomplish all the above.
10.	d	378	These are examples of a supplementary service to the core service: airline transportation.
11.	e	380	Convenience via location is the key issue in the distribution decision for services.
12.	c	381	The note is an example of something tangible to an otherwise intangible product.
13.	b	381	A focus on matching supply and demand by varying prices to ensure maximum use of productive capacity at any specific point in time is an operations-oriented pricing objective for service firms.
14.	c	382	Building social bonds includes staying in though with customers, learning about their needs, and designing the service to meet those needs.
15.	c	384	Computer microchip manufacturing is product-, rather than service-oriented. Additionally, this industry is at a disadvantage compared to Asian products.
16.	e	384	An understanding and use of marketing offers a framework for decision making, tools for effective communication, and increased support from customers and interested groups.
17.	c	386	Success in a nonprofit organization is measured by how well service goals are met and how well it serves the community.
18.	d	386	Sources of donations are likely to be apathetic or opposed.

19.	b	387	The organization could not clearly explain the content of its services and, therefore, had inadequately defined the product.
20.	e	388	Donating time and materials to a nonprofit firm will result in intangible and long-term benefits, but not immediate financial gain.
21.	a	388	Media offer nonpaid advertising, called public service advertising, to nonprofit organizations.

ESSAY QUESTION SOLUTIONS

1. **Intangibility:** Services are intangible; that is, they cannot be touched, seen, tasted, heard, or felt in the same manner in which goods can be sensed. A restaurant provides food preparation and cleanup services. Although there are many aspects that are tangible (such as the food itself, the tables and chairs, and the staff), the service benefit of prepared, served food is intangible. Marketers should rely on tangible cues to communicate a service's nature and quality. This includes creation of the right environment and atmosphere at the restaurant.

   Inseparability: Services are often sold and then produced and consumed at the same time. For a restaurant, first the meal is ordered. Then the food is produced and consumed in the same service encounter. Customers are involved in production and interact with service staff and other customers. Therefore, care must be taken that the staff is courteous and comfort is provided for all customers. Additionally, inseparability implies that services cannot be produced in a centralized location and distributed at decentralized locations convenient to the consumer. Service production facilities (the restaurant) should be located in areas most convenient to the consumer.

   Heterogeneity: Consistency and quality control are often difficult to achieve in a service because services are dependent on their labor force, and services are produced and consumed at the same time. In a restaurant a customer cannot get exactly the same service experience from meal to meal. The length of the preparation time, seating comfort, wait-staff friendliness, food, neighboring diners, and many other factors will vary. Standardization and training help increase consistency and reliability. Another way to increase consistency is to mechanize the service process by having some meals automatically prepared.

   Perishability: Services cannot be stored, warehoused, or inventoried. An empty restaurant seat produces no revenue and cannot be saved for the next meal time. Marketers must synchronize supply with demand. This may be accomplished by using differential pricing to encourage demand during nonpeak periods.

   (text pp. 373-374)

2. **Product:** Four important issues regarding the service process are (1) what is being processed, (2) core and supplementary services, (3) customization versus standardization, and (4) the service mix. Three types of service processing occur: people, possession, and information processing. The service product consists of the basic, or core service, and additional supplementary service. Customized services are more flexible and responsive, and also can command a higher price. Standardized services are more efficient and may have lower costs. Firms must also consider their service mix--the portfolio of services they offer.

   Place: Convenience is the key factor influencing the selection of a service provider by a customer. The first step in developing a service distribution strategy is setting objectives, including intensity of distribution. Next, management must decide whether to distribute services directly to end users or indirectly through other firms. Management must select locations, and this choice most clearly reveals the relationship between a firm's target market strategy and distribution strategy. Scheduling of service provision must also be determined.

   Promotion: Four promotional strategies for services include (1) stressing tangible cues, (2) using personal information sources such as celebrity endorsements, (3) creating a strong organizational image by managing the physical environment, and (4) using postpurchase communication with customers.

   Price: Intensive price competition in the service industry indicates that close scrutiny should be paid to the pricing of each service offered. Three categories of pricing objectives have been suggested for services, including (1) revenue-oriented pricing, (2) operations-oriented pricing, and (3) patronage-oriented pricing. Service firms often use a combination of these objectives.

   (text, pp. 378-382)

3. **Product:**

   - The benefits of quitting smoking may be complex, long-term and intangible. Teen-agers see themselves as "invincible" and may not react to fear tactics.

- The benefits of quitting smoking may also be weak. Many teen-agers live for today, and the addiction to tobacco and the need to boost self-image may be more important in the short-run.
- Involvement is very high in this cause ("stop smoking") and requires very effective promotional messages in order to persuade the target audience to take action. Resistance to the persuasive attempts will be strong.

Place (distribution): Unless the nonsmoking campaign has facilities that help the target market to quit smoking, distribution will probably not be relevant.

Promotion:

- Professional volunteers may be needed to "spread the word" and to start and operate a grass-roots campaign aimed at stopping smoking among teen-agers.
- Sales promotion techniques can be used to provide incentives to those who stop smoking for a certain period of time.
- Public service announcements can be used since you are representing a nonprofit organization. This will be especially important, given that budgets will likely be very limited.
- Licensing of clever logos or slogans can be used to promote the cause.

Price: Price will be less relevant to the nonsmoking campaign than for other non-profit causes. Your organization will likely be funded by the state or by other organizations, and you will be operating a cost center and managing budgets.

(text pp. 387-388)

CHAPTER 13 Marketing Channels and Logistics Decisions

PRETEST SOLUTIONS

1. A marketing channel is a set of interdependent organizations that ease the transfer of ownership as products move from producer to business user or consumer.

   (text p. 406)

2. Three types of channel intermediaries are:
   - **Retailers**: Firms that sell directly to consumers;
   - **Merchant wholesalers**: Organizations that facilitate the movement of products and services from manufacturer to producers, resellers, etc., and who take title to the products;
   - **Agents and brokers**: Professional salespeople who represent a manufacturer, wholesaler, or retailer and who sell the product.

   (text p. 409)

3. Three intensities of distribution are:
   - **Intensive**: Achieve mass market selling by distributing the good everywhere possible.
   - **Selective**: Work closely with selected intermediaries who meet certain criteria.
   - **Exclusive**: Work with a single intermediary for products that require special resources.

   (text pp. 418-419)

4. Channel conflict is a clash of goals or methods to achieve goals among channel members.

Channel leadership occurs when one member of a marketing channel exercises authority and power over other members.

Channel partnering is the joint effort of all channel members to create a supply chain that serves customers and creates a competitive advantage.

(text pp. 420-422)

5. Six major functions of the supply chain are:

- Procuring supplies and raw materials;
- Scheduling production;
- Processing orders;
- Managing inventories of raw materials and finished goods;
- Warehousing and materials-handling; and
- Selecting modes of transportation.

(text p. 430)

VOCABULARY PRACTICE SOLUTIONS

1. marketing channel (channel of distribution)
2. discrepancy of quantity, discrepancy of assortment, temporal discrepancy, spatial discrepancy
3. agents and brokers, merchant wholesalers, retailers
4. direct channel
5. dual distribution (multiple distribution), adaptive channel, strategic channel alliance
6. intensive distribution, selective distribution, exclusive distribution
7. channel power, channel control, channel leader
8. channel conflict, horizontal conflict, vertical conflict
9. channel partnering
10. logistics, supply chain, supply chain management
11. logistics service
12. logistics information system, supply chain team
13. mass customization, just-in-time production
14. order processing system, electronic data interchange (EDI)
15. inventory control system, materials requirement planning (MRP), distribution resource planning (DRP), materials-handling system
16. electronic distribution, outsourcing

TRUE/FALSE SOLUTIONS

(question number / correct answer / text page reference / answer rationale)

1.	F	407	Middleman can actually help save costs.
2.	F	407	The discrepancy between the amount of product produced and the amount the end user wants to buy is a discrepancy of quantity.
3.	T	408	
4.	F	409	Although the three functions are correct, their subdescriptions are not. Transactional functions include contacting and promoting, negotiating, and risk taking. Logistical functions include physical distribution and sorting. Facilitating functions include research and financing.
5.	T	412	

6.	T	413	
7.	F	414	This is an example of an adaptive channel. Strategic channel alliances occur when a manufacturer uses another manufacturer's already established channel.
8.	T	418	
9.	F	418	Intensive distribution indicates that the product is available in every outlet where the potential customer might want to buy it, not just in a few retailers.
10.	T	418	
11.	T	420	
12.	F	422	Since both stores are retailers and at the same level in the channel, they are part of a horizontal conflict.
13.	F	428	The most important factor in physical distribution service is on-time delivery and pickup, coupled with minimizing costs.
14.	F	430	Wholesalers are part of the supply chain team, and transportation is an important function of the team.
15.	T	431	
16.	T	432	
17.	F	433	The ordering of merchandise using computer technology would be an example of electronic data interchange (EDI).
18.	T	438	
19.	F	439	Logistics activities of services include managing quality, convenience, cost, and people.
20.	F	441	Uncertainty regarding shipping usually tops the list of reasons why smaller companies resist international markets.

AGREE/DISAGREE SOLUTIONS

(question number / sample answers)

1. Reason(s) to agree: By going directly from manufacturer to retailer, products may be less costly, will take less time to get to the retailer, and will be in transit less time. Savings can be passed on to consumers.

   Reason(s) to disagree: The direct route is not always the best route. Sometimes wholesalers can actually save costs by purchasing in bulk quantities (something small retailers may not be able to do) and sending out assortments of goods tailored to specific retailer's requests. Wholesalers can add value to the marketing channel by providing marketing services and transportation services.

2. Reason(s) to agree: Channel leaders exert power and influence over other channel members, and many channel members might not react well to this.

   Reason(s) to disagree: If the channel leader manages the relationships well among the channel members, it can have a positive influence that will avoid conflict.

3. Reason(s) to agree: Intensive distribution allows manufacturers or wholesalers to distribute their products through the largest number of outlets possible. Their products will gain more exposure, and consumers are more likely to buy them. Even computer companies are going from selective distribution to a slightly more intensive distribution by selling computers in deep discount stores.

   Reason(s) to disagree: Quantity does not mean quality. Selling a prestige product through intensive distribution will NOT necessarily increase revenue; it may even decrease revenue because it could destroy the mystique and image of the product.

4. Reason(s) to agree: Services are intangible, so distribution through marketing channels is impossible. Many of the functions of logistics—materials-handling, inventory management, warehousing—are not appropriate for intangible products.

   Reason(s) to disagree: All services have tangible aspects. Banking services are "distributed" in a bank, a retail location where customers are served. Airlines companies have ticketing agencies and counters at airports. Even electrical companies have distribution—through electrical lines right to consumers' homes.

MULTIPLE-CHOICE SOLUTIONS

(question number / correct answer / text page reference / answer rationale)

1.	b	407	Economies of scale are not one of the responsibilities of marketing channels; it is a responsibility of a manufacturer.
2.	a	407	A manufacturer may only produce one product, yet consumers may need a variety of products to be satisfied. This is a discrepancy of assortment.
3.	b	407	A temporal discrepancy is created when a product is produced, but a consumer is not ready to purchase it.
4.	d	409	By taking title to goods and reselling them to retailers (drugstores), McKesson is a merchant wholesaler.
5.	e	409	An agent is a professional sales person or groups of sales people who represent a manufacturer, a wholesaler, or a retailer.
6.	a	409	Since Lands' End sells directly to consumers, it is a retailer.
7.	e	412	For customized and highly technical business products, the most common channel structure is the direct channel because of the amount of interaction and direct communication that is required.
8.	e	413	Some producers select two or more different channels to distribute the same products to target markets, a practice called dual or multiple distribution.
9.	d	418	Intensive distribution is distribution aimed at maximum market coverage. This distribution is used for many convenience goods and attempts to make the product available in every outlet where the potential customer might want to buy it.
10.	a	418	Selective distribution is achieved by screening dealers to eliminate all but a few in any single geographic area. With fewer retailers, the product must be one that consumers are willing to search for, such as shopping goods and some specialty products.
11.	e	419	The most restrictive form of distribution, exclusive distribution, entails establishing one or a very few dealers within a given geographic area. Because buyers are willing to search or travel to acquire the product, this form of distribution is limited to consumer specialty or luxury goods and major industrial equipment.
12.	c	420	Because of its influence and dominance in the channel, Wal-Mart is a channel leader.
13.	a	422	Since all the parties are retailers, the conflict is horizontal.
14.	d	422	The conflict is in raising service levels, which costs money, and simultaneously minimizing operating costs.
15.	c	430	Stages of the product life cycle may affect the size of the warehouse but should not affect the location. The other four items clearly affect the location decision.
16.	d	435	The materials-handling system moves and handles inventory into and out of the warehousing subsystem.
17.	d	431	Tailor-made, or built-to-order, computers is an example of mass customization.
18.	d	432	Just-in-time is based on a reduction of inventory by receiving the parts from suppliers to the assembly line at the time of installation. This reduces capital tied up in inventory.
19.	b	433	Electronic data interchange is the direct electronic transmission, from computer to computer, of standard business forms between two organizations. When this technique is used orders can become virtually paperless and information about the order is available to both firms.
20.	a	437	Rail transportation often involves rough handling and lack of speed, making raw materials more suitable for rail transport.
21.	c	437	Motor carriers are not as fast as airplanes, but they are much less expensive, particularly when dealing with a bulky product like meat.
22.	b	437	Limited handling and speed, available with airway transportation, mean less risk because there is diminished opportunity for damage and spoilage.

23.	a	437	Water carriers are the only type of carrier that can cross the ocean from Europe, carry heavy products, and perform product services while in transit.
24.	e	438	In contract logistics, a manufacturer or supplier turns over the entire function of buying and managing transportation to a third party. Contract warehousing is a growing trend in the area of contract logistics.
25.	e	439	Service industries are customer-oriented and must manage intangible services by minimizing wait times, managing service capacity, and providing delivery through distribution channels.
26.	c	440	If a product is fairly standardized, does not require much service, and is being distributed in a culturally dissimilar country, independent foreign intermediaries are typical, It is unlikely that the domestic firm could act as a channel captain in a VMS.

ESSAY QUESTION SOLUTIONS

1. **Specialization and division of labor:** Specialization and division of labor maintains that breaking down a complex task into smaller, simpler ones and allocating them to specialists will result in much greater efficiency. Marketing channels achieve economies of scale through specialization and division of labor. Some producers do not have the interest, financing, or expertise to market directly to end users or consumers. These producers hire channel members to perform functions and activities that the producers are not equipped to perform or that these intermediaries are better prepared to perform. Channel members can perform some functions and activities more efficiently than producers, and they enhance the overall performance of the channel because of their specialized expertise.

   Overcoming discrepancies: Channel members help bridge the gap that several discrepancies create:

   - **Discrepancy of quantity:** Large quantities produced to achieve low unit costs create quantity discrepancies (the amount of product produced compared to the amount an end user wants to buy). Marketing channels overcome quantity discrepancies by making products available in the quantities that buyers desire.
   - **Discrepancy of assortment:** This discrepancy occurs when mass production does not allow a firm to produce all the items necessary for buyers to receive full satisfaction from products. Marketing channels overcome discrepancies of assortment by assembling in one place assortments of products that buyers want.
   - **Temporal discrepancy:** This is created when a product is produced, but the consumer is not ready to purchase it. Marketing channels overcome temporal discrepancies by maintaining inventories in anticipation of demand.
   - **Spatial discrepancy:** Mass production requires a large number of potential purchasers, so markets are usually scattered over large geographic regions. Marketing channels overcome spatial discrepancies by making products available in locations convenient to consumers and business buyers.

   Contact efficiency: Channels simplify distribution by reducing the number of transactions required to get products from manufacturers to consumers.

   (text, pp. 407-408)

2. **Intensive distribution** is aimed at maximum market coverage. The manufacturer tries to have the product available in every outlet where the potential customer might want to buy it. If a buyer is unwilling to search for a product, the product must be placed closer to the buyer. Assuming that the product is of low value and is frequently purchased, a lengthy channel may be required. Candy, gum, cigarettes, soft drinks, and any other type of convenience good or operating supply would be distributed intensively.

   Selective distribution is achieved by screening dealers to eliminate all but a few in any single geographic area. Since only a few retailers are selected, the consumer must be willing to seek out the product. Shopping goods such as electronic equipment and appliances and some specialty products are distributed selectively. Accessory equipment manufacturers in the business-to-business market usually follow a selective distribution strategy.

   Exclusive distribution entails establishing one or a few dealers within a given geographic area. This is the most restrictive form of market coverage. Since buyers may have to search or travel extensively to purchase the product, exclusive distribution is usually limited to consumer specialty goods, a few shopping goods, and major industrial equipment. Some products distributed exclusively include Rolls Royce automobiles, Chris Craft boats, Pettibone tower cranes, and Coors beer.

(text, pp. 418-419)

3. Channel conflict occurs when channel members have a clash of goals or methods of achieving goals. Two types of channel conflict are:

   - **Horizontal conflict**: occurs among channel members at the same level, such as two or more wholesalers, two or more retailers, or two or more manufacturers. An example is two merchant wholesalers that blame each other for predatory pricing in order to get business from retailers.
   - **Vertical conflict**: occurs among channel members at different levels, such as between a retailer and a wholesaler, a wholesaler and a manufacturer, or a retailer and a manufacturer. An example is a retailer that becomes angry when a manufacturer decides to expand its so-called "exclusive" line of products to other retail outlets.

   (text, pp. 420-422)

4. **Sourcing and procurement:** The goals of sourcing and procurement is to reduce the costs of raw materials and supplies. Companies that purchase will strategically manage suppliers in order to reduce the total cost of raw materials and services. Purchasers and suppliers often form a cooperative relationship to ensure that both entities are receiving benefits.

   Production scheduling: Production begins when product forecasts are given to the production plant. The manufacturer must then schedule the production to meet forecasts and special large orders. Two considerations must be taken in production scheduling. Manufacturers might be asked to engage in mass customization, or to build products to specifications requested by a large customer. Another consideration is that some customers are demanding just-in-time (JIT) delivery, requiring manufacturers to schedule JIT production and to require their own suppliers to deliver raw materials literally just-in-time for production. This has the benefit of lowering inventory costs.

   Order processing: Order processing is a subsystem of physical distribution that begins with order entry, continues with order handling, and ends with a filled order. As the order enters the system, management must monitor two flows: the flow of goods and the flow of information. Order processing has been affected by computer systems that allow automated order entry, customer information systems, delivery instructions, and other tasks. Without human intervention, fewer errors are made and customer service is improved.

   Inventory control: An inventory control system develops and maintains an adequate assortment of products to meet customers' demands. The objective of inventory management is to balance minimum inventory levels (to reduce costs) while maintaining an adequate supply of goods to meet customer demand. Two major decisions managers must make regarding inventory are when to buy (order timing) and how much to buy (order quantity).

   Warehousing and materials-handling: The final user may not need or want the good at the same time the manufacturer produces and wants to sell it. Warehousing allows manufacturers to hold these products until shipment to the final consumer is demanded. Key warehousing decisions include location, number and size, and type of warehouse. Alternative types include private warehouses, public warehouses, and distribution centers.

   A materials-handling system moves inventory into, within, and out of the warehousing subsystem. Activities include receiving goods into the warehouse or distribution center; identifying, sorting, and labeling the goods; dispatching the goods into a temporary storage area; and recalling, selecting, or picking the goods for shipment. Packaging and bar coding are also possible functions. The goal of an effective materials-handling system is to move items quickly with minimal handling.

   Transportation: The transportation subsystem allows physical distribution managers to select from a variety of different transportation modes that best fit their criteria. The subsystem is responsible for the actual movement of goods through the channel of distribution to the final consumer.

   (text, pp. 430-437)

5. Transportation modes include:

   Railroad: coal, farm products, minerals, sand, chemicals

   Motor carriers: clothing, food, computers, paper goods

   Pipelines: oil, coal, chemicals, water

   Water: oil, grain, sand, ores, coal, cars

Airways: technical instruments, perishable products, documents

Transportation criteria include:

Cost: The total amount it will cost to use a specific carrier to move the product from the point of origin to the destination. Airways have the highest cost, while water is the cheapest mode.

Transit time: The total time a carrier has possession of goods. Water has the highest transit time, and air is the fastest mode.

Reliability: The consistency of the service provided by the carrier to deliver goods on time and in an acceptable condition. Pipelines are the most reliable, while water transport is the least reliable.

Capability: Ability of the carrier to provide the appropriate equipment and conditions for moving specific kinds of goods. Water has the most capability, while pipelines have the least.

Accessibility: The carrier's ability to move goods over a specific route or network. Trucks have the best accessibility, and pipelines have the most limited accessibility.

Traceability: Relative ease with which a shipment can be located and transferred. Air is the best mode, and pipeline is the worst.

(text, p. 437)

CHAPTER 14 Retailing

PRETEST SOLUTIONS

1. Retailing is all the activities directly related to the sale of goods and services to the ultimate consumer for personal, nonbusiness use.

   (text p. 454)

1. Ten types of stores are:

   - Department stores
   - Specialty stores
   - Supermarkets
   - Convenience stores
   - Drugstores
   - Full-line discount stores
   - Discount specialty stores
   - Warehouse clubs
   - Off-price retailers
   - Restaurants

   (text p. 456)

1. Four types of nonstore retailing are:

   - Automatic vending: the use of machines to offer goods for sale.
   - Direct retailing: selling products door-to-door, office-to-office, or at home sales parties.
   - Direct marketing: getting consumers to make a purchase from their home, office, or other non-retail setting.
   - Electronic retailing: use of TV networks and online retailing.

   (text pp. 465-470)

1. The six Ps of retailing are:
   - Product: product offering.
   - Place: the proper location.
   - Price: the appropriate price.
   - Promotion: value attached to product offerings.
   - Presentation: atmosphere and other qualities of store.
   - Personnel: people that work in the store who offer customer service.

   (text p. 473)

VOCABULARY PRACTICE SOLUTIONS

1. retailing
2. independent retailers, chain stores
3. gross margin
4. department store, buyer
5. specialty store
6. supermarket, scrambled merchandising, convenience store, drugstore
7. discount store, full-line discounters, mass merchandising
8. hypermarket, supercenter
9. specialty discount store, category killer
10. warehouse membership clubs
11. off-price retailer, factory outlet
12. nonstore retailing, automatic vending
13. direct retailing, direct marketing or direct-response marketing, telemarketing
14. franchise, franchiser, franchisee
15. retailing mix, product offering, private label brand
16. atmosphere

TRUE/FALSE SOLUTIONS

(question number / correct answer / text page reference / answer rationale)

1.	F	454	Because Lands' End sells directly to consumers who buy the products for nonbusiness use, Land's End is considered a retailer.
2.	F	457	Revenue minus cost of goods sold will result in gross margin, not net income.
3.	T	457	
4.	F	461	Mass-merchandising shopping chains have greater sales volume and number of stores than department stores.
5.	T	462	
6.	F	459	This would be an example of scrambled merchandising because it involves offering nontraditional goods for a pharmacy.
7.	F	462	Discount specialty stores have earned this nickname.
8.	T	466	
9.	T	470	
10.	F	473	The six Ps of retailing are product, price, place, promotion, presentation, and personnel.

11. T 487

AGREE/DISAGREE SOLUTIONS

(question number / sample answers)

1. Reason(s) to agree: Millions of consumers already have access to anything they want to buy—food, clothing, house wares, toys, flowers—through nonstore retailing. Even high-involvement purchases, such as expensive antiques, are available on-line. As computer technology becomes even more sophisticated, and as retailers become more savvy in providing customer service without face-to-face contact, customers will have little reason to leave their homes and can spend more leisure time without having to shop.

   Reason(s) to disagree: There will always be a need for retail stores. Customers cannot simply order everything on-line or through the telephone. Many customers enjoy the "human" side of store retail shopping: shopping with friends, socializing with store personnel, and enjoying the atmosphere that only a store retail environment can provide. Technology can have a "dehumanizing" effect on retailing.

2. Reason(s) to agree: Large retail chains—from Starbucks Coffee to Barnes & Noble bookstores to discount stores such as Wal-Mart—are taking over the world. These chains are able to buy in huge quantity and to offer merchandise at lower costs to consumers, thus putting the small independents out of business. They have even entered small towns, where independent retailers have always thrived.

   Reason(s) to disagree: Small retailers will always have a purpose. As their large counterparts are offering homogenized merchandise throughout the country or even throughout the world, small retailers can tailor their merchandise to the local market more easily. They can differentiate themselves with unique merchandise and outstanding customer service, because these elements are more easily controlled by smaller business.

3. Reason(s) to agree: By adapting their merchandise offerings to foreign markets, U.S. retailers can ensure that they are meeting customer demand..

   Reason(s) to disagree: Adaptation is not always the correct strategy. Adaptation may not be required (such as in electronics, where customer preferences are similar throughout the world) and may not even be preferred (some foreign markets want to purchase the American image).

MULTIPLE-CHOICE SOLUTIONS

(question number / correct answer / text page reference /. answer rationale)

1. b 454 Large retail operations represent about 50 percent of the nation's retail sales.
2. e 457 Gross margin is the percentage of sales after cost of goods has been subtracted from net sales.
3. d 458 A specialty store is not only a type of store, but also a method of retail operations, specializing in a given type of merchandise with a deep but narrow assortment.
4. e 459 This is the definition for a supermarket. The key factors are the emphasis on food and the size of the store.
5. c 459 In many cases supermarkets offer a wide variety of nontraditional goods and services under one roof, a strategy called scrambled merchandising.
6. a 461 These strategies are used by mass merchandisers.
7. c 461 A hypermarket combines both a supermarket and discount department store and is huge in size. Hypermarkets require enormous sales volume.
8. d 461 Discount specialty stores are single-line stores offering merchandise such as sporting goods, electronics, auto parts, office supplies, or toys. These stores offer a nearly complete selection of one line of merchandise and use self-service, discount prices, high volume, and high-turnover merchandise to their advantage.
9. c 463 Warehouse membership clubs sell a limited selection of brand name appliances, household items, and groceries, usually in bulk on a cash-and-carry basis to members only.
10. a 463 Off-price discount retailers purchase goods at cost or less from manufacturers' overruns, bankruptcies, irregular stock supplies, and unsold end-of-season output. The other four categories may be current customers and would not have a use for the overruns and out-of-season stock.

11.	d	464	A factory outlet is a type of off-price retailer that is owned and operated by a manufacturer and carries one line of merchandise--its own.
12.	c	464	By selling sauce in a restaurant, the chef is conducting store retailing.
13.	b	466	Direct marketing refers to a variety of techniques such as telephone selling, direct mail, and catalogs.
14.	d	469	Use of a cable television channel to display goods that are then sold over the phone is the shop-at-home network format.
15.	c	4271	The ties to the franchisor's policies may be very restrictive for an innovative businessperson because certain products and procedures must be adhered to.
16.	a	473	The six Ps of retailing are product, price, place, promotion, personnel, and presentation, not packaging.
17.	b	474	A freestanding store has the advantage of low site costs and will be a benefit if consumers are willing to seek it out.
18.	d	478	Malls have unified images, convenient parking, and expensive leases. It is unlikely that a small, specialty store would be an anchor. The mall atmosphere and other stores will attract shoppers.
19.	a	478	Anchors are the large stores at the ends of shopping malls and may sell a variety of products, not just expensive ones.
20.	d	479	The predominant aspect of the store's presentation is the atmosphere--how the store's physical layout, decor, and surroundings convey an overall impression.
21.	a	481	Browsing patterns are not a consideration when setting a service level. They are a consideration for store layout.
22.	d	483	Retail operations have not shown a large trend toward merging with retailers in other countries.
23.	a	484	Franchises are well-received internationally, and governments are making franchising more attractive.

ESSAY QUESTION SOLUTIONS

1.

Type of Retailer	Service Level	Assortment	Price Level
Department store	high	broad	moderately high
Specialty store	high	narrow	high
Supermarket	low	broad	moderate
Convenience store	low	medium/narrow	moderately high
Discount store	moderate/low	medium/broad	moderately low
Warehouse club	low	broad	low/very low
Off-price retailer	low	medium/narrow	low

(text, p. 456)

2. **Product:** The product offering, or merchandising mix, must satisfy the target customers' desires. This level of satisfaction is often determined as a level of width and depth of the product assortment. Inventory management and physical distribution is also a key issue when dealing with products at the retail level.

Promotion: Retail promotional strategy includes advertising, public relations, publicity, and sales promotion. The objective of the promotional strategy is to help position the store relative to competitors in consumers' minds. Retail promotion is often done on a local basis.

Place: This involves selecting a proper site or location for the store. Managers must decide on a community, a specific site, and store type, including whether to be a freestanding store or part of a shopping center or mall.

Price: The right price is a critical element in retailing strategy. Price is a key element in a retail store's positioning strategy and classification.

Presentation: The presentation of a retail store to its customers helps determine the store's image. The predominant aspect of this is atmosphere, which includes employee type and density, merchandise type and density, fixture type and density, sound, odors, and visual factors.

Personnel: People are a unique aspect of retailing because most retail sales involve a customer-salesperson relationship. Personal selling issues and setting the quality level of customer service are two important personnel issues.

(text, pp. 473-483)

3. You could compete by using the following three trends:

- **Entertainment**: Though providing entertainment is not new to gourmet coffee retailers, you could differentiate yourself by providing unique entertainment. These could include magicians, comedy poetry readings, stand-up comics, or unusual music.
- **Convenience and efficiency**: Most gourmet coffee shops do not have drive-through windows. By incorporating one, you could provide a fast-food approach to an otherwise upscale product. In addition, you could gain much business by opening the drive-through early for early morning commuters.
- **Customer management**: Most coffee shops already offer frequency cards (whereby a customer can get a free coffee after the purchase of a certain number). You could go one step beyond and take vital information about a customer (such as birthdays, typical orders, etc.) and sent direct mailings, offering the customer a special coupon for their favorite latte on their birthday or telling the customer about a favorite entertainer who will make an appearance. This one-on-one marketing approach will turn your customer into a loyal one, a strategy that will be hard to match by the large coffee chains.

(text, pp. 487-489)

CHAPTER 15 Marketing Communication and Personal Selling

PRETEST SOLUTIONS

1. The four elements of the promotion mix are:

- **Personal selling**: a situation in which two people communication in an attempt to influence each other in a purchase situation.
- **Advertising**: any form of paid communication in which the sponsor or company is identified.
- **Sales promotion**: consists of all marketing activities that stimulate consumer purchasing and deal effectiveness.
- **Public relations**: evaluates public attitudes, identifies areas within the organization that the public may be interested in, and executes a program of action to earn public acceptance.

(text pp. 510-512)

2. The four steps of the communication process are:

- **Sender and encoding**: the sender, or originator of the message, encodes the message to a second party. Encoding is the conversion of the sender's ideas and thoughts into a message.
- **Message transmission**: sender transmits the message through a channel, such as through a medium (radio, TV, etc.) or by his/her own voice.
- **Receiver and decoding**: the receiver of the message decodes, or interprets, the message.

- **Feedback**: the receiver send a response to the original message.

(text pp. 514-516)

3. AIDA is a model used by advertisers to get the target market to buy a product. AIDA stands for:
   - **Awareness**: the advertiser must achieve awareness of the message with the target market.
   - **Interest**: the advertiser must create interest in the product through the message.
   - **Desire**: the advertiser must make the target market want the product.
   - **Action**: the advertisement should be effective enough to entice the target market to actually purchase the product.

   (text pp. 519-520)

4. The seven steps of the selling process are:
   - Generating leads
   - Qualifying leads
   - Approaching the customer and probing needs
   - Developing the proposing solutions
   - Handling objections
   - Closing the sale
   - Following up

   (text p. 526)

VOCABULARY PRACTICE SOLUTIONS

1. promotion, promotional strategy, differential advantage
2. promotional mix, advertising, personal selling, sales promotion, public relations, publicity
3. communication, interpersonal communication, mass communication
4. sender, encoding, channel, receiver, decoding, feedback, noise
5. integrated marketing communications
6. AIDA concept
7. push strategy, pull strategy
8. relationship selling
9. sales process
10. lead generation, cold calling, referral, networking
11. lead qualification
12. preapproach, needs assessment
13. sales proposal, sales presentation
14. negotiation
15. follow-up
16. quota
17. straight commission, straight salary

TRUE/FALSE SOLUTIONS

(question number / correct answer / text page reference / answer rationale)

1. F 509 This describes promotion, not advertising (which is communicating through mass media).
2. T 510

3. T 515

4. F 515 Even though a message is received, it will not necessarily be properly decoded; receivers interpret messages based on idiosyncratic frames of reference. Noise may also be a problem.

5. T 518

6. T 520

7. T 520

8. T 522

9. F 525 Totter Toys is using a push strategy.

10. T 531

11. T 536

12. F 536 A salesperson's responsibilities do not end with a sale; a follow-up is critical to encouraging repeat business.

13. T 539

14. F 540 This is the combination system.

AGREE/DISAGREE SOLUTIONS

(question number / sample answers)

1. Reason(s) to agree: It is advertising that is most effective at generating awareness of a product. Without awareness, the other elements (interest, desire, action) will not take place.

   Reason(s) to disagree: Advertising may be the most important promotional tool in some situations, such as selling many consumer products, but it is not as effective in selling business products. In this selling situation, tools such as personal selling may be much more important.

2. Reason(s) to agree: While advertising may be effective at the first two steps in AIDA (generating awareness and interest in the product), it is not as effective at generating desire and action (purchase). It take other promotional elements, such as personal selling or sales promotion, to achieve these.

   Reason(s) to disagree: An effective advertisement can make potential customers go through all the steps. A good advertisement will create desire for the product and in some cases even action. Without advertising, the target market could never buy the product because it would not be aware of it.

3. Reason(s) to agree: Personal selling involves pushing a product to a target market. Not very many salespeople will walk away from a potential customer, even if the customer does not appear to have a "need" for the product at first. A good example is a tenacious telemarketer who does not take "no" for a first answer and persists on speaking to the lead.

   Reason(s) to disagree: Personal selling is a part of true marketing. A savvy salesperson will always conduct a needs assessment to determine if the prospect is a good one, and what the prospect truly needs.

4. Reason(s) to agree: The selling process does not apply to certain situations, such as fund raising for a nonprofit organization.

   Reason(s) to disagree: Even some nonprofit organizations could use the seven-step process. Fund raisers can (1) generate leads (potential donors); (2) qualify leads (determine who is likely to give); (3) approach the donor and ask questions; (4) develop and propose solutions (present how the donors can contribute by allowing deductions from payroll); (5) handle objections (why the donor cannot or will not give); (6) close the sale (get the commitment from the donor); and (7) follow-up (a thank you letter).

MULTIPLE-CHOICE SOLUTIONS

(question number / correct answer / text page reference / answer rationale)

1. d 509 A differential advantage is a set of unique features of a company and its products that are perceived by the target market as significant and superior to the competition.

2. b 510 Advertising is a form of impersonal, sponsor-paid, one-way mass communication.

3. a 512 Public relations is the marketing function that executes a program of action to gain publicity.

4.	d	510	Oral, face-to-face presentation in a conversation with one or more prospective purchasers for the purpose of making sales constitutes personal selling.
5.	c	511	Sales promotion includes marketing activities that stimulate consumer purchasing such as coupons, contests, free samples, and trade shows.
6.	c	513	Communication to large audiences, usually through a medium such as television or a newspaper, is called mass communication.
7.	d	515	Noise is anything that interferes with, distorts, or slows down the transmission of information.
8.	e	515	Receivers are the people who decode the message.
9.	a	515	For successful communication to occur, the sender and receiver need to have overlapping frames of reference to have common understanding.
10.	a	518	Informative promotion is a necessary ingredient for a highly technical product category.
11.	b	519	This promotion is aimed at persuading consumers to change their attitudes and try the snack.
12.	b	519	Reminder promotion is used to keep a familiar brand name in the public's mind.
13.	a	519	Reminder promotion is prevalent during the maturity stage of the product life cycle.
14.	c	520	AIDA stands for awareness, interest, desire, action.
15.	a	521	Public relations is best at building awareness.
16.	e	521	Sales promotion is best at inciting consumer action (purchase).
17.	d	522	The packaging should not matter because it will not affect which medium is used or how complex the message needs to be.
18.	d	525	The use of aggressive personal selling and trade advertising by a manufacturer to convince a wholesaler and/or retailer to carry and sell its merchandise is a push strategy.
19.	b	525	The manufacturer using the pull strategy focuses its promotional efforts on the consumer.
20.	a	526	Personal selling is the direct communication that a representative engages in with one or more prospective purchasers for the purpose of making a sale.
21.	c	529	Brown's is selling a complex, high-value product to a few customers in a small area.
22.	e	531	Lead generation is the identification of those firms and people most likely to buy the seller's offering.
23.	a	532	Networking uses friends and acquaintances as a means of meeting potential clients.
24.	c	535	The sales presentation often involves the face-to-face presentation of the product or its benefits to the customer.
25.	b	535	The need-satisfaction approach emphasizes that people buy products to satisfy needs and solve problems.
26.	e	536	Negotiation involves offering special concessions during the closing of a sale.
27.	a	540	In a straight commission plan, the salesperson receives no pay until a sale is made.
28.	c	540	Because of all the nonselling activities, these salespeople will need a guaranteed straight salary.
29.	a	540	Combination plans include both a guaranteed salary and an incentive or commission. This provides control plus motivation.

ESSAY QUESTION SOLUTIONS

1. **Advertising** is a form of impersonal, one-way mass communication paid for by the sponsor. Advertising is transmitted by different media, including television, radio, newspapers, magazines, books, direct mail, billboards, and transit cards.

   Public relations is the marketing function that evaluates public attitudes, identifies areas within the organization that the public may be interested in, and executes programs to earn public understanding and acceptance. A solid public relations program can generate favorable publicity. A firm can generate publicity in the form of news items, feature articles, or sporting event sponsorship.

   Personal selling involves a planned face-to-face presentation to one or more prospective purchasers for the purpose of making sales. Personal selling is more prevalent in the industrial goods field because of the complex, technical nature of many industrial products. Personal selling can also be used to encourage wholesalers and retailers to carry and resell a product. One form of personal selling is telemarketing.

Sales promotion includes a wide variety of activities for stimulating consumer purchasing and dealer effectiveness. Examples include free samples, contests, bonuses, trade shows, and coupons.

(text, pp. 461-463)

2.

1. The sender encodes ideas into a promotional message.

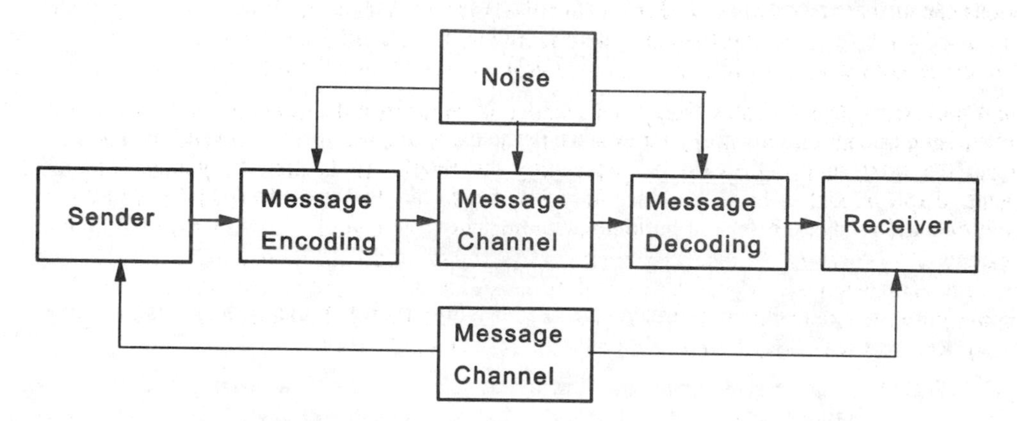

2. The message is transmitted through a channel, or communication medium.
3. The message may or may not be received.
4. The receiver decodes the message. Messages may or may not be properly decoded.
5. The receiver's response to a message is feedback to the source. In promotions, feedback is indirect rather than direct.
6. Any phase of this process may be hindered by noise.

(text, pp. 464-468)

3. The AIDA model steps are Attention-Interest-Desire-Action.

Attention: A salesperson would attract a potential customer's attention by a greeting and approach.

Interest: A good sales presentation, demonstration, or promotional literature handed out during the sales presentation would create interest in the product.

Desire: The salesperson could create desire by illustrating how the product's features will satisfy the consumer's needs.

Action: The salesperson could make a special offer or a strong closing sales pitch to obtain purchase action.

(text, pp. 474-477)

4. **Nature of the product:** Industrial products are usually promoted with more personal selling than advertising. Consumer products are promoted primarily through advertising. Sales promotion, branding, and packaging are more important for consumer goods than industrial goods. When the costs or risks of a product's use increases, personal selling becomes more important.

Stage in product life cycle: During the introduction stage, emphasis is placed on advertising and public relations as well as some sales promotion and personal selling. During growth, sales promotion efforts are reduced, while advertising and public relations continue. At maturity, sales promotion and advertising become the focus. All promotion is reduced during the decline stage.

Target market characteristics: Widely scattered customers, highly informed buyers, and brand-loyal repeat purchasers generally require a blend of more advertising and sales promotion and less personal selling. Personal selling is required for industrial installations, even if buyers are extremely competent. Print advertising can be used when potential customers are difficult to locate.

Type of buying decision: A routine, low-involvement buying decision requires sales promotion and reminder advertising. Advertising and public relations can be used for a new purchase situation, while personal selling is most effective for complex buying decisions.

Available funds: A firm with limited funds can rely heavily on free publicity if the product is unique. If personal selling is necessary, the firm may use manufacturers' agents who work on a commission basis. Some sales promotions can also be inexpensive. Although advertising is very expensive, it has low cost per contact, which may be necessary for a large potential market. There is usually a trade-off among the funds available, the number of people in the target market, the quality of communication needed, and the relative of the promotional elements.

Push and pull strategies: A push strategy is a situation in which manufacturers use aggressive personal selling and trade advertising to convince a wholesaler or a retailer to carry and sell their merchandise. The wholesaler, in turn, must push the merchandise forward by persuading the retailer to handle the goods. The retailer then uses advertising, displays, and so on to convince the consumer to buy the pushed products. A pull strategy stimulates consumer demand to obtain product distribution. In this case, the manufacturer focuses its promotional efforts on end consumers. As consumers demand the product, the retailer orders the merchandise from the wholesaler. As the wholesaler is confronted with rising demand, it places orders for the pulled merchandise from the manufacturer. Stimulating consumer demand pulls the product through the distribution channel. Heavy sampling, consumer advertising, cents-off campaigns, and couponing are part of a pull strategy.

(text, pp. 522-525)

5. **Sales lead generation:** You can get sales leads to sell computers from your own organization's database, by networking at industry trade shows, or by doing research on firms in your local area.

Sales lead qualification: From your sales leads, you should screen out those firms that are not likely to buy from those who are likely to buy. This can be determined by initial interest in the computers.

Doing a needs assessment: You should find out customers and their needs, the competition, and the industry. Are customers ready to make a decision about buying new computers? What kinds of features do they seek?

Developing and proposing solutions: Once you determine customer needs, you should develop a sales proposal that is relevant to the needs of the customer and make the presentation about the computer products that you have. The presentation should include the types of computers, the service contract, and pricing.

Handling objections: During the presentation, the customer is likely to object to certain elements of the proposal. Know in advance what the customer might object to, and know how to handle these objections during the presentation.

Closing the sale: At the end of a sales presentation, you can attempt to close the sale if the prospect's objections are handled properly. Whenever the customer makes a commitment to buy, sale closing and order processing should begin. However, if the commitment to buy is not forthcoming, a number of techniques can be used to attempt to close the sale, including negotiation on the computers.

Following up: Once the customer has purchased the computers, you should follow up. This should including calling or visiting the customer to determine if there were any problems with the computers and service.

(text, pp. 529-536)

CHAPTER 16 Advertising, Sales Promotion and Public Relations

PRETEST SOLUTIONS

1. Three types of product advertising are:
   - **Pioneering advertising**: stimulates primary demand for a new product or product category.
   - **Competitive advertising**: influences demand for a specific brand.
   - **Comparative advertising**: directly or indirectly compares two or more competing brands on one or more specific attributes.

   (text, p. 556)
2. The steps to creating an advertising campaign are:
   - Determine advertising objectives
   - Make creative decisions
   - Make media decisions
   - Evaluate the campaign

   (text pp. 557-561)
3. Seven types of media are:
   - Newspapers
   - Magazines
   - Radio
   - Television
   - Outdoor media
   - The Internet and World Wide Web
   - Alternative media, such as computer screen savers, interactive kiosks, and video shopping carts

   (text pp. 563-567)
4. Six types of consumer sales promotion tools are:
   - Coupons and rebates
   - Premiums
   - Loyalty marketing programs
   - Contests and sweepstakes
   - Sampling
   - Point-of-purchase promotion

   Six types of trade sales promotion tools are:
   - Trade allowances
   - Push money
   - Training
   - Free merchandise
   - Store demonstrations
   - Business meetings, conventions, and trade shows

(text pp. 571-576)

5. Six public relations tools include:
   - New product publicity
   - Product placement
   - Consumer education
   - Event sponsorship
   - Issue sponsorship
   - Internet Websites

   (text pp. 577-580)

VOCABULARY PRACTICE SOLUTIONS

1. advertising response function
2. institutional advertising, advocacy advertising, product advertising, pioneering advertising, competitive advertising, comparative advertising
3. advertising campaign, advertising objectives
4. medium
5. advertising appeal, unique selling proposition
6. cooperative advertising
7. infomercial
8. cost per contact, media mix, reach, frequency, audience selectivity
6. media schedule, continuous media schedule, flighted media schedule, pulsing media schedule, seasonal media schedule
10. consumer sales promotion, trade sales promotion
11. coupon, premium, loyalty marketing program, frequent-buyer program, sampling, point-of-purchase display
12. trade allowance, push money
13. crisis management

TRUE/FALSE SOLUTIONS

(question number / correct answer / text page reference / answer rationale)

1.	T	554	
2.	T	556	
3.	F	556	This is comparative advertising.
4.	F	558	This ad sells the product's attributes.
5.	T	560	
6.	T	563	
7.	F	563	The college is calculating cost per contact.
8.	T	569	
9.	F	569	This is an example of a seasonal schedule.
10.	F	571	Providing price-off packages to loyal customers is a waste of money; the airlines should consider a loyalty program, such as frequent flyer.
11.	T	573	
12.	T	576	
13.	T	576	

AGREE/DISAGREE SOLUTIONS

(question number / sample answers)

1. Reason(s) to agree: Many other variables may affect a product's sales—such as sales promotions, selling efforts, or a change in price—so that advertising efforts cannot be isolated. Sometimes a product's sales will even decrease, even when advertising is in place.

   Reason(s) to disagree: An advertiser could prove the effect of its advertising by isolating advertising efforts somehow. If the advertiser, for instance, decides to stop advertising for awhile without changing anything else in the marketing mix, it can determine what impact this will have on sales.

2. Reason(s) to agree: Advertising has the ultimate goal of selling a product or of creating a perception of an institution in the minds of consumers. As such, advertising messages may not necessarily lie but may withhold certain information known to the advertiser that may be negative about the product or company.

   Reason(s) to disagree: Advertising can be quite truthful, especially when product claims are backed by reputable and scientific research. There are a number of antitrust laws that protect the consumer against false and misleading advertising.

3. Reason(s) to agree: Sales promotion cannot stand alone in a promotion mix. Coupons, rebates, discounts, sweepstakes, etc., are only effective when there is awareness of the product, which can only be accomplished through advertising.

   Reason(s) to disagree: As a communication tool, sales promotion can actually work in the place of advertising when limited spending is required.

4. Reason(s) to agree: After advertising, personal selling, and sales promotion, public relations is simply an addition to the entire promotion mix. Advertising generally covers much of what public relations wants to achieve—shifting of attitudes and perceptions about a company or its products.

   Reason(s) to disagree: Public relations—and its product, publicity—is a key component of the promotion mix. Whereas consumers recognize advertising as "biased," publicity is generated by journalists who are deemed to be objective in their writing. Some businesses rely heavily on publicity when advertising budgets are tight—such as local restaurants, whose success may be based on a single food critic's positive review.

MULTIPLE-CHOICE SOLUTIONS

(question number / correct answer / text page reference / answer rationale)

1. a 552 In general, toy sellers have a large advertising-to-sales ratio. This is common for this industry and is not a sign of inefficiency.
2. b 554 There is a saturation point for advertising.
3. c 554 The lack of fat and cholesterol are product attributes. Benefits are good health and possible weight loss.
4. b 555 The company is trying to enhance its image to the public in the form of institutional advertising.
5. e 556 Pioneering advertising is intended to stimulate primary demand for a new product or product category and is heavily used g the introductory stage of the product life cycle.
6. a 556 By comparing its customer service to that of its competitor, the company is conducting comparative advertising.
7. c 557 The first step in the process is to decide what you want to accomplish in by determining objectives.
8. d 559 By showing the typical busy life of professional parents, this company is using a lifestyle execution.
9. d 560 A product symbol (such as the animated scrubbing bubbles) is one executional style that uses a character to represent the product.
10. b 563 Cooperative advertising is an arrangement under which a manufacturer pays a percentage of the advertising cost that a retailer places for the manufacturer's brand.
11. e 565 A relatively new form of direct-response advertising, the infomercial, is in the form of a 30-minute advertisement that resembles a TV talk show.
12. c 563 The media mix is the combination of advertising that will be conducted among the various media vehicles.
13. a 568 Cost per contact is the cost of reaching one member of the market. Comparing the cost of contacting the same amount of people using different media mixes is a common aid in selecting media.

14. a 564 Some media, such as magazines, have a long life span with the consumer, and advertisements may be seen repeatedly. Ads in other media such as radio and television are gone instantly.

15. c 569 The media schedule designates when and where advertising will appear.

16. b 569 The flighting media scheduling strategy schedules ads heavily for a period, then drops them for a period, and then repeats them.

17. e 570 Sales promotion is an activity in which a short-term incentive is offered to induce the purchase of a particular good or service.

18. c 573 A premium is an item offered, usually with proof of purchase, to the consumer.

19. d 574 A sample allows the consumer to try the product without investing any money or having a full size container. It is an expensive strategy but works well if the trial product is significantly different and better than other products on the market.

20. d 575 A trade allowance is a price reduction offered by manufacturers to intermediaries in exchange for performance of specified functions or purchasing during special periods.

21. d 576 This is nonpaid information about the firm, directed to an interested public.

22. e 577 Publicity is the media activity that results from public relations.

ESSAY QUESTION SOLUTIONS

1. **Institutional advertising** is used when the goal of the campaign is to establish, change, or maintain the image of a product or service, the company, or the industry. Institutional advertising has four important audiences: (1) the public, which includes legislators, businesspeople and opinion leaders; (2) the investment community, which is mainly comprised of stockholders; (3) customers; and (4) the company's employees.

   Advocacy advertising is a special form of institutional advertising that allows a corporation to express its views on controversial issues. Most advocacy campaigns react to unfair criticism or media attacks. Other campaigns may attempt to ward off impending regulatory threats.

   Product advertising touts the benefits of a specific product or service. It is used if the advertiser wishes to enhance the sales of a specific product, brand, or service. Product advertising can take three forms: pioneering advertising, competitive advertising, and comparative advertising.

   Pioneering advertising is intended to stimulate primary demand for a new product or product category. It is used during the introductory stage of the product life cycle to offer information about product class benefits.

   Competitive advertising is used to influence demand for a specific brand of a good or service. This advertising emphasizes the building of brand name recall and favorable brand attitudes. This type of advertising is often necessary during the growth stage of the product life cycle, when competitive entry eliminates the need to stimulate product category demand. This type of advertising stresses subtle differences between brands such as target market or price.

   Comparative advertising compares two or more specifically named or shown competitive brands on one or more specific product attributes. Advertisers may make taste, price, and preference claims often at the expense of the competing brand. The FTC has fostered the growth of comparative advertising, although new laws prohibit advertisers from falsely describing competitors' products.

   (text, pp. 555-557)

2. Advertising appeals include:

   Profit motive: The low price of Less-U will save consumers money.

   Concern for health: Less-U will give body-conscious consumers the trim, healthy body necessary for a long, healthy life.

   Love/romance: Less-U will create a new, sexy body that is necessary for attracting romantic partners.

   Fear: Overweight consumers should be embarrassed at their present state; without Less-U, these consumers will be unattractive and die early.

   Admiration: An attractive celebrity spokesperson could promote Less-U.

   Convenience: Less-U will save food preparation time and is available at all convenient outlets.

   Fun and pleasure: While slender, the consumer can enjoy fun sports.

Vanity and egotism: Consumers can be shown admiring their new, shapely figures.

Environmental consciousness/considerate of others: Less-U could be promoted as having environmentally-friendly packaging, or showing how slender consumers take up less room and is more considerate of others.

Advertising executional styles include:

Slice-of-life: Family members in a normal household setting such as a dinner table or living room could discuss the merits of Less-U.

Lifestyle: The ad could not only show how important a healthy body is for family members, but also show how much the family members enjoy baked desserts and how Less-U fits into this lifestyle.

Spokesperson/testimonial: A typical family member or celebrity could make a testimonial or endorse Less-U.

Fantasy: This ad could build a fantasy for a family member who cooks--after using Less-U, the person wins baking contests, accolades from friends and relatives, and other recognition.

Humor: Famous comedians could make light fun of "fat" and "skinny" cookies.

Real or animated product symbols: This ad could show an animated Less-U character being baked into cakes and cookies.

Mood or image: This ad would build a mood or image around Less-U--perhaps one of rewards and pleasure because people could eat more desserts prepared with Less-U.

Demonstration: Chefs from cooking shows could demonstrate the use of Less-U and its benefits in an ad.

Musical: The benefits of Less-U could be demonstrated through an MTV-style music video.

Scientific: Research or scientific evidence could be used to demonstrate Less-U's superiority over butter, margarine, and shortening.

(text, pp. 559-560)

3. **TRADITIONAL**

Newspapers are generally a mass-market medium. The largest source of newspaper ad revenue stems from the local retail sector through cooperative advertising between retailers and manufacturers. Advantages include (1) geographic selectivity and flexibility, (2) short-term advertiser commitments, (3) news value and immediacy, (4) advertising permanence, (5) stable readership, (6) high individual market coverage, (7) co-op and local tie-in availability, and (8) short lead time. Disadvantages include (1) little demographic selectivity, (2) limited color capabilities, (3) different local and national rates, (4) low pass-along rate, (5) may be expensive, and (6) noise from competing ads and news stories.

Magazine advertising has increased in recent years because of segmented niche marketing. Advantages include (1) good color reproduction, (2) message longevity, (3) demographic selectivity, (4) regional and local market selectivity, (5) long life, and (6) high pass-along rate. Disadvantages include (1) long-term advertiser commitments, (2) slow audience buildup, (3) limited demonstration capacity, (4) lack of urgency, (5) long lead times, and (6) high total cost.

Radio is another medium that lends itself well to cooperative advertising. Local advertising accounts for 77 percent of radio ad volume. Advantages include (1) low cost, (2) high frequency, (3) immediate message, (4) short-notice rescheduling, (5) stable audience, (6) portable medium, (7) negotiable costs, (8) short-term advertiser commitments, (9) entertainment carryover (10) audience selectivity, (11) geographical selectivity, and (12) low production costs. Disadvantages include (1) no visuals, (2) short message life, (3) background sound, and (4) commercial clutter.

Television can be divided into three basic types: network television (ABC, CBS, NBC, and Fox), independent stations, and cable television. The largest growth market is cable television. Advantages include (1) wide reach, (2) creative demonstration opportunities, (3) immediate messages, and (4) entertainment carryover. Disadvantages include (1) little demographic selectivity, (2) short message life, (3) consumer skepticism toward claims, and (4) high cost.

Outdoor advertising is a flexible, low-cost medium that may take a variety of forms. Examples include billboards, skywriting, bus shelters, taxicabs, giant inflatables, construction site fences, minibillboards in malls, lighted moving signs in bus terminals and airports, and ads painted on the sides of cars and trucks. Outdoor advertising reaches a broad and diverse market. Therefore, it is normally limited to promoting convenience products and select shopping products. Advertisers usually base billboard use on census tract data. Advantages include (1) repetition, (2) moderate cost, and (3) flexibility. Disadvantages include (1) short message, (2) lack of demographic selectivity, and (3) high noise.

NEW FORMS

Fax machines can be used to electronically deliver direct mail, advertisements, menus, and other solicitations.

Video shopping carts can provide shoppers with product information and advertisements triggered by the cart's location in the store.

Electronic place-based media include video monitors placed throughout stores that broadcast appropriate (often custom) programming and advertisements.

Interactive advertising uses personal computers to transmit immediate, personalized advertisements to consumers.

Minibillboards are used in high schools and college campuses with ads personalized to each school.

Cinema/video: Advertisements are now being placed at the beginning of movies in theaters and on rented videocassettes.

Infomercials are a relatively new form of direct-response advertising that take the form of 30-minute talk-show format commercials.

Computer: screen savers, CD-ROMs, online ads on the web, home pages, etc.

(text, pp. 563-567)

4. **Coupons** are certificates given to consumers entitling them to an immediate price reduction when they purchase the item. Coupons for Steri-Flor could be mailed to households or placed in home magazines.

   Premiums are items offered to the consumer, usually requiring proof of purchase. The premium should be related to the product in some way. Premiums could be offered to purchasers of Steri-Flor, with proof of purchase required. Related premiums might be floor mops, sponges, buckets, or other items related to floor cleaning.

   Frequent-buyer programs reward brand-loyal customers for repeat purchases of products or services.

   Contests are promotions in which participants compete for prizes based on some skill or ability. Entrants could write an essay or poem about Steri-Flor or complete a puzzle about Steri-Flor's attributes.

   Sweepstakes allow anyone to participate and are characterized by chance drawings for prizes. Steri-Flor could offer a sweepstakes for people to win free housecleaning services, trips, cash prizes, or a year's supply of Steri-Flor.

   Sampling refers to free samples and trial sizes. Sampling allows consumers to try products with minimal risk. Trial sizes reduce the risk of trying new products. They also eliminate the problem of being stuck with a large quantity of a disliked product. Steri-Flor could be sent in small samples to households with instructions for use or comparison against the consumer's regular floor cleaner. Trial sizes could also be sold at a minimal price at grocery stores.

   Point-of-purchase displays are special displays set up at retail locations to build traffic, advertise the product, or induce impulse buying. Displays could be designed in the shape of the Steri-Flor packaging, or a display could show comparison floor samples. The displays would encourage consumers to stock up on Steri-Flor.

   (text, pp. 571-575)

5. **Trade allowances** are price reductions offered by manufacturers to intermediaries. The price reduction or rebate is in exchange for performance of specified functions or purchasing during special periods. Trade allowances could be offered to wholesalers in the form of price reductions or rebates. In exchange, wholesalers would promise to purchase during the holiday season or perform marketing functions for Steri-Flor.

   Push money is a bonus that intermediaries receive for pushing the manufacturer's brand. The push money is often directed toward the retailer's salespeople. Money or trips could be offered as a bonus to the wholesaler's sales force for pushing the Steri-Flor brand.

   Training programs may be provided for an intermediary's personnel if the product is complex. Because Steri-Flor may not be very complex, training may be in the form of teaching sales techniques or effective communications and demonstrations.

   Free merchandise may be offered in lieu of quantity discounts. It may also be used as payment for trade allowances provided though other sales promotions. Steri-Flor could offer one free case of the product for every ten cases purchased, or free cases could be offered in exchange for promotional functions performed by the intermediary.

Store demonstrations can be performed at the retail establishment for customers. Customers can then sample products or see how they are used. Demonstrators could set up booths in stores to show comparisons of Steri-Flor cleaned floors versus floors cleaned by competing products.

Business meetings, conventions, and trade shows are all ways to meet other vendors and potential customers of floor-cleaning products.

(text, pp. 575-576)

6. Public relations tools include:

Press relations. Placing newsworthy information in the news media to attract attention to a person, product, or service.

Product publicity. Publicizing specific products.

Corporate communications. Creating internal and external communications to promote understanding of the firm or institution.

Public Affairs. Building and maintaining national or local community relations.

Lobbying. Dealing with legislators and government officials to promote or defeat legislation and regulation.

Employee and investor relations. Maintaining positive relationships with employees, shareholders, and others in the financial community.

Counseling. Advising management about public issues and company positions and image.

Crisis management. Responding to unfavorable publicity or a negative event.

During a crisis, the public relations team should:

- Get professional public relations help for handling news media relations.
- React quickly and start early. The worst damage to a company's or product's reputation tends to occur immediately after the problem becomes public knowledge.
- Avoid the "no comment" response and do not ignore the situation.
- Make a team effort. Rely on senior management, public relations professionals, attorneys, quality control experts, manufacturing employees, and marketing personnel.
- Provide lots of communication and squelch incorrect rumors.
- Do not try to shift the blame away from the company. Take responsibility for the results of the problem.

(text, pp. 576-578)

CHAPTER 17 Pricing Concepts

PRETEST SOLUTIONS

1. Three categories of pricing objectives include:
   - Profit-oriented objectives: Profit maximization, satisfactory profits, target return on investment
   - Sales-oriented objectives: Market share, sales maximization.
   - Status quo pricing objectives

   (text pp. 601-604)

2. **Elastic demand**: is a situation in which consumer demand is sensitive to changes in price. When demand is elastic, marketers must be very careful about changing price or the demand will be impacted significantly.

Inelastic demand: is a situation in which an increase or a decrease in price will not significantly affect demand for the product. When demand is inelastic, the marketer has more leeway in increasing prices to maximize profit.

Unitary elasticity: means that an increase in sales exactly offsets a decrease in prices so that total revenue remains the same.

(text p. 607)

3. Three pricing methods using cost as a determinant are:
   - **Markup pricing**: cost of buying the product from the producer plus amounts for profit and for expenses not otherwise accounted for.
   - **Profit maximization pricing**: occurs when marginal revenue equals cost.
   - **Break-even pricing**: pricing based on determining what sales volume must be reached before total revenue equals total costs.

   (text pp. 610-615)

4. Five alternative pricing determinants are:
   - Stage in the product life cycle
   - Competitors' prices
   - Distribution strategy
   - Promotion strategy
   - Relationship of price to quality

   (text pp. 615-620)

VOCABULARY PRACTICE SOLUTIONS

1. price
2. revenue, profit
3. return on investment (ROI), market share, status quo pricing
4. demand, supply, price equilibrium
5. elasticity of demand, elastic demand, inelastic demand, unitary elasticity
6. variable cost, fixed cost, average variable cost (AVC), average total cost (ATC), marginal cost (MC)
7. markup pricing, keystoning
8. marginal revenue (MR), profit maximization
9. break-even analysis
10. selling against the brand
11. prestige pricing

TRUE/FALSE SOLUTIONS

(question number / correct answer / text page reference / answer rationale)

1. F 600 Price is not just monetary; it is what the customer gives up in exchange for a product or service, such as time or energy.
2. T 602
3. F 604 Maximization of cash should never be a long-run objective.
4. T 604
5. F 607 Unitary elasticity means that any price change will exactly offset sales change so that revenue stays the same.
6. T 607

7. T 610

8. T 613

9. F 616 Firms in such situations must usually choose between pricing below the market or at the market level. Firms should price above the market only if there is a clear, demonstrable advantage over competing, existing products.

10. F 620 Consumers use price as a quality indicator when there is a substantial degree of uncertainty involved in the purchase decision.

AGREE/DISAGREE SOLUTIONS

(question number / sample answers)

1. Reason(s) to agree: Most firms have both a short- and long-term goals of maximizing profit. Therefore, pricing objectives must be consistent with this goal.

   Reason(s) to disagree: Some firms may have a short-term goal of maximizing market share, which may be done at the expense of profit. To maximize share, a firm may have to price low in order to gain quick sales volume. This practice has been used by many Japanese firms entering a foreign market.

2. Reason(s) to agree: It would be a real coincidence if an increase (or decrease) in price directly offset a decrease (or increase) in sales.

   Reason(s) to disagree: Theoretical or not, unitary elasticity could occur. Products that are discounted for a short period of time during a sales promotion may experience unitary elasticity, whereby the decrease in price is offset in the short-term by an increase in sales volume. The objective in this case is to get more products into the hands of consumers, who try the product and may repurchase even when the price is raised to its normal level.

3. Reason(s) to agree: Since most firms have a profit goal, pricing must at least cover costs.

   Reason(s) to disagree: Sometimes pricing is not based on costs directly. When competition is fierce, pricing may actually fall below costs. Supermarkets follow a strategy called "loss leadership," whereby they price an item below their cost in order to bring customers into their stores to purchase other, more profitable products.

4. Reason(s) to agree: Prestige products, such as Rolex or Rolls Royce, enjoy a high quality perception, partially due to the high price that these products command.

   Reason(s) to disagree: In many product categories, price is not the only determinant of quality. If a product's quality does not deserve its high price, the high pricing strategy will not work in the long run.

MULTIPLE-CHOICE SOLUTIONS

(question number / correct answer / text page reference / answer rationale)

1. b 600 The company is not losing money; in fact, it is making $100,000 in profit.

2. e 601 Profit maximization means setting prices so that total revenue is as large as possible relative to total costs.

3. a 601 The objective of satisfactory profits is characterized by seeking a level of profits that is satisfactory to management and shareholders.

4. d 602 ROI is net profits after taxes divided by total assets: $50,000/500,000 = 10 percent.

5. a 602 By pricing low to gain short-term unit sales volume, the firm is trying to gain market share.

6. e 604 Status quo pricing is best described as meeting the competition.

7. d 607 Since the total revenue dropped when the price was doubled, the demand is elastic.

8. d 607 When demand and supply are approximately equal, price equilibrium is reached.

9. c 607 Inelastic demand is characterized by a decrease in price that leads to an increase in demand, but the increase in demand does not offset the decrease in revenue caused by the price decrease.

10. a 607 Under unitary elasticity the increase in demand exactly offsets the decrease in price. In this case, $6 \times 300 = 1800$ and $5 \times 360 = 1800$.

11. e 610 Variable costs vary with the level of output.

12. a 610 The payment on leased equipment remains the same, no matter how many pizzas are produced, and therefore is a fixed cost.

13. c 610 Marginal cost is the change in total costs associated with a one-unit change in output.

14. b 611 Keystoning doubles costs to set prices.

15. d 610 Markup pricing is adding profit to the base cost of a product.

16. b 613 Break-even quantity is the total fixed costs ($50,000) divided by fixed cost contribution per unit ($.20).

17. e 606 The analysis only includes company costs but does not consider consumer demand.

18. b 615 In the maturity stage of the product life cycle, price promotion often occurs to maintain market share.

19. d 616 The egg producer is passing a large portion of its profit margin on to the wholesalers and retailers, not maximizing profit margin for itself.

20. c 616 Since the product is new and innovative, it has no direct competition.

21. b 620 Most consumers equate price and quality.

22. c 620 Prestige pricing strategy sets high prices to connote high product quality and exclusiveness.

23. c 620 The price needs to fit the product quality for a more cohesive image.

ESSAY QUESTION SOLUTIONS

1. **Profit-oriented pricing objectives include:**

   - Profit maximization
   - Satisfactory profits
   - Target return on investment

   Profit-oriented pricing objectives have several disadvantages. In particular, profit maximization has two problems. First, many firms do not have adequate accounting data for setting profit maximization goals. Additionally, this goal does not provide a basis for planning because it is vague and lacks focus. Target ROI may be difficult to evaluate because it should be viewed against the competitive environment, risks in the industry, and economic conditions.

   Sales-oriented pricing objectives include:

   - Market share (dollars or units)
   - Sales maximization (dollars or units)

   These demand-oriented objectives can also have disadvantages. In particular, dollar/unit sales maximization objectives ignore profits, competition, and the marketing environment as long as sales are rising. This type of sales maximization should be used on a short-term basis only.

   Status quo pricing objectives include:

   - Maintaining the existing price
   - Meeting the competition

   While these passive policies are simple, they may not be responsive to price changes required by consumers.

   (text, pp. 601-604)

2. The price established depends primarily on demand for the good or service and the cost to the seller for that good or service. Other factors that would influence price include stage in the product life cycle, the competition, distribution strategies, promotion strategies, and perceived quality.

   Demand: The quantity of a product that people demand to buy depends on its price. The higher the price, the fewer goods consumers will demand. Conversely, lower prices increase demand levels. Elasticity of demand is also an important factor. Elastic demand occurs when consumers are sensitive to price changes. Inelastic demand means that an increase or a decrease in price will not significantly affect demand for the product. Unitary elasticity exists when the increase in sales exactly offsets the decrease in price so that total revenue remains the same. Factors that affect elasticity of demand include (1) the availability of substitute goods and services, (2) price relative to a consumer's purchasing power, (3) product durability, and (4) other uses for the product.

Cost: Setting prices based solely on costs ignores demand and other important factors such as marketing mix components or consumer needs and wants. Prices determined strictly on the basis of cost may be too high for the target market, thereby reducing or eliminating sales. Cost-based prices may also be too low, causing the firm to earn a lower return than it should. Costs play an important role in price setting, however. Costs serve as a threshold guideline or a floor below which a good or service must not be priced in the long run. Cost-based pricing methods include (1) markup pricing, (2) formula pricing, (3) break-even analysis, and (4) target return pricing.

Stage in the product life cycle: Management usually sets prices high during the introduction stage, and prices begin to decline throughout the life cycle as demand for the product and competitive conditions change.

Competition: If a firm faces no competition, it may set high prices. These high prices, however, serve to attract competitors. If a firm is in a competitive industry, the firm faces a decision of whether to price below the market or at the market level.

Distribution strategy: Adequate distribution for a new product can often be attained by offering a larger than usual profit margin to intermediaries. A variation is to give dealers large trade allowances to stimulate demand at the retail level. Finally, consumers may purchase a higher-priced item if it is located at a convenient outlet.

Promotion strategy: Reduced prices are often used as a promotional tool to induce consumers to shop in a particular store. Price promotions can also take the form of discount coupons and rebates.

Relationship of price to quality: Consumers tend to rely on price as an indicator of product quality; that is, a higher price indicates higher quality in the form of better materials, more careful workmanship, or higher service levels. Conversely, lower price indicates lower quality as illustrated by the adage, "you get what you pay for." Marketers can take advantage of the price-quality phenomenon by increasing the price of the product to enhance the image of their product. This is known as a prestige pricing strategy.

(text, pp. 604-620)

3. **Introduction:** Generally, prices are set high in the introduction stage to recover development costs. However, pricing strategies followed in this stage depend on demand elasticity. If demand is inelastic, a high introductory price is warranted. If demand is elastic and consumers are price-sensitive, price should be set at the market level or lower.

   Growth: Price may stabilize at this level as competitors enter the marketplace. Price may fall somewhat as economies of scale allow lower costs to be passed on to the consumer in the form of a lower price.

   Maturity: This stage brings on further price declines as competition increases and inefficient, high-cost firms are eliminated. Remaining competitors typically offer similar prices. Price increases are cost-initiated rather than demand-initiated.

   Decline: In the final stage of the product life cycle, prices may decline even further as the few remaining competitors attempt to salvage the last vestiges of demand. If only one firm is left in the market, prices will stabilize or even rise as the product becomes a specialty good.

   Product: Price levels must be set according to the cost of the product, demand for the product, elasticity of demand for the product, and the perceived relationship of price to quality of the product.

   Place/distribution: Adequate distribution for a new product can often be attained by offering a large profit margin to wholesalers and retailers. Price can be set higher than normal if the product is distributed to outlets that are convenient to the consumer.

   Promotion: Price is often used as a promotional tool to increase consumer interest. Special low prices are often advertised as an inducement. Discount coupons, cents-off campaigns, price rebates, and other discounts are all price-promotion marketing tools.

   (text, pp. 615-620)

PROBLEM SOLUTIONS

1. The following calculation table shows the resultant unit and dollar shares. Market share should be expressed in percentage points.

Company	Units Sold	Unit Price	Total Dollars	Unit Share	Dollar Share
Xylo	1,000	$1.00	$1,000	1,000/2,000 = 50%	$1,000/5,000 = 20%
Yeti	600	$5.00	$3,000	600/2,000 = 30%	$3,000/5,000 = 60%
Zeta	400	$2.50	$1,000	400/2,000 = 20%	$1,000/5,000 = 20%
	2,000		$5,000		

(text, pp. 603-603)

2. Markup percentages for the retailer are stated in terms of the final selling price. The dollar markup is calculated as selling price minus cost, and percentage markup can be calculated by dividing dollar markup by selling price.

a) Dollar markup ÷ selling price = percent markup

$\$5 \div \$25 = 20\%$

b) Selling price = (dollar markup + cost)

Dollar markup ÷ selling price = percent markup

$\$6 \div (\$6 + \$4) = 60\%$

c) Dollar markup = (selling price - cost)

(Selling price - cost) ÷ selling price = percent markup

$(S - \$15) \div S = .25$

$(S - \$15) = .25S$

$\$15 = .75S$

Selling price = $20

d) (Selling price - cost) ÷ selling price = percent markup

$(\$12 - C) \div \$12 = .75$

$(\$12 - C) = (.75 \times \$12)$

$C = \$12 - \$9 = \$3$

(text, p. 610)

3. Using the break-even formula indicates that TV-Terry must sell 1,500 televisions to break even.

Fixed cost contribution = selling price - variable cost

FCC = $500 - $300 = $200

Break-even quantity = total fixed costs ÷ fixed cost contribution

BEQ = $300,000 ÷ $200 = 1,500 units

(text, p. 613)

CHAPTER 18 Setting the Right Price

PRETEST SOLUTIONS

1. The five steps in setting the right price are:
   - Establish pricing goals
   - Estimate demand cost and profits
   - Choose a price strategy to help determine a base price
   - Fine tune the base with pricing tactics
   - Results can lead to the right price

   (text p. 630)
2. Four illegal issues regarding price are:
   - **Unfair trade practices:** Selling a product below cost.
   - **Price fixing:** An agreement between two or more firms on the price they will charge for a product.
   - **Price discrimination:** Charging different customers different prices for no reason.
   - **Predatory pricing:** Charging a very low price for a product with the intent of driving competitors out of business or out of a market.

   (text pp. 663-635)
3. Five types of discounts, rebates, and allowances that can be used are:
   - Quantity discounts
   - Cash discounts
   - Functional discounts
   - Seasonal discounts
   - Promotional allowances

   (text p. 636)
4. Nine types of special pricing tactics are:
   - Single-pricing
   - Flexible pricing
   - Professional services pricing
   - Price lining
   - Leader, or loss leader, pricing
   - Bait pricing
   - Odd-even pricing
   - Price bundling
   - Two-part pricing

   (text pp. 640-643)

VOCABULARY PRACTICE SOLUTIONS

1. price strategy
2. price skimming, penetration pricing
3. unfair trade practice acts, price fixing, predatory pricing

4. base price
5. cash discount, quantity discount, cumulative quantity discount, noncumulative quantity discount
6. functional discount (or trade discount), trade loading, everyday low prices (EDLP)
7. seasonal discount, promotional allowance, rebate
8. FOB origin pricing, uniform delivered pricing, zone pricing, freight absorption pricing, basing-point pricing
9. single-price tactic, flexible pricing (or variable pricing), price lining
10. leader pricing (or loss-leader pricing), bait pricing, odd-even pricing (or psychological pricing)
11. price bundling, unbundling, two-part pricing
12. product line pricing, joint costs
13. delayed-quotation pricing, escalator pricing, price shading

TRUE/FALSE SOLUTIONS

(question number / correct answer / text page reference / answer rationale)

1. F 630 The first step in setting price is to establish pricing goals.
2. F 631 If a firm introduces an item similar to others already on the market, its pricing freedom will be restricted.
3. T 632
4. T 633
5. F 634 The presidents are engaging in price fixing.
6. T 636
7. F 637 Rebates involve a cash refund for the purchase of a product during a specific period.
8. T 643
9. F 643 Achieving goals for an entire product line rather than for any individual product is an example of product line pricing.
10. F 644 Complementary products are those that are consumed together, not produced together. The sale of one complement causes an increase in the sale of the related complement.
11. T 645
12. T 647

AGREE/DISAGREE SOLUTIONS

(question number / sample answers)

1. Reason(s) to agree: Some firms, such as Procter & Gamble and Wal-Mart, have greatly reduced the number of discounts given to customers and have simplified their pricing strategies altogether. By using EDLP, manufacturers can eliminate "trade loading" (which happens when retail or wholesale customers buy products in bulk when discounted). In addition, retailers can advertise less and maintain a consistent pricing strategy.

   Reason(s) to disagree: In some industries, discounted pricing and "specials" may be too ingrained to eliminate without losing business. Products that are discounted for sales promotion and that are advertised as such can bring customer traffic into a store, which can produce more profits in the long run.

2. Reason(s) to agree: Leader pricing indicates that retailers are pricing certain items at or below cost in order to bring traffic into the store. If customers buy nothing else at the store, the store can lose money very quickly.

   Reason(s) to disagree: Supermarkets have been successful at leader pricing because customers will rarely go to the supermarket to buy just one item (the promoted item) and will pick up other, more profitable products.

3. Reason(s) to agree: Pricing at product at $9.99 instead of $10.00 is an insult to consumers. They realize that there is only a 1 cent difference. Retailers should just round up their pricing to make it easier for customers to pay at the cash register.

Reason(s) to disagree: Many consumers perceive that odd-numbered prices (such as $9.99) indicates that a product is discounted. Because of this, they may be more likely to pick up the product.

4. Reason(s) to agree: Since price is the single most important criteria for the purchase of most products, prices should be reduced during tough economic times.

   Reason(s) to disagree: Customers do not buy just on price; they purchase value. By bundling product together or by introducing new "value-priced" product to complement a more expensive line, marketers can hold their market share during a recession.

MULTIPLE-CHOICE SOLUTIONS

(question number / correct answer / text page reference / answer rationale)

1. e 630 Different individuals in an organization may have pricing objectives that are not mutually compatible and will involve trade-offs.
2. a 632 Price skimming is pricing high to capitalize on a product's uniqueness.
3. c 632 Penetration pricing is pricing low in order to gain volume market share.
4. b 633 Status quo pricing is simply meeting the competition.
5. b 634 Publishing and circulating minimum fee schedules is an example of price fixing.
6. d 635 Price discrimination is illegally offering two or more customers different prices which can cause unfair competition.
7. d 635 Price discrimination violates the Robinson-Patman Act.
8. c 636 A cash discount is offered to those who make immediate payment upon delivery.
9. a 636 A functional discount is the customary discount from list price that is offered to intermediaries in recognition of their functions that are performed in the selling of the product.
10. a 636 A seasonal discount is a price reduction for buying merchandise out of season.
11. b 636 A promotional allowance may be used to offer free goods or displays to a retailer in return for promoting a manufacturer's products, or it may pay for some or all of the advertising costs.
12. a 637 A rebate is a cash refund given for the purchase of a product during a specific period of time.
13. b 640 Because title will pass to the purchaser at the time of shipment with FOB origin pricing, the risks and costs will also pass to the purchaser at that point.
14. b 640 With uniform delivered pricing, all customers will pay the same price regardless of their location.
15. d 640 With zone pricing, the freight prices are set according to geographic areas.
16. a 640 With freight absorption pricing, the seller pays all or part of the freight costs and does not pass them on to the purchaser, keeping the purchase price low.
17. b 640 With basing-point pricing, customers pay freight from a set base point, regardless of the location from which the goods are shipped.
18. c 640 The single price tactic offers all goods and services at the same price.
19. b 641 Flexible pricing is defined as selling essentially the same product to different customers for different prices.
20. a 641 Price lines allow a retailer to appeal to several different target markets. It is not an uncommon strategy, and The Sports Stop's competitors probably use it also. It should not affect inventory overall and will not force a price-quality comparison.
21. b 642 Leader pricing involves selling a product near or even below cost to attract business.
22. e 643 Marketing two or more products in a single package for a special price is called price bundling.
23. d 643 Two-part pricing involves two separate charges to consume a single product or service.
24. d 644 The manager is trying to determine the relationships between the various products in the line and is looking at the products, not the buyer.
25. e 645 Delayed-quotation pricing delays the setting of the final price.

26. a 645 Like delayed-quotation pricing, escalator pricing allows for price increases and delays the setting of the final price.

27. b 645 These activities fit the description of price shading.

ESSAY QUESTION SOLUTIONS

1. The four-step process is:

   Establish price goals: You must first decide what you want to achieve through pricing. An example would be to gain quick market share and to position the shoes as a complement to other athletic shoes (not as a replacement to the pricier sport-oriented shoes).

   Estimate demand, costs, and profit: You must conduct market research to get an idea of what kind of demand exists for these trendy shoes. If the demand is elastic, for example, this will dictate lower pricing. Working with your manufacturing facilities, you should get an idea of all costs involved in producing and marketing the shoes, as well as determine what profit is acceptable to your company.

   Choose a price strategy: You should then determine a long-term strategy that will be used for the introduction of the shoes and for the other phases of the product life cycle. For example, you could use a penetration pricing policy that would price products lower than other name brand athletic shoes so that consumers would buy your shoes in addition to more sports-oriented shoes.

   Fine tune the base with pricing tactics: You should plan other pricing tactics for the introduction of the shoes. This could include providing introductory promotional allowances to entice retailers to feature the new shoes, functional discounts to wholesalers and retailers who stock the shoes, quantity discounts to those that purchase several cases at once, and geographic pricing.

   (text pp. 630-633)

2. There are three basic methods for setting a price on a new good or service: price skimming, price penetration, and status quo pricing.

   Price skimming: With this method a high introductory price is charged that skims the top off a market in which there is inelastic demand. The high introductory price attracts a smaller market share but recoups costs quickly. Price-skimming advantages include (1) quick recovery of product development or educational costs, (2) pricing flexibility that allows subsequent lowering of price, and (3) the ability to market prestige products successfully. Disadvantages include encouragement of competitive entry into the market.

   Penetration pricing: With this method a firm introduces a product at a relatively low price, hoping to reach the mass market in the early stages of the product life cycle. The low price allows the product to penetrate a large portion of the market, resulting in large market share and lower production costs. Penetration pricing advantages include (1) a tendency to discourage competitive entry, (2) large market share due to high volume sold, and (3) lower production costs resulting from economies of scale. Disadvantages include (1) lower profits per unit, (2) higher volume required to reach the break-even point, (3) slow recovery of development costs, and (4) inability to later raise prices.

   Status quo pricing: With this method the price charged is identical or close to that of the competition. This strategy may be used more often by small firms for survival. Status quo policies have the advantage of simplicity. Disadvantages include ignoring demand or cost.

   (text, pp. 632-633)

3. **FOB origin pricing:** This price tactic requires the purchaser to pay for the cost of transportation from the shipping point. A manager would choose to use FOB origin pricing if he or she is not concerned about total costs varying among the firms's clients or if freight charges are not a significant pricing variable. Industrial products such as hydraulic cranes and power plants are shipped FOB factory.

   Uniform delivered pricing: With this price tactic, the seller pays the actual freight charges, but bills every buyer with an identical, flat freight charge. This equalizes the total cost of the product for all buyers, regardless of location. A manager would select this policy if the firm is trying to maintain a nationally advertised price or when transportation charges are a minor part of total costs. The tactic also reduces price competition among buyers. This pricing method could be used for most consumer food and drug items or small industrial parts.

Zone pricing: This price tactic is a modification of uniform delivered pricing in which the geographic selling area is divided into zones. A flat freight rate is charged to all customers in a given zone, but different rates will apply to each zone. A marketing manager would use this strategy to equalize total costs among purchasers within large geographic areas. The U.S. Parcel Post service uses this structure. This structure may also be used by building suppliers.

Freight absorption pricing: With this price tactic, the seller pays all or part of the actual freight charges and does not pass these charges along to the customer. A manager would choose this tactic if competition is extremely intense, if the firm is trying to break into new market areas, or if greater economies of scale are a company goal. Media direct marketers often use this tactic for records, tapes, kitchenware, and other consumer items.

Basing-point pricing: This method requires the seller to designate a location as a basing point and charge all purchasers with the freight cost from that point (regardless of the point from which the goods are actually shipped). This tactic might be used for firms that sell relatively homogeneous products and for which transportation costs are an important component of total costs. Basing-point pricing has been prevalent in the steel, cement, lead, corn oil, wood pulp, sugar, gypsum board, and plywood industries.

(text, p. 640)

4. **Single-price tactic:** In this case, all goods and services are offered at the same price (or perhaps two or three prices). Examples of retailers employing this tactic include One Price Clothing Stores, Dre$$ to the Nine$, Your $10 Store, and Fashions $9.99. Advantages include (1) removal of price comparisons from the buyer's decision-making process, (2) a simplified pricing system, and (3) minimization of clerical errors. Disadvantages include (1) continually rising costs and (2) necessity for frequent revisions of the selling price.

   Flexible pricing: With this pricing tactic, different customers pay different prices for essentially the same merchandise bought in equal quantities. This policy is often found in the sale of shopping goods, specialty merchandise, and industrial goods (except for supply items). Automobile dealers and appliance retailers commonly follow this practice. Advantages include (1) allowance for competitive adjustments for meeting or beating another seller's price, (2) ability for the seller to close a sale with price-conscious consumers, and (3) the ability to procure business from a potential high-volume shopper. Disadvantages include (1) the lack of consistent profit margins, (2) the potential ill will of high-paying purchasers, (3) the tendency for salespeople to automatically lower the price to make a sale, and (4) the possibility of a price war among sellers.

   Professional services pricing: This is used by people with lengthy experience, training, and often certification by a licensing board. Fees may be based on the solution of a problem or performing an act. Flat rate pricing may also be used. Examples include lawyers, physicians, or family counselors. Advantages include (1) prices justified according to the education and experience of the service provider, and (2) the simplicity of flat-rate pricing. Disadvantages include (2) difficulty in attaching dollar amounts to experience, education, or certifications, and (2) a temptation to charge "all the traffic will bear" in an inelastic demand situation.

   Price lining: When a seller establishes a series of prices for a type of merchandise, it creates a price line. Price lining is offering a product line with several items placed in the line at specific price points. Examples include Hon offering file cabinets at $125, $250, and $400, and The Limited offering dresses at $40, $70, and $100. Advantages include (1) reduction of confusion for salespeople and consumers, (2) a wider variety of merchandise offered to the buyer at each price, (3) the ability of the seller to reach several market segments, and (4) smaller total inventories for the seller. Disadvantages include (1) rising costs that force confusing changes in price line prices and (2) difficulty in determining where to place the prices within a line.

   Leader pricing: This is an attempt by the marketing manager to induce store patronage through selling a product near or below cost. This type of pricing is common in supermarkets and is also used at department and specialty stores. Advantages include (1) increase in store patronage, (2) potential higher volume of sales per customer, and (3) inducement of store switching. Disadvantages include (1) potential of consumers to stock up on only the leader items and (2) lack of response because of competition with other stores offering similar bargains.

   Bait pricing: This is a deceptive tactic that tries to get consumers into a store though false or misleading advertising. Once in the store, high-pressure sales tactics are used to persuade the consumer to buy more expensive merchandise. Advantages may include (1) increase in store patronage, (2) potential higher volume of sales per customer, and (3) inducement of store switching. The main disadvantage is that the practice is illegal.

   Odd-even pricing: This tactic establishes prices ending in odd or even numbers. Odd-numbered prices are intended to denote bargains, while even-numbered prices are used to imply quality. Retail food stores often use odd-numbered pricing, while prestige products such as perfumes, fur coats, or luxury watches are frequently sold at even-numbered prices. Advantages include (1) implied bargains (odd) or quality (even) and (2) stimulation of demand for some products. Disadvantages include (1) creation of a sawtoothed demand curve and (2) changes in the price elasticity of demand.

Price bundling: In this case, two or more goods and/or services are combined into a single package for a special price. Examples include the sale of maintenance contracts with appliances or office equipment, vacation packages, complete stereo systems, and options on automobiles. Service industries often use bundling. Advantages include (1) stimulation of demand for the bundled items if the consumers perceive the price as a good value, (2) better coverage of constant fixed costs (especially in service industries), and (3) assistance in selling the maximum number of options (on a car, for example). Disadvantages include (1) customers' resistance if one of the bundled items is not wanted and (2) consumers' incorrect value perceptions.

Unbundling: With unbundling, services are split off and charged for. For example, a hotel may charge for parking, or department stores may charge for gift wrapping. Advantages include keeping costs down. A possible disadvantage is that customers may not want to pay "extra" for items that have typically been bundled.

Two-part pricing: This involves two separate charges to consume a single product or service. There is usually some type of base fee, plus a charge per use. Examples include health clubs, amusement parks, and telephone companies. Advantages include (1) consumers' preference of two-part pricing when they are unsure of utilization, (2) high-use consumers paying a higher total price, and (3) possible increase in revenue for the seller by attracting low-use consumers. Disadvantages include (1) difficulty in establishing pricing levels from usage estimates, and (2) resistance by high-use consumers.

(text, pp. 640-643)

5. During a recession, you could use:

   Value Pricing: You could stress the long life of the car as a value to the consumer.

   Bundling: You could offer greater service with the purchase of the car (longer warranty, special services such as a special valet service for cars that have problems).

   (text, pp. 646-647)

6. The E-Lam Corporation should continue to manufacture and sell all three types of machines. An investigation of overall figures shows that a $60,000 profit was earned on the three items in the line:

	Portable	**Vending**	**Desktop**	**Total**
Sales	$40,000	$80,000	$90,000	$210,000
Less: Cost of goods sold	50,000	50,000	50,000	150,000
Gross margin	($10,000)	$30,000	$40,000	$ 60,000

The portable line should not be dropped just because it is currently showing a loss; the joint costs would have to be allocated to the remaining two lines:

	Vending	**Desktop**	**Total**
Sales	$80,000	$90,000	$170,000
Less: Cost of goods sold	75,000	75,000	150,000
Gross margin	$ 5,000	$15,000	$ 20,000

Equal allocation of joint costs may not be the right way to distribute the costs. Other allocation bases that may be used include weighting, market value, or quantity sold. Other allocation methods would change the figures for each machine type, but not overall figures.

(text, p. 644)

CHAPTER 19 Marketing and the Internet

Please refer to www.swpco.com

PRETEST SOLUTIONS

1. Four ways in which companies can communicate with customers over the Internet are:
   - **Send e-mail:** Companies can send and receive e-mail message directly with customers.
   - **Create a newsgroup or discussion list:** Companies can crate a discussion group among customers to help publicize and share information about products.
   - **Launch a home page:** Companies can communicate with the mass market in the form of a home page that provides much information about the company and its products.
   - **Sponsor an established group, list, or site:** Companies can provide a sponsorship to groups in order to publicize themselves.
2. Five ways to conduct marketing research online are: online information search engines, trade magazine and journal archives, electronic news services, electronic surveys, and industry newsgroups and discussion lists.
3 Three privacy issues with online commerce are:
   - **Electronic cash**: payment for products or services online.
   - **Firewalls**: are special computer programs that check all incoming and outgoing information streams for proper identification and authorization.
   - **Spamming**: broadcasting commercial messages to a vast number of people.

VOCABULARY PRACTICE SOLUTIONS

1. Internet, World Wide Web
2. multimedia, hypertext
3. surfer
4. spamming
5. servers
6. download
7. electronic cash
8. cookies
9. firewalls
10. intelligent agents

TRUE/FALSE SOLUTIONS

(question number / correct answer / answer rationale)

1. T
2. F Today's surfer could live anywhere in the world, has an average age of 33, and includes many females.
3. T
4. F E-mail communication is the number-one reason for using the Internet.
5. F Newsgroups are often moderated so that unwanted commercial messages are not forwarded to group members.
6. F Electronic surveys are much less expensive than mail or phone surveys.

7. F Spamming is not illegal but may be considered an invasion of privacy to some receivers.

8. T

9. F Intelligent agents are software programs that comparison shop for low prices.

10. T

AGREE/DISAGREE SOLUTIONS

(question number / sample answers)

1. Reason(s) to agree: The Internet has increased the number of ways in which a company can reach its customers. By using e-mail, home pages, newsgroups and other Internet-based communication vehicles, companies can now reach their customers 24 hours a day.

   Reason(s) to disagree: The Internet has made marketing much more complex. In order to keep up with competition, a company must now use all Internet vehicles to reach its target markets. Communication technology has made it easier to communicate, but now communication is more frequent. Staff must be trained to handle customer communication online.

2. Reason(s) to agree: Though Internet usage has become mainstream, samples of people responding to surveys over the Internet may still be biased. For instance, if a marketer is trying to conduct a survey to people aged 60 or above, the sample responding to the survey may be biased (more active, more technologically driven than the typical 60-year-old).

   Reason(s) to disagree: Whether a marketing research sample is biased depends on the target market it is supposed to represent. Some target markets are well represented through the Internet, which can be an inexpensive and quick way to get information. There are techniques that can be used to take the bias out of the sample.

3. Reason(s) to agree: The Internet has changed all four Ps: product (by the type of service offered on a home page); price (firms can offer either offer lower prices because of lowered costs or higher prices because of offering more convenient shopping); promotion (the Internet has increased the variety of ways in which companies can communicate with their markets); and distribution (certain types of services can actually even be distributed online).

   Reason(s) to disagree: The four Ps still hold true for the Internet. Technology has just provided another means of reaching target markets.

MULTIPLE-CHOICE SOLUTIONS

(question number / correct answer / answer rationale)

1. b The U.S. military originally developed the Internet to provide uninterrupted communication in the event of a sabotage or attack.

2. a Multimedia refers to any combination of text, graphics, sound, or other data formats in a single document.

3. c Most users of the Internet are North American with an average age of 35.

4. b E-mail is used more often than any other Internet activity.

5. c Spamming is sending unsolicited commercial e-mail messages.

6. e One can launch one's own Web site, often at low costs.

7. b The two-way interaction indicates the highest level of involvement.

8. a The Internet population is not quite yet "average."

9. d Firewalls restrict access to a computer network, among other security measures.

10. e Although costs of standardized products are lower, consumers still seek unique goods that suit their individual tastes.

ESSAY QUESTION SOLUTIONS

1. Four benefits of conducting electronic surveys are:

   - **Interactive contact** with respondents to surveys
   - **Ease of sending and receiving** surveys
   - **Completion** of items in the surveys
   - **Inexpensive** compared to phone surveys, in-person surveys (such as mall intercept), and mail surveys

2. Guidelines for the web site include:

   - Include the Web address in print advertisements and stationery. Make sure your company's phone number is also available.
   - Register with online search engines such as Infoseek, Yahoo, and Lycos along with key words so that visitors can get to your home page instantly.
   - Advertise through an offline subscription service. Several companies can automatically load changing Web site content, including advertising, onto subscribers' computers.
   - Frequently change the Web site's content. This way, retrieval services and people have a reason to frequent the site.
   - Pay consumers to visit the site.

   Other options to a web site include:

   Send e-mail. Sending and retrieving electronic messages is the most frequently performed Internet activity. The method is quick, inexpensive, and selective. However, marketers should be careful to avoid spamming.

   Create a newsgroup or discussion list. These Internet services allow people to participate in online discussions about specific topics of interest.

   Sponsor an established group, list, or site. Sponsoring established Internet communication groups, discussion lists, or Web sites instantly gives businesses a targeted audience.

3. You should cover these topics regarding the marketing mix:

   Product. The focus is on branded commodities: the lowest worldwide price for well-known goods. Pre-technology focuses were on availability of delivered goods, and on manufacturer's image, reputation, and product quality.

   Place. The focus is on shipping from the home or office via desktop computer. Previously, products had to be bought from local vendors, or transportation could be used to broaden the geographic availability of products.

   Promotion. Modern promotion is via Internet-based Web sites. Ads can be interactive, targeted, and personalized.

   Price. Today, intelligent software agents located the lowest worldwide price. In the past, price could be set by local sellers until some price competition generated lower prices.

CHAPTER 20 One-to-One Marketing

PRETEST SOLUTIONS

1. Six forces are:
   - A more diverse society
   - More demanding and time-poor consumers
   - A decline in brand loyalty
   - The explosion of new media alternatives
   - Changing channels of distribution
   - Demand for marketing accountability

   (text pp. 675-676)

2. In one-to-one marketing, the communications process differs in that there is less "noise" in the process of sending a message, messages are personalized to the customer, and feedback from the customers is more direct (and often less costly to obtain).

   (text pp. 677-679)

3. Five advantages are:
   - The ability to identify the most profitable and least profitable customers
   - The ability to create long-term relationships with customers
   - The ability to target marketing efforts only to those people most likely to be interested
   - The ability to offer varied messages to different consumers
   - Increased knowledge about customers and prospects

   Disadvantages are:
   - The cost and time involved in creating a one-to-one marketing database
   - Increasing privacy concerns by consumers

   (text pp. 679-680)

4. Eight common applications are:
   - Identifying the best customers
   - Retaining loyal customers
   - Cross-selling other products or services
   - Designing targeted marketing communications
   - Reinforcing consumer purchase decisions
   - Inducing product trial by new customers
   - Increasing the effectiveness of distribution channel marketing
   - Maintaining control over brand equity

   (text pp. 680-684)

5. Three levels of one-to-one marketing are:
   - CIS (customer information system)
   - MIS (marketing intelligence system)
   - DSS (decision support system)

(text pp. 694-696)

VOCABULARY PRACTICE SOLUTIONS

1. one-to-one marketing
2. one-to-one marketing communications process
3. database marketing, database, marketing database
4. response list, compiled list
5. database enhancement, modeled data, custom data
6. recency, frequency, and monetary (RFM) analysis; lifetime value (LTV) analysis; predictive modeling; data mining
7. customer information system (CIS), marketing intelligence system (MIS), data warehouse, data mart

TRUE/FALSE SOLUTIONS

(question number / correct answer / text page reference / answer rationale)

1. T 672
2. F 675 *Declining* brand loyalty is an influence on the emergence of one-to-one marketing.
3. T 676
4. T 678
5. F 680 The upfront costs of developing a database are generally high.
6. F 680 Privacy is a major concern to most customers.
7. F 682 This is an example of cross selling other products.
8. T 683
9. T 686
10. F 695 What is described is actually a marketing intelligence system (MIS).

AGREE/DISAGREE SOLUTIONS

(question number / sample answers)

1. Reason(s) to agree: Database technology is making is easier for marketers to personalize and individualize products and services to customers. Once one-to-one marketing becomes more common, no consumer would want to be part of a mass marketing effort.

   Reason(s) to disagree: Some products and industries still lend themselves better to mass marketing. Though products such as Coca-Cola can target special promotions on a one-to-one basis, its product offerings would still be better to market on a mass level.

2. Reason(s) to agree: Business marketing involves larger and more concentrated customers. Business marketing has always included aspects of one-to-one marketing because of the importance of each customer to a firm's sales and profits.

   Reason(s) to disagree: One-to-one marketing can be as effective on consumer markets as on business markets. Because consumer markets are more fragmented and more difficult to reach, one-to-one marketing may be more difficult. However, database technology has allowed marketers to overcome these issues and to personalize messages to an otherwise anonymous audience.

3. Reason(s) to agree: One-to-one marketing requires a database full of personal information about consumers—such as demographic information, their purchasing patterns, and their interests. Without these vital data, one-to-one marketing would not be feasible.

   Reason(s) to disagree: Marketers with "lower tech" methods can still use one-to-one marketing. They can still create lists by collecting information by mail or through response cards (such as product registration). The process may be cumbersome but is quite feasible.

4. Reason(s) to agree: Consumers are being inundated with personalized messages from marketers. Eventually, consumer lobby groups will force more legislation to protect the privacy of all consumers.

   Reason(s) to disagree: Even if legislation stifles the effort of some one-to-one marketers, companies may use the legislation to their benefit. By honoring privacy and sending messages only to those who are interested, marketers can better identify which customers are most likely to purchase their products.

MULTIPLE-CHOICE SOLUTIONS

(question number / correct answer / text page reference / answer rationale)

1. b 671 One-to-one marketing does not focus on share of market.
2. a 675 Lower production costs have not affected the emergence of one-to-one marketing.
3. c 675 Sales promotions, increased power of retailers, and the proliferation of brands have all contributed to decreased loyalty.
4. b 678 Because the communication occurs directly from the marketer to the customer, there is less distraction from competing products.
5. c 678 Reading and interpreting a message is considered decoding.
6. e 680 One-to-one marketing is relatively expensive to set up.
7. a 681 One-to-one marketing can minimize wasted time and money on focusing on the wrong people.
8. b 681 Repeat customers are a sign of loyalty, which is a result of creating long-term relationships.
9. c 681 A limited number of marketing segments is *not* a disadvantage of one-to-one marketing.
10. d 680 Providing sales leads is not a common application of one-to-one marketing.
11. a 681 Providing points to encourage customers to continue choosing the hotel is an example of retaining loyal customers.
12. b 683 Sending out a team to congratulate the customer is an example of reinforcing a purchase decision.
13. c 682 Suggesting other titles is an example of cross-selling other products.
14. b 684 The compilation of names, addresses, and other information about customers is a marketing database.
15. e 686 A response list is created from people who have responded to an offer, such as a rebate.
16. c 686 A compiled list is created from names and addresses gleaned from directories or membership rosters.
17. d 688 Modeled data are information that has already been sorted into distinct groups or clusters or customers.
18. b 689 Custom data are enhancement information acquired by the marketer through customer surveys, customer participation programs, product registration, warranty cards, or loyalty marketing programs.
19. e 691 Recency-frequency-monetary analysis identifies those customers who bought most recently, most frequently, or spent a specified amount of money.
20. c 692 Lifetime value analysis projects the future value of the customer over a period of years.
21. c 694 A customer information system (CIS) is used to track data captured from purchase transactions and past marketing activity.
22. d 695 A marketing intelligence system (MIS) builds on information captured by a CIS by integrating it with a greater array of data, such as customer profiles and lifestyles, in order to predict future response.
23. a 696 A marketing decision support system is the most sophisticated level of marketing database.
24. e 696 What is described is a data warehouse.

ESSAY QUESTION SOLUTIONS

1. Five advantages of one-to-one marketing are:

- **The ability to identify the most profitable and least profitable customers**: The retailers can identify which customers bought the most merchandise most recently so that it can target special promotions and loyalty programs to these customers.
- **The ability to create long-term relationships with customers**: The retailer likely has customers who visit the store frequently and should try to hold onto them. By conducting one-to-one marketing, the retailer can make their customers even more loyal.
- **The ability to target marketing efforts only to those people most likely to be interested**: The retailer can save time and money by focusing their efforts to people who enjoy the outdoors and therefore more likely to purchase their merchandise.
- **The ability to offer varied messages to different consumers**: The retailer may find through its marketing efforts that certain customers buy certain merchandise, based on their sports interests. For example, if a customer has purchased merchandise for camping, the retailer can target promotional messages to this customer based on camping merchandise.
- **Increased knowledge about customers and prospects**: The more the retailer understands its customers, the more it can target special marketing programs to them. The customer can also become a valuable marketing tool by referring the store to other prospects.

(text pp. 679-680)

2. Eight applications for this database are:

- **Identify the best customers**: The resort can identify which customers have enough income, the correct demographic profile, and the lifestyle and interests that would be suited to the service that it offers.
- **Retain loyal customers**: The resort can use its list of past customers to target loyalty programs, such as awarding points that can be used toward free stays or upgrades in rooms.
- **Cross-sell other products**: The resort can use the database to sell other services, such as vacation or tour packages.
- **Design targeted marketing communications**: The resort can identify which features certain customers enjoyed the most and provide promotional offers regarding these features (such as a free dinner to customers who enjoyed one of the resort's restaurants).
- **Reinforce consumer purchase decisions**: The resort can use the database to send warm greetings to new customers who have just made a decision about coming to the resort but who have not yet arrived. This will give the new customer a better feeling about the purchase decision.
- **Induce product trial by new customers**: The database can contain names and addresses of people who dined at one of the restaurants at the resort but who did not stay at the resort. The resort can then send special messages or promotional offers to these customers to encourage them to stay at the resort.
- **Increase the effectiveness of distribution channel marketing**: The database can also contain the names and addresses of key travel agencies that recommend the resort most often. Special offers can then be provided these agencies to further encourage these recommendations. In addition, information about customers can be given to travel agents so that they understand the kind of people enjoy the resort the most.
- **Maintain control over brand equity**: The resort can also protect the brand name "Island Breeze" and to provide consumers with a good overall feeling about the resort.

(text pp. 680-684)

3. Five data manipulation techniques are:

- **Customer segmentation:** You can take the data and divide it into different customer segments using demographic, geographic, psychographic, or buying behavior information. For example, you can develop customer profiles for each of the models of cars you sell.
- **Recency-frequency-monetary analysis (RFM):** You can determine which customers have bought recently and most frequently and how much they have spent to date. This information can be used to build loyalty programs and to target specific messages about new car models that are coming out.

- **Lifetime value analysis (LVA):** You can conduct an LVA to determine how much your most loyal customer is worth in actual dollars. You can then select the most loyal and create long-term relationships with them.
- **Predictive modeling:** You can develop certain criteria, such as demographic or lifestyle profiles, and determine how likely certain customers will buy another vehicle from you.
- **Data mining:** You can discover other independent and nonobvious variables that lead to purchase, such as occupation or personality, so that you can better understand how to target future marketing efforts.

(text pp. 691-693)